Amino Acids and Peptides

Volume 17

A Specialist Periodical Report

Amino Acids and Peptides

Volume 17

A Review of the Literature Published during 1984

Senior Reporter
J. H. Jones, *University of Oxford*

Reporters
G. C. Barrett, *Oxford Polytechnic*
J. Brennan, *UMIST*
J. S. Davies, *University College of Swansea*
I. J. Galpin, *University of Liverpool*
P. M. Hardy, *University of Exeter*
R. W. Hay, *University of Stirling*
K. B. Nolan, *University of Surrey*

The Royal Society of Chemistry
Burlington House, London W1V 0BN

ISBN 0-85186-154-7
ISSN 0269-7521

Set by Unicus Graphics Ltd., Horsham, West Sussex
and printed in Great Britain
by Whitstable Litho Ltd., Whitstable, Kent

Preface

It became apparent as the fifteenth and sixteenth volumes of 'Amino Acids, Peptides, and Proteins' were being assembled that full coverage of the syllabus originally set in the sixties had become impossible without self-defeating compression or selectivity. An exhaustive review of a year's output in the whole field of amino acids, peptides, and proteins could no longer be contained in a single volume. The proverbial quart would not go into our pint pot. It was therefore decided to contract the scope and shift the emphasis of the title, which is now renamed as the Specialist Periodical Report on 'Amino Acids and Peptides'. The biological edges have been trimmed, and proteins have been excised altogether. Painful surgery, but essential if the title was to avoid unreasonable compression, expense, and delays in publication. Even with this great reduction in scope, the present volume has about 10% more literature citations in it than the first, which covered the publications of 1968. On the credit side, furthermore, the coverage of β-lactam chemistry is fuller than before and has a separate section for the first time.

1984 saw the passing of one peptide chemistry Nobel Laureate (V. du Vigneaud, died 31st July) and the creation of another (R. B. Merrifield). Bruce Merrifield developed his exquisitely simple concept of solid-phase peptide synthesis in the early sixties, and the only surprise about the Nobel Prize is that it was not awarded sooner. Asked where the original idea for solid-phase synthesis came from, he replied:[1]

> 'Making peptides by the solution method was difficult, time-consuming, and it was difficult to purify the intermediates. I could identify a need for some improvement. The general idea of the solid-phase way to do it was influenced, no doubt, by chromatography. We were doing chromatography in isolations, so I guess I jumped from there to the idea of doing the synthesis on the support ... After arriving at the basic idea, the second step was to demonstrate that you could make anything, *e.g.* make a dipeptide. That was the hardest part, getting the first positive result. The next phase was to elaborate on that, modify and try to improve as we went along. That's what we've done ever since and we're still doing it.'

The low-key modest style is typical of the man, and we salute his genius the more because of it.

1984 was also remarkable for the publication of an unusually large number of important books and symposium proceedings in the field. Of the latter we have not only the proceedings of specialized symposia on opioids[2] and β-lactam anti-

[1] 'Chemalog Hi-lites', Chemical Dynamics Corp., 1985, Vol. 9, No. 2, p. 7.

[2] 'Opioids. Past, Present and Future', Proc. Symp. held in honour H. W. Kosterlitz, Cambridge, U.K., 1983, ed. J. Hughes, H. O. J. Collier, M. J. Rance, and M. B. Tyers, Taylor and Frances, London, 1984.

biotics[3] but also the proceedings of the Eighth American[4] and Eighteenth European[5] Peptide Symposia, as well as those of the Fourth U.S.S.R–F.R.G.[6] and Twenty-second Japanese[7] Peptide Symposia. Japanese peptide chemistry is still not as fully appreciated in the world at large as it ought to be. One suspects that a good deal more hullabaloo would have accompanied the total classical synthesis of ribonuclease – the very Everest of peptide synthesis and indeed of organic synthesis in general – if it had been achieved in Europe or America. This is partly to be ascribed to geographical and language barriers, but also to the diffidence with which our Japanese colleagues present themselves. Of recent years, however, the Japanese Peptide Symposia, which were originally established for domestic purposes, have acquired an international flavour, and their proceedings have been published in English since 1976, opening up Japanese peptide chemistry to the rest of the world in a most welcome manner.

Graham Barrett has contributed the annual chapter on amino acids to these Reports now for twelve successive years, and is uniquely qualified by his experience and exhaustive study of recent literature to edit an authoritative treatise on the subject. This he has now done,[8] personally contributing key chapters on synthesis, resolution, and chemical reactions. The book has twenty-two chapters in all, covering chemical, biochemical, analytical, and structural aspects of all kinds of amino acids. It gives convenient rapid access to the literature of the last two decades, and is an indispensable sequel to Greenstein and Winitz (1961).[9]

Bodanszky and Ondetti (1966)[10] and its successor Bodanszky, Klausner, and Ondetti (1976)[11] have done much to shape the chemical thinking behind many of the achievements of peptide synthesis in the last twenty years. They are now superseded by M. Bodanszky's 'Principles of Peptide Synthesis' (1984),[13] an admirable account, although the critical reader will find a few details of fact and interpretation to quibble over. Its companion volume can be accorded a less

[3] 'Recent Advances in the Chemistry of β-Lactam Antibiotics', Proc. 3rd Int. Symp., Cambridge, U.K., 1984, ed. A. G. Brown and S. M. Roberts, Special Publication No. 52, The Royal Society of Chemistry, London, 1985.

[4] 'Peptides, Structure and Function', Proc. 8th Am. Pept. Symp., Tucson, Arizona, 1983, ed. V. J. Hruby and D. H. Rich, Pierce, Rockford, Illinois, 1983.

[5] 'Peptides 1984', Proc. 18th Eur. Pept. Symp., Djurönäset, Sweden, 1984, ed. U. Ragnarsson, Almqvist and Wiksell International, Stockholm, 1984.

[6] 'Chemistry of Peptides and Proteins', Vol. 2, Proc. 4th U.S.S.R.–F.R.G. Pept. Symp., Tübingen, West Germany, ed. W. Voelter, E. Bayer, Yu. A. Ovchinnikov, and E. Wünsch, Walter de Gruyter, Berlin, West Germany, 1984.

[7] 'Peptide Chemistry 1984', Proc. 22nd Jpn. Pept. Symp., ed. N. Izumiya, Peptide Institute Inc., Osaka, 1984.

[8] 'Chemistry and Biochemistry of the Amino Acids', ed. G. C. Barrett, Chapman and Hall, London, 1985.

[9] J. P. Greenstein and M. Winitz, 'Chemistry of the Amino Acids', Wiley, New York, 1961.

[10] M. Bodanszky and M. A. Ondetti, 'Peptide Synthesis', 1st Edn., Interscience, New York, 1966.

[11] M. Bodanszky, Y. S. Klausner, and M. A. Ondetti, 'Peptide Synthesis', 2nd Edn., Wiley-Interscience, New York, 1976.

[12] M. Bodanszky, 'Principles of Peptide Synthesis', Springer-Verlag, Berlin, West Germany, 1984.

restrained recommendation: 'The Practice of Peptide Synthesis' (1984),[13] compiled by M. and A. Bodanszky, in which all the currently standard reactions of classical peptide synthesis (and a few more besides) are illustrated by carefully selected examples from the literature, with full experimental detail and comment. It is, however, still the case that solid-phase synthesis is better served than the classical approach as far as an experimentalists' handbook is concerned. This was already so by 1969 with the publication of Stewart and Young,[14] which has now been revised and brought out as a second edition.[15] Essential for anyone working with solid-phase methods, those engaged in classical solution work will also find it a very useful book to keep close at hand, because its coverage extends to peptide chromatography, sources of materials, and so on.

Balliol College, Oxford JOHN JONES
September 1985

13 M. Bodanszky and A. Bodanszky, 'The Practice of Peptide Synthesis', Springer-Verlag, Berlin, West Germany, 1984.
14 J. M. Stewart and J. D. Young, 'Solid Phase Peptide Synthesis', 1st Edn., Freeman, San Francisco, California, 1969.
15 J. M. Stewart and J. D. Young, 'Solid Phase Peptide Synthesis', 2nd Edn., Pierce, Rockford, Illinois, 1984.

Contents

Abbreviations

Abbreviations for amino acids and their use in the formulation of derivatives follow, with rare exceptions, the 1983 Recommendations of the I.U.P.A.C.-I.U.B. Joint Commission on Biochemical Nomenclature, which are reprinted as an Appendix in Volume 16 of 'Amino Acids, Peptides, and Proteins'. Exceptions and additions are defined in the text as they occur.

1

Amino Acids

BY G. C. BARRETT

1 Introduction

Thorough coverage centred on the 1984 literature, though omitting routine biological applications and reports of the distribution of well-known amino acids, is the intention for this chapter. There is therefore continuity with preceding volumes of this *Specialist Periodical Report* (to which reference is occasionally made in order to help the reader put into context some recent progress reported here for an on-going topic of study).

2 Textbooks and Reviews

The 1983 recommendations for amino acid nomenclature are only a library distant, since the I.U.P.A.C.-I.U.B. Newsletter (1984) has been reproduced in major journals.[1a] The recommendations have been quickly followed by nomenclature for amino acid amides (1984).[1b]

Important compilations providing support of research work with amino acids represent the latest outputs from sources already well known for similar recent monographs.[2] Other reviews, much less readily accessible, deal with various facets of medium- to large-scale production of amino acids.[3]

[1] (*a*) 'Nomenclature and Symbolism for Amino Acids and Peptides (Recommendations 1983)', *Eur. J. Biochem.*, 1984, **138**, 9; *Arch. Biochem. Biophys.*, 1984, **229**, 399; *Int. J. Pept. Protein Res.*, 1984, **24** (1); *Hoppe-Seyler's Z. Physiol. Chem.*, 1984, **365**, 1; *Can. J. Biochem. Cell Biol.*, 1984, **62**, viii; *Pure Appl. Chem.*, 1984, **56**, 595; *J. Biol. Chem.*, 1985, **260**, 14; (*b*) *Biochem. J.*, 1985, **225**, I.

[2] *Methods Enzymol.*, 1984, **106**: H. L. Cooper, M. H. Park, and J. E. Folk, p. 344 (hypusine), R. L. Sass and M. E. Marsh, p. 351 (N^{π}- and N^{τ}-histidinoalanine), A. N. Glazer, p. 359 [S^{β}-(bilin)cysteine derivatives], K. Lerch, p. 355 [S^{β}-(2-histidyl)cystine]; *Methods Enzymol.*, 1984, **107**: R. Amado, R. Aeschbach, and H. Neukom, p. 377 (dityrosine), S. C. Fry, p. 388 (iodotyrosine), H. J. Waite and C. V. Benedict, p. 397 (dopa), S. Hunt, p. 413 (halogenated tyrosines), G. L. Nelsestuen, p. 503 (γ-carboxyglutamic acid), T. H. Koch, M. R. Christy, R. M. Barkley, R. Sluski, D. Bohemier, J. J. Van Buskirk, and W. M. Kirsch, p. 563 (synthesis of β-carboxyaspartic acid and identification of γ-carboxyglutamic acid), M. X. Sliwkowski, p. 620 (*Se*-adenosyl selenomethionine); 'CRC Handbook of H.P.L.C. for the Separation of Amino-acids, Peptides, and Proteins', ed. W. S. Hancock, Chemical Rubber Company, Boca Raton, Florida, U.S.A., 1984, Vols. 1 and 2 [*e.g.* S. Ishimitsu, S. Fujimoto, and A. Ohara, p. 275 (*o*-, *m*-, and *p*-tyrosine)].

[3] K. Yokozeki, C. Aguchi, and Y. Hirose, *Ann. N. Y. Acad. Sci.*, 1983, **413**, 551; K. Matsumoto and I. Chibata, *Hakko to Kogyo*, 1983, **41**, 834 (*Chem. Abstr.*, 1984, **100**, 175 219); B. Hoppe and J. Martens, *Chem. Unserer Zeit*, 1984, **18**, 73; 'Preparation of Optically-active Amino-acids: Theoretical Principles, Problems, Horizons', I. Yu. Galaev and Yu. V. Galaev, *Izd. Sarat. Univ., Saratov, U.S.S.R.*, 1983 (*Chem. Abstr.*, 1984, **101**, 171 747).

3 Naturally Occurring Amino Acids

Occurrence of Known Amino Acids. – Points of interest to appeal to a cross-section of readers are found in the location of D-2-aminopimelic acid and its *trans*-3,4-dehydro analogue in *Asplenium unilaterale*,[4] of five γ-carboxyglutamic acid residues within a novel heptadecapeptide toxin in the venom of a fish-hunting cone snail, *Conus geographus*,[5] and of the crosslinking amino acid residues lysinoalanine (in alkali-treated partial hydrolysates of β-casein and broad-bean protein)[6] and 3-hydroxypyridinium-containing moieties (in cartilage).[7] Other heteroaromatic and aromatic moieties of familiar types feature in a useful study of optimum conditions for protein hydrolysis in which tryptophan degradation is largely avoided (92% recovery using 3M mercaptoethane sulphonic acid at 166 °C for 25 min)[8] and in structure elucidation of the common aglycone moiety of the actaplanin antibiotics (made up of hydroxylated phenylalanine and phenylglycine units condensed into a tetracyclic peptide array).[9]

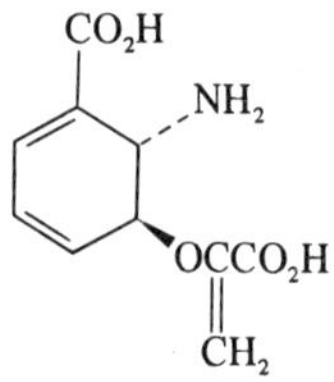

(1)

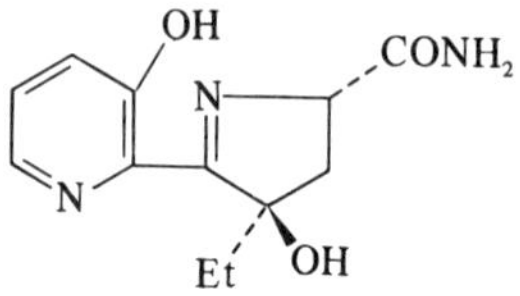

(2)

New Natural Amino Acids. – First findings reported in this section range from free amino acids [*trans*-4-hydroxy-*N*-methyl-L-proline in the red alga *Chondria coerulescens*[10] and an intermediate (1) in the transformation of chorismic acid to anthranilic acid by anthranilate synthase I from *Serratia marcescens*][11] to simple derivatives histargin (a new carboxypeptidase B inhibitor from *Streptomyces roseoviridis*, in which arginine and histidine are linked *via* carboxy groups by 1,2-diaminoethane),[12] siderochelin C (2) from an *Actinomycete*,[13] and an

4 N. Murakami and S. Hatanaka, *Phytochemistry*, 1983, **22**, 2735.
5 J. M. McIntosh, B. M. Olivera, L. J. Cruz, and W. R. Gray, *J. Biol. Chem.*, 1984, **259**, 14 343.
6 H. Noetzold, H. Winkler, B. Wiedemann, and E. Ludwig, *Nahrung*, 1984, **28**, 299 (*Chem. Abstr.*, 1984, **101**, 67 984).
7 J. J. Wu and D. R. Eyre, *Biochemistry*, 1984, **23**, 1850.
8 K. Maeda, J. J. Scheffler, and A. Tsugita, *Hoppe-Seyler's Z. Physiol. Chem.*, 1984, **365**, 1183.
9 A. H. Hunt, T. K. Elsey, K. E. Merkel, and M. Debono, *J. Org. Chem.*, 1984, **49**, 641.
10 S. Sciuto, R. Chillemi, M. Piattelli, and G. Impellizzeri, *Phytochemistry*, 1983, **22**, 2311.
11 P. P. Policastro, K. G. Au, C. T. Walsh, and G. A. Berchtold, *J. Am. Chem. Soc.*, 1984, **106**, 2443.
12 H. Umezawa, T. Aoyagi, K. Ogawa, H. Iiunima, H. Naganawa, M. Hamada, and T. Takeuchi, *J. Antibiot.*, 1984, **37**, 1088.
13 L. A. Mitscher, T. Hogberg, S. D. Drake, A. W. Burgstahler, M. Jackson, B. Lee, R. I. Sheldon, H. E. Gracey, W. Kohl, and R. J. Theriault, *J. Antibiot.*, 1984, **37**, 1260.

unusual deoxynucleotide, α-*N*-(9-β-D-2′-deoxyribofuranosylpurin-6-yl)glycinamide, specified by bacteriophage Mu.[14]

New Amino Acids from Hydrolysates. – This section continues to record unsuspected and unlikely (but real) protein amino acids, with a spectacular 'first', the location of aminomalonic acid, $H_3\overset{+}{N}CH(CO_2^-)CO_2H$, in *Escherichia coli* and atherosclerotic plaque proteins (the latter also contain β-carboxyaspartic and γ-carboxyglutamic acids).[15]

Although both *cis* and *trans* isomers of 3- and 4-hydroxyproline appear in collagen hydrolysates, the *cis* isomers are formed during the hydrolysis procedure.[16]

The presence of ε-(γ-glutamyl)lysine in protein hydrolysates has been established through sensitive h.p.l.c. methods.[17]

4 Chemical Synthesis and Resolution of Amino Acids

General Methods of Synthesis. – Amination of simple substrates is represented in reactions of sodium chloroacetate with secondary amines in tetrahydrofuran[18] and of aliphatic aldehydes with $CHCl_3$ and NH_3 in CH_2Cl_2–H_2O containing a phase-transfer agent[19] and in reductive amination of keto acids using sodium cyanoborohydride and an ammonium salt.[20] Analogous carboxylation processes are represented in electroreduction of Schiff bases $PhCR^1{=}NCHR^2Ph$ in the presence of CO_2[21] and in hydrocarbonylation of *N*-vinyl- and -allyl-phthalimides catalysed by Rh or Pd complexes.[22]

Standard procedures are employed in alkylation of diethyl acetamidomalonate (*e.g.* 2,6-dihalotyrosines[23]), formation of α-aminonitriles ($RCHO + Me_3SiCN$ catalysed by $ZnI_2 \rightarrow NCCHROSiMe_3$, which is reacted with a secondary amine in MeOH),[24] more conventional Strecker synthesis of alicyclic α-amino acids from corresponding ketones and $PhCH_2NH_2$ with KCN,[25] azlactone syn-

14 D. Swinton, S. Haltmann, P. F. Crain, C. S. Cheng, D. L. Smith, and J. A. McCloskey, *Proc. Natl. Acad. Sci. U.S.A.*, 1983, **80**, 7400.

15 J. J. Van Buskirk, W. M. Kirsch, D. L. Kleyer, R. M. Barkley, and T. H. Koch, *Proc. Natl. Acad. Sci. U.S.A.*, 1984, **81**, 722.

16 G. Bellon, R. Berg, F. Charstang, A. Malgras, and J. P. Borel, *Anal. Biochem.*, 1984, **137**, 151.

17 M. Griffin and J. Wilson, *Mol. Cell. Biochem.*, 1984, **58**, 37.

18 F. M'Henni and Z. Mighri, *J. Soc. Chim. Tunis*, 1984, **11**, 3 (*Chem. Abstr.*, 1984, **101**, 171 679).

19 X. Sun, Y. Shi, H. Zhu, Z. Zhou, and C. Lin, *Nanjing Daxue Xuebao, Ziran Kexue*, 1983, 658 (*Chem. Abstr.*, 1984, **100**, 210 384).

20 S. P. Reid and P. J. Reeds, *Anal. Biochem.*, 1984, **142**, 24 (*cf.* R. F. Borsch, M. D. Bernstein, and H. D. Durst, *J. Am. Chem. Soc.*, 1971, **93**, 2897).

21 T. Iwasaki and K. Harada, *Bull. Inst. Chem. Res., Kyoto Univ.*, 1983, **61**, 72 (*Chem. Abstr.*, 1984, **100**, 121 561).

22 G. Delogu, G. Faedda, and S. Gladiali, *J. Organomet. Chem.*, 1984, **268**, 167.

23 R. A. Pascal and Y. C. J. Chen, *J. Org. Chem.*, 1985, **50**, 408.

24 K. Mai and G. Patil, *Tetrahedron Lett.*, 1984, **25**, 4583.

25 W. J. Layton, S. L. Smith, P. A. Crooks, T. Deeks, and R. D. Waigh, *J. Chem. Soc., Perkin Trans. 1*, 1984, 1283.

thesis,[26-28,150] alkylation of isocyanoacetic esters[29] and glycine derivatives [*e.g.* the Schiff base $Ph_2C{=}NCH_2CO_2Et$[30] and $PhSCH_2NMeCH_2CO_2Et$,[31] the latter with NaH undergoing cycloaddition with $PhCH{=}C(CO_2Me)_2$ in HMPA–dimethoxyethane to yield *N*-methylproline derivatives[31]], and an example of the Ugi four-component condensation, leading to compound (3).[32]

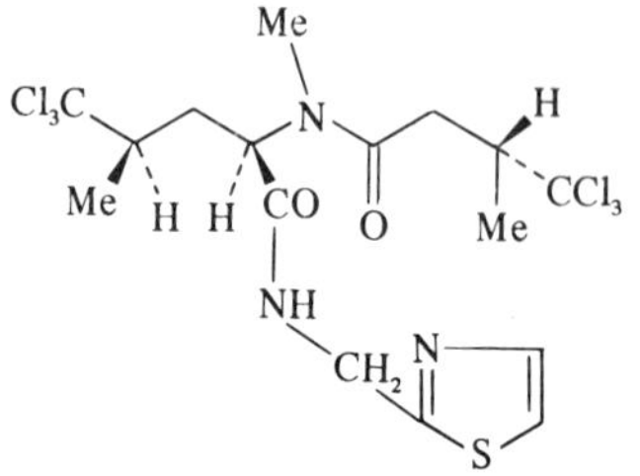

(3)

A new synthesis has been reported, based on the rearrangement of acetimidates $R^1CH{=}CHCHR^2OC({=}NH)CCl_3$ derived from allylic alcohols.[33,93] Overnight refluxing in xylene followed by treatment of the resulting amide CCl_3CONH-$CHR^2CH{=}CHR^1$ with $NaIO_4$–RuO_3 then hydrolysis in aqueous HCl gives the amino acid $H_3\overset{+}{N}CHR^2CO_2^-$. The potential of this method is limited by both the accessibility of the allylic alcohol and the compatibility of the eventual amino acid side chain R^2 with the reaction conditions (the conversion of an alcohol into the acetimidate requires NaH and CCl_3CN as reagents).

Asymmetric Synthesis. – Many of the recent papers on this topic cover what could be described as established general methods, since many of them extend studies that have featured in this section in preceding volumes. One of the longest-established of these, the asymmetric hydrogenation of amino-acrylates,[34]

26 A. K. Sen and S. Mukhopadhyay, *Indian J. Chem., Sect. B*, 1983, **22**, 939 (*Chem. Abstr.*, 1984, **100**, 210 368).
27 P. K. Tripathy and A. K. Mukerjee, *Synthesis*, 1984, 418.
28 M. Ali and N. H. Khan, *Indian J. Chem., Sect. B*, 1984, **23**, 868 (*Chem. Abstr.*, 1984, **101**, 230 986).
29 C. Herdeis and U. Nagel, *Heterocycles*, 1983, **20**, 2163.
30 M. J. O'Donnell, W. Bruder, K. Wojciechowski, L. Ghosez, M. Navarro, F. Sainte, and J. P. Antoine in 'Peptides: Structure and Function', Proceedings of the 8th American Peptide Symposium, ed. V. J. Hruby and D. H. Rich, Pierce Chemical Company, Rockford, Illinois, U.S.A., 1983, p. 151.
31 N. Imai, Y. Terao, K. Achiwa, and M. Sekiya, *Tetrahedron Lett.*, 1984, **25**, 1579.
32 S. E. De Laszlo and P. G. Williard, *J. Am. Chem. Soc.*, 1985, **107**, 199.
33 S. Takano, M. Akiyama, and K. Ogasawara, *J. Chem. Soc., Chem. Commun.*, 1984, 770.
34 I. Ojima, *Pure Appl. Chem.*, 1984, **56**, 99; K. Harada and M. Takasaki, *Bull. Chem. Soc. Jpn.*, 1984, **57**, 1427; T. Yamagishi, M. Yatagai, H. Hatakeyama, and M. Hida, *ibid.*, p. 1897; M. Inoue, K. Ohta, N. Ishizuka, and S. Enomoto, *Chem. Pharm. Bull.*, 1983, **31**, 3371; L. O. Nindakova, F. K. Shmidt, E. I. Klabunovskii, and V. A. Pavlov, *Izv. Akad. Nauk S.S.S.R., Ser. Khim.*, 1984, 720.

azlactones,[35] and Schiff bases,[36] is represented in familiar forms, employing chiral phosphine–Rh or –Co catalysis[34] or the incorporation of a chiral moiety into the substrate.[34,36] The protonation of amino-acrylic acid itself occurs with modest (15–20%) enantiomeric excess during its conversion into alanine catalysed by the *Pseudomonas striata* amino acid racemase.[37]

The reason for the continuing flow of papers is the incomplete understanding of the relationship between stereoselectivity and structure in this area of asymmetric synthesis. This uncertainty also applies to asymmetric transaminations of α-keto acids using a chiral pyridoxamine[38] and aminolysis of azlactones by chiral amines.[39] The crop of papers in which asymmetric alkylation processes are extended generally describe high stereoselectivity, however; a synthesis of α-methylated (*S*)-amino acid esters $H_2NCRMeCO_2Me$ through alkylation of the Schiff base of methyl L-alaninate with an alkyl bromide RBr after lithiation with $LiNPr^i_2$, where the Schiff base is formed with the aldehyde formed from 1,2,3,4-protected D-galactose (*cf.* precedent work, Vol. 14, p. 11),[40] gives enantiomeric excesses of 44–85%.[40] L-Serine undergoes α-alkylation with retention of its configuration through reaction of the derived lithium enolate (4) with electrophiles.[41] The same principle also applies to the asymmetric alkylation of nickel(II) complexes of (*N*-benzyl-L-prolyl)-*o*-aminobenzaldehyde with acetaldehyde, leading to L-threonine and its *allo* isomer in enantiomeric yields of 86 and 76%, respectively.[42] Amides formed between DL-amino acids and (*S*)-prolinol methyl ether can be lithiated and re-protonated with up to 92% diastereoselectivity;[43] much less selectivity is seen in the alkylation of DL-amino acids esterified either with (*S*)-prolinol or with (–)-menthol, since diastereoisomeric excesses range from 5 to 46%.[43]

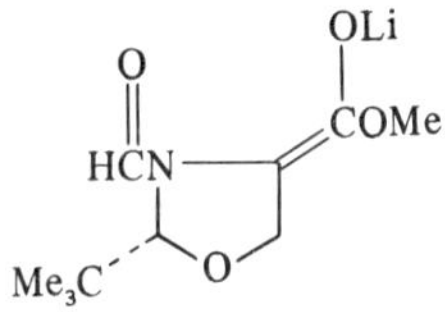

(4)

[35] E. I. Karpeiskaya, L. F. Godunova, E. S. Levitina, M. R. Lyubeznova, E. I. Klabunovskii, E. D. Lubuzh, and A. I. Lutsenko, *Izv. Akad. Nauk S.S.S.R., Ser. Khim.*, 1984, 85.

[36] K. Harada and S. Shioni, *Bull. Chem. Soc. Jpn.*, 1984, **57**, 1367.

[37] B. Badet, K. Lee, H. G. Floss, and C. T. Walsh, *J. Chem. Soc., Chem. Commun.*, 1984, 838.

[38] Y. Tachibana, M. Ando, and H. Kuzuhara, *Bull. Chem. Soc. Jpn.*, 1983, **56**, 3652; K. Bernauer, R. Deschenaux, and T. Taura, *Helv. Chim. Acta*, 1983, **66**, 2049; R. Deschenaux and K. Bernauer, *ibid.*, 1984, **67**, 373.

[39] L. N. Kaigorodova, E. S. Levitina, E. I. Karpeiskaya, and E. I. Klabunovskii, *Izv. Akad. Nauk S.S.S.R., Ser. Khim.*, 1984, 813.

[40] I. Hoppe, U. Schöllkopf, and R. Toelle, *Synthesis*, 1983, 789.

[41] D. Seebach and J. D. Aebi, *Tetrahedron Lett.*, 1984, **25**, 2545.

[42] Yu. N. Belokon, N. I. Chernoglazova, K. A. Kochetkov, N. S. Garbalinskaya, M. G. Ryzhov, V. I. Bakhmutov, M. B. Saporovskaya, E. A. Paskonova, and V. I. Maleev, *Izv. Akad. Nauk S.S.S.R., Ser. Khim.*, 1984, 804.

[43] K. G. Davenport, D. T. Mao, C. M. Richmond, D. E. Bergbreiter, and M. Newcomb, *J. Chem. Res. (S)*, 1984, 148.

Continuing studies[44] with alkylation of lithiated 2,5-dimethoxy-3,6-dihydropyrazines (see also refs. 42 and 76) confirm the high levels of enantiomer purity that can be achieved (see Vol. 16, p. 6; for a review see ref. 45). For example,[44] condensation of the (*S*)-3-isopropyl compound with $Me_3CSiMe_2CR^1R^2CHO$ after lithiation gives $R^1CR^2{=}CRCH(\overset{+}{N}H_3)CO_2^-$ in better than 95% enantiomer excess.

A novel transfer-of-chirality operation applied to the synthesis of *N*-benzyloxycarbonyl-D-amino acids in 78–84% enantiomer excess is based on [2,3]-sigmatropic rearrangement of a vinyl selenide (Scheme 1).[46]

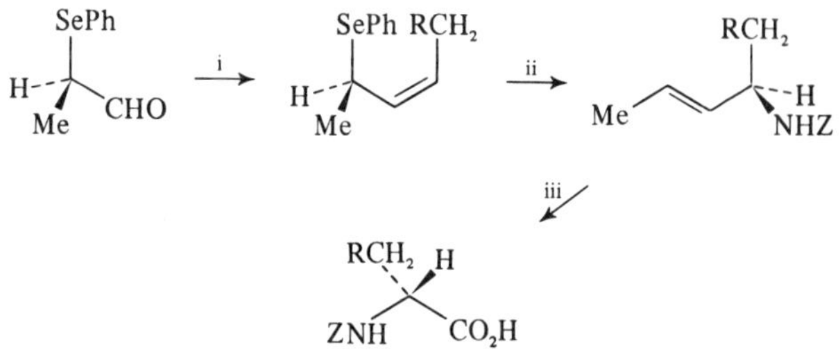

Reagents: i, Wittig synthesis; ii, $PhCH_2OCONH_2$; iii, O_3, Jones oxidation

Scheme 1

Prebiotic Synthesis Models. – The ripples continue to spread out from the original 'electric discharge-CH_4/H_2O/N_2 or NH_3' experiment, and as in recent years (see Vol. 16, p. 6) one of the original authors has again reappeared on the expanding wavefront with a comparison of relative yields of C_3–C_6 amino acids in such a system when the NH_3 concentration varies and when other simple alkanes are used in place of methane.[47] Shock-wave compression (amplitude 10 GPa) converts ammonium salts of acrylic, crotonic, cinnamic, and fumaric acids into *β*-alanine, *β*-aminobutyric acid, phenylalanine, and aspartic acid, respectively, in yields of up to 10%.[48]

Extended reaction times convert glycine-formaldehyde or -acetaldehyde mixtures at pH 3.5 and at 60–80 °C not only into the expected serine and threonine but also into alanine, glutamic acid, aspartic acid, norvaline, isoleucine, and four other protein amino acids.[49]

[44] U. Schöllkopf, J. Nozulak, and U. Groth, *Tetrahedron*, 1984, **40**, 1409; Y. Jiang, U. Groth, and U. Schöllkopf, *Huaxue Xuebao*, 1984, **42**, 86 (*Chem. Abstr.*, 1984, **100**, 210 372).

[45] U. Schöllkopf, *Pure Appl. Chem.*, 1983, **55**, 1799.

[46] J. N. Fitzner, R. G. Shea, J. E. Fankhauser, and P. B. Hopkins, *J. Org. Chem.*, 1985, **50**, 417.

[47] D. Ring and S. L. Miller, *Origins Life*, 1984, **15**, 7.

[48] A. A. Zharov, G. A. Adadurov, A. G. Kazakevich, V. M. Zulin, and I. I. Zhukuvleva, *Izv. Akad. Nauk S.S.S.R., Ser. Khim.*, 1984, 1199.

[49] Ch. Ivano, O. Ivanov, R. Simeonova, and G. Mirkova, *Origins Life*, 1983, **13**, 97.

Synthesis of Protein Amino Acids and Other Naturally Occurring Amino Acids. – It would be inappropriate to ignore the burgeoning literature, but necessary to give representative citations, only, of fermentative production of amino acids. Enzymic synthesis of amino acids from β-chloroalanine has been reviewed;[50] biosynthetic studies include formation of L-isoleucine by methanogenic bacteria,[51] enhanced L-proline production from L-glutamic acid by a barley mutant,[52] and conversion of L-aspartic acid into L-alanine.[53] The production of L-dopa (*Mucuna pruriens*)[54] and L-tryptophan[55] has been given detailed attention.

Protein amino acids are frequently the objective of exploratory studies with new or modified general syntheses, and examples of this type appear elsewhere in this chapter. The protein amino acids themselves are starting points for the synthesis of other natural products (see Section 6) including amino acids. The starting protein L-amino acid may appear as such in the synthetic target, as in the synthesis of histopine (5) *via* reductive alkylation of L-histidine with pyruvic acid in the presence of $NaBH_3CN$ and separation of the resulting mixture of diastereoisomers.[56] An alternative synthesis based on the established route to this general class of crown-gall tumour metabolites, in this case using L-histidine and (*R*)- or (*S*)-α-bromopropionic acid, was also explored in this study.[56]

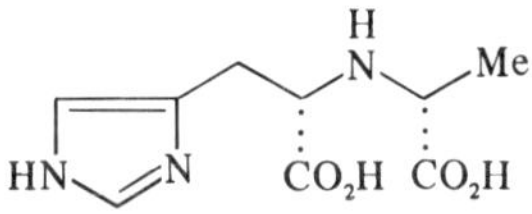

(5)

L-Glutamic acid was the starting point for differently conceived syntheses of N^{δ}-hydroxy-L-ornithine,[57,58] both sketched in Scheme 2. In one of these studies[57] alternative approaches were thwarted by the propensity of the urethane nitrogen atom in N^{α}-Boc-L-glutamic semialdehyde to undergo intramolecular reaction and also by transamidation rearrangements that occurred on attempted reduction of the side-chain carboxy group in certain glutamic acid α-hydroxamate derivatives. However, when the nitrogen atom is enclosed in an oxazolidone ring, this problem is avoided.[58] L-Serine has been used for the synthesis of L-2,3-diaminopimelic acid through application of the Mitsunobo reaction (Z-Ser-OMe with Ph_3P and diethyl or di-isopropyl azodicarboxylate to give the corresponding α-azido-alanine, subjected to H_2S–py reduction).[59] The

[50] T. Nagasawa and H. Yamada, *Kagaku, Zokan (Tokyo)*, 1984, 107.
[51] I. Ekiel, I. C. P. Smith, and G. D. Sprott, *Biochemistry*, 1984, **23**, 1683.
[52] J. S. H. Kueh, J. M. Hill, S. J. Smith, and S. W. J. Bright, *Phytochemistry*, 1984, **23**, 2207.
[53] M. C. Fusee and J. E. Weber, *Appl. Environ. Microbiol.*, 1984, **48**, 694.
[54] H. J. Huizing and H. J. Wichers, *Prog. Ind. Microbiol.*, 1984, **20**, 217.
[55] S. Takao, A. Yokota, and M. Tanida, *J. Ferment. Technol.*, 1984, **62**, 329.
[56] H. A. Bates, A. Kaushal, P. N. Deng, and D. Sciaky, *Biochemistry*, 1984, **23**, 3287.
[57] R. K. Olsen, K. Ramasamy, and T. Emery, *J. Org. Chem.*, 1984, **49**, 3527.
[58] (*a*) G. Benz, *Liebigs Ann. Chem.*, 1984, 1424; (*b*) B. H. Lee and M. J. Miller, *Tetrahedron Lett.*, 1984, **25**, 927.
[59] B. T. Golding and C. Howes, *J. Chem. Res. (S)*, 1984, 1.

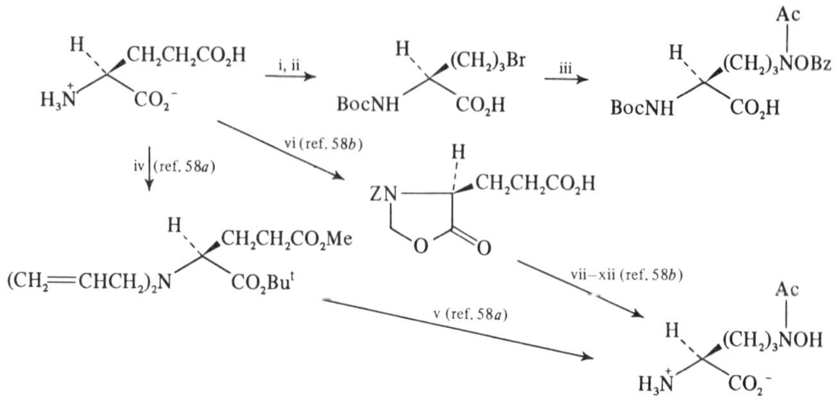

Reagents: $BocN_3$; ii, established route; iii, *O*-benzyl acetohydroxamate, NaH; iv, esterify, CH_2=$CHCH_2Br$; v, $(Ph_3P)_3RhCl$, CF_3CO_2H, hydrogenolysis; vi, ZCl, CH_2O/TsOH; vii, $SOCl_2$; viii, Bu_3SnH or $Li(OBu^t)_3AlH$; ix, $PhCH_2ONH_2$; x, $NaBH_3CN$; xi, OH^-; xii, HBr

Scheme 2

conversion of L-serine into D-α-amino acids[60] involves *N*-benzenesulphonyl-L-serine lithium salt for aminoacylation of a Grignard reagent, the resulting side-chain carbonyl group being converted into a methylene group through Raney nickel reduction of the derived dithioketal; oxidation ($-CH_2OH \rightarrow -CO_2H$) was achieved using O_2/PtO_2, leading to excellent yields of D-amino acids $H_3\overset{+}{N}CH(CH_2R)CO_2HBr^-$ after cleavage of the N-protecting group with 48% HBr.[60]

Another example in which a chiral natural product, this time (*R*,*R*)-tartaric acid, serves as starting material for a natural amino acid is the 25-stage synthesis of (2*S*,3*R*,4*R*,6*E*)-3-hydroxy-4-methylamino-6-octenoic acid (a constituent of cyclosporin A).[61]

trans-α-(Carboxycyclopropyl)glycine, a constituent of ackee seed, has been prepared through cyclopropanation of (*E*)-EtO_2CCH=$CHCH(OEt)_2$, conventional amino acid synthesis through the masked aldehyde group involving the Strecker route.[62] Stammer's group (see Vol. 15, p. 12) continue their studies on synthesis of cyclopropane-based amino acids by diazoalkane cyclopropanation of amino-acrylates with examples including coronamic acid.[63]

Synthesis of β- and Higher Homologous Amino Acids. – The large range of examples, many of them represented in natural products, that are covered by the title of this section is matched by a constant stream of papers. There are relatively few general methods specific to each class of ω-amino acid, and text-book methods of synthesis of amines are used, needing little particular discussion.

[60] P. J. Mauer, H. Takahata, and H. Rapoport, *J. Am. Chem. Soc.*, 1984, **106**, 1095.
[61] R. M. Wenger, *Helv. Chim. Acta*, 1983, **66**, 2308.
[62] S. R. Landor, P. D. Landor, and M. Kalli, *J. Chem. Soc., Perkin Trans. 1*, 1983, 2921.
[63] J. M. Bland, C. H. Stammer, and K. I. Varughese, *J. Org. Chem.*, 1984, **49**, 1634; M. Suzuki, E. E. Gooch, and C. H. Stammer, *Tetrahedron Lett.*, 1983, **24**, 3839.

3-Ketoglutaric acid is converted through some conventional steps but including a notable use of *Arthrobacter* for stereoselective formation of ethyl (*S*)-3-hydroxyglutarate, into either L-carnitine, [$Me_3\overset{+}{N}CH_2CH(OH)CH_2CO_2^-$], or the 4-amino-3-hydroxybutanoic acid itself.[64] Other γ-amino acids reached through enantiospecific synthesis are (3*S*,4*S*)-statine,[65] as its *N*-Boc ester, $Me_2CHCH_2CH(NHBoc)CH(OH)CH_2CO_2Me$, starting with *N*-Boc-L-leucinal, and (−)-gabaculine (6), starting with benzoic acid and including a notable role for the $Fe(CO)_3$ moiety in enabling enantiospecific introduction of ^2H as well as the correct location of the amino group.[66] Detoxinine has been synthesized from (*S*)-$CH_2{=}CHCH_2CH(\overset{+}{N}H_3)CO_2^-$ through a stereoselective route[67] that competes with an alternative route described in Vol. 16, p. 10. The synthesis of (+)-galantinic acid (7) starting with (*R*)-$CH_2{=}CHCH(\overset{+}{N}H_3)CO_2^-$ mimics the detoxinine synthesis in some respects, involving stereospecific epoxidation and regiospecific ring cleavage with $Li_2(CN)Cu(CH{=}CHCH_2OSiMe_2CMe_3)_2$.[67,68] L-Glutamic acid serves as starting material for (*S*)-4-amino-4,5-dihydrofuran-2-carboxylic acid, found to be a potent γ-aminobutyric acid transaminase inhibitor;[69] so also is (*S*)-4-amino-5-hexenoic acid, prepared from (*S*)-5-vinyl-2-pyrrollidone, available from L-glutamic acid through straightforward elaboration.[70]

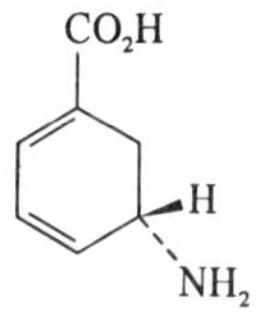

(6)

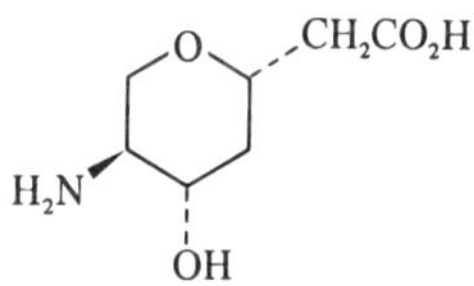

(7)

Synthesis of α-Alkyl Analogues of Protein Amino Acids. – Alkylation reactions continue to gain favour for this purpose as reaction procedures become optimized, in comparison with total synthesis by standard general methods employing ketones. In addition to examples described elsewhere in this chapter,[30,75] alkylation of Schiff bases $R^1R^2C{=}NCHR^3CO_2R^4$ is easily accomplished using an alkyl halide in refluxing MeCN in the presence of K_2CO_3 and $Bu_4N^+Br^-$.[71] α-Alkylated L-leucines are accessible through the use of lithiated 2,5-dimethoxy-3,6-diisobutyl-3,6-dihydropyrazine (see also refs. 44 and 45) as chiral substrate;

64 A. S. Gopalan and C. J. Sih, *Tetrahedron Lett.*, 1984, **25**, 5235.

65 B. Rague, J. A. Fehrentz, R. Guegan, Y. Chapleur, and B. Castro, *Bull. Soc. Chim. Fr.*, 1983, 230.

66 B. M. R. Bandara, A. J. Birch, and L. F. Kelly, *J. Org. Chem.*, 1984, **49**, 2496.

67 Y. Ohfune, N. Kurokawa, H. Nishio, and M. Matsunaga, *Tennen Yuki Kagobutsu Toronkai Koen Yoshishu*, 1983, 500 (*Chem. Abstr.*, 1984, **100**, 210 365).

68 Y. Ohfune and N. Kurokawa, *Tetrahedron Lett.*, 1984, **25**, 1587.

69 J. P. Burkhart, G. W. Hobert, and B. W. Metcalf, *Tetrahedron Lett.*, 1984, **25**, 5267.

70 W. Frieben and F. Gerhart, Brit. U.K. Pat. Appl. GB 2,133,002 (*Chem. Abstr.*, 1984, **101**, 231 027).

71 M. J. O'Donnell, K. Wojciechowski, L. Ghosez, M. Navarro, F. Sainte, and J. P. Antoine, *Synthesis*, 1984, 313.

alkylation is followed by hydrolysis to give the α-alkyl-leucine methyl ester.[72] Similar use of the chiral oxazinone formed between DL-2-(2-furyl)glycine and (*S*)-$Pr^iCH(OAc)COCl$ or (*S*)-$Bu^tCH(OAc)COCl$ after conversion into its potassium salt leads to (*S*)-α-alkyl-α-(2-furyl)glycines with asymmetric induction levels of 50–95%.[73]

Synthesis of Other Aliphatic, Alicyclic, and Saturated Heterocyclic α-Amino Acids. – The use of 1,2-diaminoethane in the Strecker synthesis yields 'bis-α-amino acids' *via* the amino nitrile:[74]

$$EtCHO + KCN + H_2NCH_2CH_2NH_2 \rightarrow NCCHEtNHCH_2CH_2NHCHEtCN$$

Hydrolysis to the amino acid succeeded only after benzoylation of both secondary amino groups. No such problem arose in the hydrolysis of the nitrile groups in alkylation products of the Schiff base $Ph_2C{=}NCH_2CN$ in a study of bis-alkylation by 1,ω-dibromoalkanes $Br(CH_2)_nBr$ leading to 1-aminocyclopropane-1-carboxylic acid and 'cycloleucine' ($n = 4$) and to 2,6-diaminopimelic acid ($n = 3$).[75]

The acetamidomalonate synthesis in its half-nitrile version has been used in the synthesis of tetaine analogues, using 3-bromomethylenecyclohexene as alkylating agent.[76] β-(2,3-Epoxycyclohexyl)alanine was formed from the hydrolysis–decarboxylation product by standard methods.

LL-3-Amino-2-piperidine-6-carboxylate (9) has been obtained in better than 99.5% chiral efficiency in an intramolecular alkylation of 2-bromoethyl-3,6-dihydropyrazine (8) followed by acid hydrolysis.[77] This astonishing selectivity may be restricted to this particular example, since higher ω-bromoalkyl homologues would presumably react with competing inter- and intra-molecular alkylation pathways.

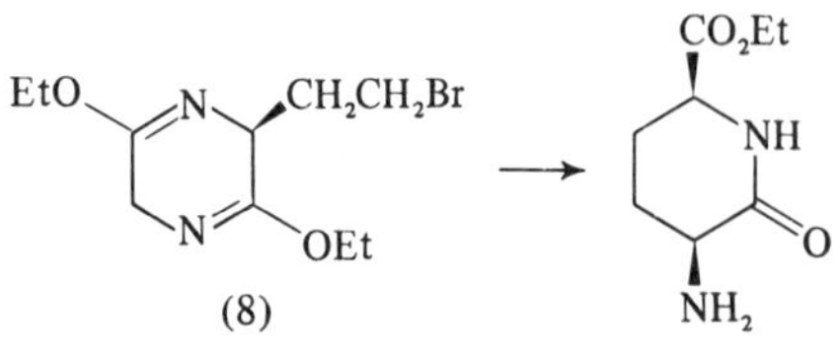

(8) (9)

Ring opening of lactones (10) derived from 2-amino-5-oxoalkanoic acids by treatment with NH_3 is followed by closure to give pyrrolines (11) after removal of the N-protecting group; further steps lead to *cis*-5-alkylprolines in high optical

[72] U. Schöllkopf, U. Busse, R. Kilger, and P. Lehr, *Synthesis*, 1984, 271.

[73] U. Schöllkopf and R. Scheuer, *Liebigs Ann. Chem.*, 1984, 939.

[74] R. Iyer, S. S. Lawate, and M. S. Sonaseth, *Indian J. Chem., Sect. B*, 1984, **23**, 3 (*Chem. Abstr.*, 1984, **101**, 192 413).

[75] M. J. O'Donnell, W. A. Brude, T. M. Eckrich, D. F. Shullenberger, and G. S. Staten, *Synthesis*, 1984, 127.

[76] M. Smulkoski, M. Dzieduszycka, and E. Borowski, *Pol. J. Chem.*, 1982, **56**, 699 (*Chem. Abstr.*, 1984, **100**, 86 098).

[77] D. S. Kemp and P. E. McNamara, *J. Org. Chem.*, 1984, **49**, 2286.

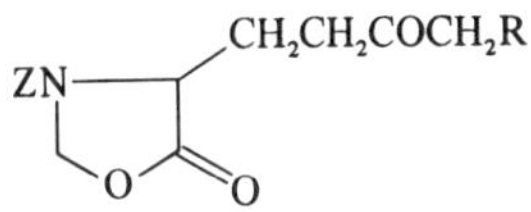

(10)

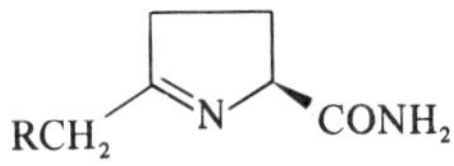

(11)

yield.[78] The use of one amino acid to synthesize another is also the basis of a route to long-chain α-amino acids starting from L-$BrCH_2CH_2CH(NHBoc)CO_2Me$; reaction with an organocuprate R_2CuLi involves no racemization[79] (see also ref. 300).

Synthesis of ω-Alkoxy-α-Amino Acids. – Anodic methoxylation of methyl *N*-formylprolinate in MeOH at a Pt anode gives the corresponding 5-methoxy compound. The group introduced in this way can be substituted by nucleophiles (dimethyl malonate, or 1,3,5-trimethylbenzene in the presence of $AlCl_3$) to give 5-substituted prolines.[80]

Synthesis of Halogenoalkyl Amino Acids. – Introduction of fluorine atoms into simple amino acids continues to offer considerable chemical interest, and the products are important as potential enzyme inhibitors. Replacement of the hydroxy group of serine using SF_4 has hitherto been difficult to perform in satisfactory yields, now[81] seen to be due to competition by a reaction in which serine and SF_4 combine in a 2:1 ratio. High dilution then became the simple cure for the problem. The other papers selected for review here deal with total synthesis by standard methods: Strecker procedure with 3-fluoro-2-hydroxynitriles and amines *via* 2-amino-3-fluoronitriles (inversion of configuration is notable)[82] and reductive amination of 3-fluoropyruvates *p*-$RC_6H_4CHFCOCO_2Na$ (to give the *erythro* isomer predominantly).[83]

Synthesis of Aliphatic α-Amino Acids Carrying Side-chain Hydroxy Groups. – The preparation of the simplest sort under this heading, the aldol-type addition reaction of an aldehyde (or reaction with an acyl chloride, then $NaBH_4$ reduction) with a glycine derivative, continues to provide points of interest. Lithium enolates of di-*N*-benzylglycine esters yield *threo*-β-hydroxy-α-amino acid derivatives $(PhCH_2)_2NCH(CHR^1OH)CO_2R^2$ with remarkably high diastereoselectivity.[84]

[78] T. L. Ho, B. Gopalan, and J. J. Nestor in 'Peptides: Structure and Function', Proceedings of the 8th American Peptide Symposium, ed. V. J. Hruby and D. H. Rich, Pierce Chemical Company, Rockford, Illinois, U.S.A., 1983, p. 147.

[79] J. A. Bajgrowicz, A. El Hallaoui, R. Jacquier, C. Pigiere, and P. Viallefont, *Tetrahedron Lett.*, 1984, **25**, 2231.

[80] M. Malmberg and K. Nyberg, *Acta Chem. Scand., Ser. B*, 1984, **38**, 85.

[81] A. W. Douglas and P. J. Reider, *Tetrahedron Lett.*, 1984, **25**, 2851.

[82] A. I. Ayi and R. Guedj, *J. Fluorine Chem.*, 1984, **24**, 137.

[83] T. Tsushima, K. Kawada, J. Nishikawa, T. Sato, K. Tori, T. Tsuji, and S. Misaki, *J. Org. Chem.*, 1984, **49**, 1163.

[84] G. Guanti, L. Banfi, E. Narisano, and C. Scolastico, *Tetrahedron Lett.*, 1984, **25**, 4693.

Hydroxyglutamic acid derivatives have been prepared from glutamic acid itself (*threo*-4-hydroxy-D-glutamic acid, from *N*-phthaloyl 4-bromo-D-glutamic acid dimethyl ester, showed high glutamine synthetase-inhibiting activity[85]) and from L-serine (*threo*-β-hydroxy-L-glutamic acid) starting with a stereoselective addition of the serinal derivative (12) to 1-methoxy-3-(trimethylsilyloxy)-1,3-butadiene (Scheme 3).[86]

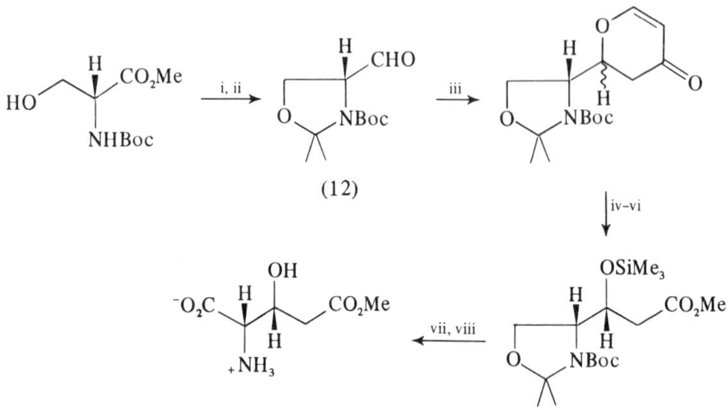

Reagents: i, DMP, TsOH; ii, di-isobutylaluminium hydride, −78 °C; iii, MeOCH=CHC$(OSiMe_3)$=CH_2; iv, $NaIO_4$–RuO_2; v, Et_2NSiMe_3; vi, MeOH, H^+; vii, $KMnO_4$; viii, $H_3O^+Cl^-$

Scheme 3

Synthesis of Aliphatic α-Amino Acids with Unsaturated Side Chains. – This section covers the logical progression from 'dehydroamino acids' (alias 2-amino-2-alkenoic acids) to vinyl and allenic amino acids, and analogues of well known α-amino acids in which an unsaturated grouping has been introduced.

Serine Schiff bases ArCH=NCH(CH_2OH)CO_2Me yield imidazolylcarbonyl derivatives with carbonyldi-imidazole, which suffer elimination in the presence of Et_3N to give the corresponding dehydroalanine Schiff bases.[87] Oxidative β-decarboxylation of *N*-acyl aspartate α-esters by sodium hypochlorite gives the corresponding *N*-acyl dehydroalanines through *N*-chlorination followed by dehydrochlorination.[88] α-Azidocarboxylic esters treated with Ac_2O–Re_2S_7 give *N*-mono- or -di-acetyl dehydroamino acids,[89] while Ac_2O–$NaReO_4$ at 90 °C was appropriate for the similar conversion of α-azidocarboxylic acid amides.[90] Direct introduction of an alkylidene group into a glycine derivative is a way of

[85] V. P. Krasnov, L. V. Alekseeva, N. A. Firsova, I. K. Kodess, and N. L. Burde, *Khim.-Farm. Zh.*, 1984, **18**, 655.

[86] P. Garner, *Tetrahedron Lett.*, 1984, **25**, 5855.

[87] G. Wulff and H. Boehnke, *Angew. Chem.*, 1984, **96**, 362.

[88] M. Seki, T. Moriya, and K. Matsumoto, *Agric. Biol. Chem.*, 1984, **48**, 1251.

[89] F. Effenberger and T. Beisswenger, *Chem. Ber.*, 1984, **117**, 1497.

[90] T. Beisswenger and F. Effenberger, *Chem. Ber.*, 1984, **117**, 1513.

describing the well known azlactone synthesis of amino acids, referred to in an earlier section of this chapter; a novel variant of the approach from a substituted glycine with an aldehyde has been described, employing $(R^3O)_2P(O)CH(NHR^1)$-COR^2, a series of *N*-acyl-2-(dialkyloxyphosphinyl)glycine esters and amides.[91] Condensation of carbonyl compounds with ethyl isocyanoacetate gives 'dehydro-isocyano acids'.[92]

Vinylglycine continues to challenge the application of synthetic methodology and has become available in racemic form from (*Z*)-$HOCH_2CH{=}CHCH_2OH$ through imidation with CCl_3CN followed by [2,3]-sigmatropic rearrangement in refluxing t-butylbenzene then oxidation ($CH_2OH \rightarrow CO_2H$) and hydrolysis[93] (see also ref. 33). The L-enantiomer has been synthesized from *N*-benzyloxycarbonyl-L-glutamic acid α-methyl ester through $Cu(OAc)_2$-catalysed decarboxylative elimination by lead tetra-acetate, followed by deprotection.[94] β-Methyleneaspartic acid has been obtained from 1,1,2-tricarbethoxyprop-1-ene (from diethyl malonate and ethyl pyruvate) through amination with chloramine and ester hydrolysis accompanied by decarboxylation.[95]

α-Allenic amino acids are potent inhibitors of bacterial tyrosine decarboxylase,[96] mammalian 4-aminobutyrate–2-oxoglutarate aminotransferase, and bacterial ornithine decarboxylase.[97] As such, this class of protein amino acid analogue is certain to be the subject of more attention, and serviceable syntheses have been established, either through Claisen rearrangement of α-benzamido-propargylic esters into 2-phenyl-4-allenic oxazolinones followed by methanolysis and hydrolysis[96,97] or from α-acetylenic amino acids (see Vol. 16, p. 14) with formaldehyde in the presence of CuBr and di-isopropylamine.[97]

Synthesis of Aromatic and Heteroaromatic Amino Acids. – The usual theme for this section, synthesis of close analogues of a number of natural aromatic and heteroaromatic α-amino acids, applies again this year.

Standard approaches have been successful for the synthesis of β-(4-aminophenyl)alanine[98] (azlactone synthesis *via* *p*-nitrobenzylidene-oxazolinones), methyldopa[99] (veratraldehyde from vanillin, thence to 3,4-dimethoxyphenylacetone *via* Darzens condensation), and arsanilazo and sulphanilazo derivatives of tyrosine and histidine[100] (electrophilic substitution of the *N*-acetyl amino acids by diazotized arsanilic and sulphanilic acids). *p*-Boronophenylalanine, 4-$(HO)_2$-$BC_6H_4CH_2CH(\overset{+}{N}H_3)CO_2^-$, is accessible through a similar general approach.[101]

[91] U. Schmidt, A. Lieberknecht, and J. Wild, *Synthesis*, 1984, 53.
[92] C. Herdeis and A. Dimmerling, *Arch. Pharm. (Weinheim, Ger.)*, 1984, **317**, 86.
[93] D. M. Vyas, Y. Chiang, and T. W. Doyle, *J. Org. Chem.*, 1984, **49**, 2037.
[94] S. Hanessian and S. P. Sahoo, *Tetrahedron Lett.*, 1984, **25**, 1425.
[95] E. S. Hand and D. C. Baker, *Int. J. Pept. Protein Res.*, 1984, **23**, 420.
[96] A. L. Castelhano, D. H. Pliura, G. J. Taylor, K. C. Hsieh, and A. Krantz, *J. Am. Chem. Soc.*, 1984, **106**, 2734.
[97] P. Casara, K. Jund, and P. Bey, *Tetrahedron Lett.*, 1984, **25**, 1891.
[98] Y. Yuan and C. Yang, *Dalian Gongxueynan Xuebao*, 1984, **23**, 135.
[99] R. Du, X. Luo, and B. Guo, *Jinan Liyi Xuebao*, 1984, **52** (*Chem. Abstr.*, 1984, **101**, 152 292).
[100] G. J. Pielak, M. S. Urdea, K. Igi, and J. I. Legg, *Biochemistry*, 1984, **23**, 589.
[101] T. Hamada, K. Aoki, T. Kobayashi, and K. Kanda, *Annu. Rep. Res. React. Inst., Kyoto Univ.*, 1983, **16**, 112.

Hydriodic acid degradation of the red-brown polymeric pigment phaeomelanin gives (4′-amino-3′-hydroxyphenyl)alanine and (7′-hydroxybenzothiazol-4′-yl)-alanine, amongst other products, and straightforward applications of standard routes to amino acids have been used for their synthesis.[102]

As well as examples in the preceding paragraphs, the heteroaromatic area is represented by the synthesis of substituted tryptophans. 2-(Alkanethio)tryptophans are best obtained[103] *via* 2,3-dihydropyrrolo[2,3-*b*]indoles, easily prepared from L-tryptophan. Electrochemical methoxylation of *N*-methoxycarbonyl-L-proline (see Vol. 16, p. 13) and conventional indolization of the derived pyrrolines into 5-substituted L-tryptophans (after removal of protecting groups, notably removal of the methoxycarbonyl group by Me_3SiI in refluxing chloroform) have been described.[104] 1-Aryl-2-cyanoaziridines are effective electrophiles for reaction at the 3-position of indoles in the presence of BF_3–Et_2O, giving satisfactory yields of nitriles from which *N*-aryltryptophans were obtained.[105]

Synthesis of *N*-Substituted Amino Acids. – This section has a narrower coverage than the title might imply, since protected amino acids for analytical or synthetic use are excluded (some coverage of these appears in the 'Reactions' and 'Analytical Methods' sections later in this chapter).

Alkylation of the Schiff bases $4\text{-}R^3C_6H_4CH{=}NCR^1R^2CO_2Me$ or the amidines $Me_2NCH{=}NCR^1R^2CO_2Me$ by Me_2SO_4, Et_2SO_4, or methyl triflate occurs without racemization.[106] More conventional alkylation procedures have been used for the preparation of 'amino diacids' ($R^5O_2CCR^1R^2NH_2 + BrCR^3R^4CO_2Et \rightarrow R^5O_2CCR^1R^2NHCR^3R^4CO_2Et$)[107] and similar compounds (see earlier sections). The racemization-free route from *N*-benzyloxycarbonylamino acids to *N*-methylamino acids (hydrogenation of lactones formed with formaldehyde) has been applied for the synthesis of *N*-methyl-L-phenylalanine.[108]

The vaguely named *N*-methyl analogue of *S*-adenosylmethionine[109] (*i.e.* MeN in place of S^+Me) has been synthesized from 2′,3′-*O*-(1-methylethylidene)adenosine 5′-tosylate, $MeNH_2$, and $PhCH_2O_2CCH(N_3)CH_2CH_2I$, the product being elaborated by standard methods.

Synthesis of Amino Acids Containing Sulphur or Selenium. – Oxidation of penicillamine with H_2O_2 yields D-ββ-dimethylcysteic acid.[110] The other non-routine examples of syntheses of sulphur-containing amino acids also start with

[102] D. G. Patil and M. R. Chedekel, *J. Org. Chem.*, 1984, **49**, 997.
[103] M. Ohno, S. Tanaka, T. C. Shieh, and T. F. Spande, *J. Org. Chem.*, 1984, **49**, 5069.
[104] K. Irie, A. Ishida, T. Nakamura, and T. Oh-Ishi, *Chem. Pharm. Bull.*, 1984, **32**, 2126.
[105] S. Apparao, G. Singh, H. Ila, and H. Junjappa, *Indian J. Chem., Sect. B*, 1984, **23**, 15 (*Chem. Abstr.*, 1984, **101**, 171 670).
[106] M. J. O'Donnell, W. A. Bruder, B. W. Daugherty, D. Liu, and K. Wojciechowski, *Tetrahedron Lett.*, 1984, **25**, 3651.
[107] B. Garrigues, *Tetrahedron*, 1984, **40**, 1151.
[108] G. Cipens, V. A. Slavinskaya, D. Sile, V. D. Grigorova, D. Kneile, and D. Eglite, *Latv. P.S.R. Zinat. Akad. Vestis, Kim. Ser.*, 1984, 620 (*Chem. Abstr.*, 1984, **102**, 79 293).
[109] M. Davis, N. P. B. Dudman, and H. F. White, *Aust. J. Chem.*, 1983, **36**, 1623.
[110] A. Calvo, R. Faggiani, D. A. Harvey, H. E. Howard-Lock, W. F. Kean, and C. J. L. Lock, *J. Crystallogr. Spectrosc. Res.*, 1984, **14**, 59.

homocysteine sodium salt and a 5′-chloro-5′-deoxynucleoside (syntheses of *S*-adenosyl-L-homocystine and related compounds, including a first synthesis of N^6,N^6-dimethyladenosyl-L-homocysteine[111]) or with the deoxynucleoside itself, condensation to various *S*-nucleosidylhomocysteines being catalysed by *S*-adenosylhomocysteine hydrolase in intact *Alcaligenes faecalis* cells.[112]

Preparations of *m*- and *p*-methylselenylphenylalanines have been reported,[113] the starting point being the corresponding cyanoselenylbenzyl acetamidomalonates (reaction with MeMgBr and hydrolysis in refluxing 12M HCl).

Amino Acids Synthesized for the First Time. – This section is to be taken with the chapter as a whole by readers seeking coverage of newly synthesized amino acids. Amino acids of this description are merely listed here: DL-2-amino-4-(*o*-chlorophenoxy)butanoic acid,[114] N^6-monoethyl-L-arginine,[115] N^α-alkyl-N^β-phenylcarbamyl-DL-asparagines,[116] (4-bis-chloroethylaminophenyl)alanine,[26] β-(3-pyridyl)-L-alanine,[151] β-ruthenocenylalanine,[117] *m*-cyclohexyloxy-L-tyrosine,[118] (*E*)-3′-hydroxyiminomethyl-L-tyrosine,[119] *S*-[ω-(*o*-alkoxyphenyl)alkyl]homocysteines,[120] and *N,N,S*-tris(carboxymethyl)-DL-methionine.[203]

Synthesis of Labelled Amino Acids. – This section is organized around the isotopic replacements described, in order of increasing atomic mass. α-Amino acids are covered first.

Deuterium-substituted analogues of protein amino acids can be prepared with secure knowledge of the absolute configuration by straightforward methods; (2*R*,3*R*)-phenylalanine-2,3-$^2H_2$ is available through addition of $^2H_2$ in the presence of Pd to the diketopiperazine of D-alanyldehydrophenylalanine, with better than 98.8% chiral purity.[121] Efficient and specific α-deuteriation of *N*-acetyl-DL-acetylamino acids occurs through equilibration in basic ^2H_2O–Ac_2O at 40 °C; resolution using porcine kidney acylase I gave L-[α-^{2}H]phenylalanine,[122] a compound that could also be prepared using pyridoxal derivatives in basic ^2H_2O.[122]

No problems of selectivity were involved in the synthesis of (*S*)-N^τ-methyl-$^2H_3$-L-histidine by methylation with C^2H_3I of the histidine protected at its

111 K. Ramalingam and R. W. Woodard, *J. Org. Chem.*, 1984, **49**, 1291.
112 S. Shimizu, S. Shiozaki, T. Oshiro, and H. Yamada, *Agric. Biol. Chem.*, 1984, **48**, 1383.
113 C. A. Loeschorn, C. J. Kelley, R. N. Hanson, and M. A. Davis, *Tetrahedron Lett.*, 1984, **25**, 3387.
114 U. Petzold, S. Neumann, F. Jacob, and M. Strube, *Wiss. Beitr. Martin-Luther Univ., Halle-Wittenberg*, 1984, 47 (*Chem. Abstr.*, 1984, **101**, 186 015).
115 Y. B. Cho, G. Furst, and W. K. Paik, *Anal. Biochem.*, 1984, **139**, 377.
116 G. A. Zeinalova, N. S. Kyazimova, and E. A. Nagieva, *Dokl. Akad. Nauk Az.S.S.R.*, 1983, **39**, 57 (*Chem. Abstr.*, 1984, **101**, 131 050).
117 W. H. Soine, C. E. Guyer, and F. F. Knapp, *J. Med. Chem.*, 1984, **27**, 803.
118 E. Giralt, D. Andreu, R. Eritja, A. Grandas, and E. Pedroso, *An. Quim., Ser. C*, 1983, **79**, 390.
119 Z. Arnold, *Pol. J. Chem.*, 1983, **56**, 1021.
120 O. W. Lever, C. Hyman, and H. L. White, *J. Pharm. Sci.*, 1984, **73**, 1241.
121 K. Tanimura, T. Kato, M. Waki, S. Lee, Y. Kodera, and N. Izumiya, *Bull. Chem. Soc. Jpn.*, 1984, **57**, 2193.
122 H. Fujihara and R. L. Schowen, *J. Org. Chem.*, 1984, **49**, 2819.

N^{π} nitrogen atom by cyclocondensation.[123] Equilibration of phenylalanine and its *p*-OH, -OMe, -OEt, and -NO_2 analogues in ^2H_2O–K_2PtCl_4 brought about exchange of aromatic and CH_3 protons preferentially.[124]

^2H- and ^{13}C-labelled prolines have been described,[125] the [4,4-$^2H_2$]imino acid from ethyl acetamidocyanoacetate and $BrCH_2C^2H_2CH_2Br$ and the [4-^{13}C] analogue in a corresponding manner.

^{11}C-Labelled amino acids continue to excite interest, not only because of their value in imaging studies but also in the need for rapid syntheses, bearing in mind the relatively short half-life of this isotope. A total synthesis time of 46 min is reported for [1-^{11}C]dopa prepared by carboxylation of 3,4-$(MeO)_2$-$C_6H_3CH_2CHLiNC$, including 16 min for h.p.l.c. resolution.[126] [3-^{11}C]Phenylalanine and [3-^{11}C]dopa in racemic forms have been prepared through the azlactone route,[127] and an alternative route to [3-^{11}C]phenylalanine[128] uses $Ph^{11}CH_2Cl$ for the alkylation of $Ph_2C{=}NCH_2CO_2Et$. Similar alkylations [of $Me_2NCH{=}NCH(Pr^i)CO_2Me$] with $^{11}CH_3I$ gave DL-α-[^{11}C]methyl valine[129] and DL-α-[^{11}C]methyl ornithine [by methylation of the N^{α}-(*p*-nitrobenzylidene)-ornithine lactam].[129]

Strecker syntheses of aspartic acid-1-^{13}C and glutamic acid-1-^{13}C (EtO_2CCH_2-CHO and $EtO_2CCH_2CH_2CHO$, respectively, with $K^{13}CN/NH_4OH/NH_4Cl$) proceed in 45% and 85% yields, respectively.[130] Glutamic acid-3-^{13}C was prepared from $PrO_2CCH_2{}^{13}CH_2Br$ through the acetamidomalonate route.[130]

DL-[2-^{13}C,^{15}N](*p*-Hydroxyphenyl)glycine, for use in nocardicin A biosynthesis studies, has been synthesized through Vilsmeier reaction of anisole with [^{13}C]-dimethylformamide and then the use of the resulting benzaldehyde with $^{15}NH_4Cl$, NaCN, and MeOH at 40 °C in the Strecker procedure.[131] DL-Glutamine-2,5-$^{15}N_2$ has been prepared from α-ketoglutaric acid through reaction with $^{15}N_2H_4$ and hydrogenation of the resulting cyclic hydrazone in the presence of Pd. The synthesis of ^{15}N-amino acids has been reviewed briefly.[133]

6-[^{18}F]Fluoro-L-dopa is available, though in low yields, through reaction of $AcO^{18}F$ with 3-methoxy-4-hydroxy-L-phenylalanine ethyl ester hydrochloride followed by removal of protecting groups.[134]

123 S. S. Yuan and A. M. Ajami, *J. Labelled Compd. Radiopharm.*, 1984, **21**, 97.
124 M. Kanska, *J. Radioanal. Nucl. Chem.*, 1984, **87**, 95.
125 P. E. Young and D. A. Torchia in 'Peptides: Structure and Function', Proceedings of the 8th American Peptide Symposium, ed. V. J. Hruby and D. H. Rich, Pierce Chemical Company, Rockford, Illinois, U.S.A., 1983, p. 155.
126 J. M. Bolster, W. Vaalburg, W. Van Veen, T. Van Dijk, H. D. Van der Molen, H. Wynberg, and M. G. Woldring, *Int. J. Appl. Radiat. Isot.*, 1983, **34**, 1650.
127 C. Halldin and B. Laangstroem, *Int. J. Appl. Radiat. Isot.*, 1984, **35**, 779.
128 M. R. Kilbourn, D. D. Dischino, and M. J. Welch, *Int. J. Appl. Radiat. Isot.*, 1984, **35**, 603.
129 F. Oberdorfer, *Int. J. Appl. Radiat. Isot.*, 1984, **35**, 559.
130 U. Fotader and D. Cowburn, *J. Labelled Compd. Radiopharm.*, 1983, **20**, 1003.
131 C. A. Townsend and G. M. Salituro, *J. Chem. Soc., Chem. Commun.*, 1984, 1631.
132 W. M. Lagna and P. S. Callery, *J. Labelled Compd. Radiopharm.*, 1984, **21**, 337.
133 H. Engelmann, A. Dauert, H. Niclas, and H. Lueneberger, *Zfl. Mitt.*, 1983, **77**, 92 (*Chem. Abstr.*, 1984, **101**, 103 826).
134 R. Chirakal, G. Firnau, J. Couse, and E. S. Garnett, *Int. J. Appl. Radiat. Isot.*, 1984, **35**, 651.

Chirally deuteriated (*R*)-*N*-trifluoroacetyl-β-alanine-β-^{2}H has been prepared from (*S*)-$PhCH^2HCH_2OH$ *via* RuO_4 oxidation of the derived (*S*)-$PhCH^2HCH_2$-$NHCOCF_3$.[135]

Resolution of Amino Acids. – This topic has settled into regularly populated subsections, covering the various categories of the established resolution techniques.

Schiff bases formed between DL-amino acids and (+)-(*R*,*R*,*R*)- or (−)-(*S*,*S*,*S*)-2-hydroxypinan-3-one can be separated cleanly by column chromatography and converted into the amino acids by mild acid hydrolysis.[136] Alternative chiral derivatization procedures have been explored, in particular the use of 1-(4-substituted phenylsulphonyl)-L-prolyl chlorides, in conjunction with liquid-chromatographic separation.[137]

Chromatographic separation of DL-amino acids over a chiral stationary phase offers a well studied variation, and a new study of the use of cellulose chromatography for the purpose adds to a lengthy list of papers describing this method. Resolution of representative amino acids (seventeen examples) is fully described in this report.[138] Man-made rather than natural chiral solids feature in resolutions of *threo*- and *erythro*-β-hydroxyaspartic acids (L-lysine within the stationary phase gives optimal separation of enantiomers for the *threo* diastereoisomer whereas L-ornithine is more appropriate for the *erythro* isomer).[139] Bonding of an L-amino acid (best results with L-pipecolic acid) to silica gel can be accomplished by first derivatizing the silica with 3-glycidoxypropyltrimethoxysilane, as the basis for resolution of DL-amino acids on the ligand-exchange chromatography principle.[140] A very similar approach, the bonding of *N*-formyl-L-isoleucine or -L-valine to aminopropylated silica gel, has also proved effective in liquid-chromatographic resolution.[141] α-(6,7-Dimethyl-1-naphthyl)isobutylamine bonded to a low-polarity stationary phase proved satisfactory for resolution of *N*-3,4-dinitrobenzoyl-DL-amino acids and other bifunctional amines.[142]

Incorporation of a chiral solute into an otherwise conventional liquid-chromatographic system is an established variant of the principle just reviewed, but the first instance of enantioselectivity based on energy differences of intermolecular hydrogen bonds between the chiral additive and the respective enantiomers has now been claimed.[143] In this case, *N*-acetyl-L-valine t-butylamide in $CHCl_3$–n-hexane was an effective chiral mobile phase for the resolution of *N*-acetyl-DL-amino acid t-butyl esters. The complexation principle with a

135 R. K. Hill, S. R. Prakash, and T. M. Zydowsky, *J. Org. Chem.*, 1984, **49**, 1666.
136 J. A. Bajgrowicz, B. Cossec, C. Pigiere, R. Jacquier, and P. Viallefont, *Tetrahedron Lett.*, 1984, **25**, 1789.
137 C. R. Clark and J. M. Barksdale, *Anal. Chem.*, 1984, **56**, 958.
138 S. Yuasa, M. Itoh, and A. Shimada, *J. Chromatogr. Sci.*, 1984, **22**, 288.
139 S. Anpeiji, Y. Toritani, K. Kawada, S. Kondo, S. Murai, H. Okai, H. Yoshida, and H. Imai, *Bull. Chem. Soc. Jpn.*, 1983, **56**, 2994.
140 G. Guebitz, F. Juffmann, and W. Jennenz, *Chromatographia*, 1982, **16**, 103.
141 J. N. Akanya, S. M. Hitchen, and D. R. Taylor, *Chromatographia*, 1982, **16**, 224.
142 W. H. Pirkle and M. H. Hyun, *J. Org. Chem.*, 1984, **49**, 3043.
143 A. Dobashi and S. Hara, *J. Chromatogr.*, 1983, **267**, 11.

copper(II) salt of *N*-n-dodecyl-L-proline as a chiral solute in counter-current liquid–liquid chromatography has proved effective for the clean resolution of DL-isoleucine.[144]

Resolution based on differential solubility of systems formed by each enantiomer with a non-bonded chiral entity has taken interesting forms in recent years, extending the standard techniques based on diastereoisomeric salt formation. However, the success or otherwise of these newer methods is still dependent on the physical nature of the systems; thus, methylammonium salts of *N*-acetyl-DL-phenylalanine cannot be resolved by preferential crystallization (seeding with crystals of the desired enantiomer) whereas ethylammonium and 1,1,3,3-tetramethylbutylammonium homologues are shown to form racemic compounds at melting temperatures (but racemic mixtures at room temperature) and therefore can be resolved by this method.[145] Precipitates formed in aqueous L-phenylalanine and DL-valine, DL-leucine, or DL-isoleucine favour the aliphatic D-amino acids, and optical purities of 84–100% have been determined for the D-amino acids recovered from these 1 : 1 adducts.[146a] A similar approach[146b] employing the ternary (1 : 1 : 1) complex (D-alaninato)(L-isoleucinato)copper(II) led to the resolution of DL-alanine using L-isoleucine and copper(II) acetate in aqueous solution. A further example (the patent literature is not routinely scanned in preparing this chapter) is the resolution of DL-amino acids by diastereoisomeric salt formation with (+)-α-phenylethanesulphonic acid.[147]

The discovery that preferential adsorption of D-amino acids occurs from solutions of DL-amino acids in contact with clay (montmorillonite) has inspired further studies of this type (see Vol. 14, p. 16) with the use of pressure-jump relaxation methods.[148] Arginine that has intercalated the interlamellar layer of montmorillonite undergoes slow hydrolysis into ornithine and urea.[148a] Ion-exchange properties of a magnesium hydrotalcite compound were similarly enantioselective, with L-histidine intercalation being favoured at the expense of its D-isomer.[148b]

Enantioselective occlusion into centrosymmetric crystals of glycine accounts for the results obtained with aqueous solutions of DL-amino acids,[149] on the basis that the crystals float in such a way that only one face is available for occlusion and that addition of a particular enantiomer of a hydrophobic amino acid orients the floating crystals so that the face that occludes amino acids of opposite chirality is exposed to the solution. This curious involvement of an

144 T. Takeuchi, R. Horikawa, and T. Tanimura, *J. Chromatogr.*, 1984, **284**, 285.

145 T. Shiraiwa, H. Miyazaki, A. Ikawa, and H. Kurokawa, *Nippon Kagaku Kaishi*, 1984, 1425; for similar phase-diagram studies with DL-α-phenylglycine sulphate see T. Shiraiwa, A. Ikawa, K. Fujimoto, K. Iwafuji, and H. Kurokawa, *ibid.*, p. 765.

146 (*a*) T. Shiraiwa, A. Ikawa, K. Sakaguchi, and H. Kurokawa, *Chem. Lett.*, 1984, 113; *Bull. Chem. Soc. Jpn.*, 1984, **57**, 2234; (*b*) T. Shiraiwa, H. Fukuoka, M. Yoshida, and H. Kurokawa, *Bull. Chem. Soc. Jpn.*, 1984, **57**, 1675.

147 I. Chibata, S. Yamada, C. Hongo, and R. Yoshioka, Eur. Pat. Appl. EP 119,804 (*Chem. Abstr.*, 1984, **101**, 231 031).

148 (*a*) T. Ikeda and T. Yasunaga, *J. Phys. Chem.*, 1984, **88**, 1253; (*b*) T. Ikeda, H. Amoh, and T. Yasunaga, *J. Am. Chem. Soc.*, 1984, **106**, 5772.

149 I. Weissbuch, L. Addadi, Z. Berkovitch-Yellin, E. Gati, M. Lahav, and L. Lieserowitz, *Nature (London)*, 1984, **310**, 161.

air–water interface has been suggested[149] to account for the generation and amplification of optical activity in an initially racemic system.

Enzymic resolution of DL-amino acids continues to be the method of choice for many workers, particularly methods that allow the recovery of both enantiomers. The use of subtilisin Carlsberg, for example,[150] for the catalysed hydrolysis of (*R*,*S*)-*N*-acetylphenylalanine methyl ester into the (*S*)-*N*-acetylphenylalanine and unchanged (*R*)-enantiomer, and applied similarly[151] in synthesis of D- and L-β-(3-pyridyl)alanine, illustrates this principle. The hydrolysis of L-leucyl-γ-substituted glutamates by leucine aminopeptidase provides an effective means of resolution based on the same principle, with *threo*-γ-methyl- and *threo*-γ-fluoro-glutamates but not with the *erythro* isomers.[152]

Whole-cell studies include the use of *Streptomyces zaomyceticus* [catalysed hydrolysis of *N*-benzyloxycarbonyl-(*p*-hydroxyphenyl)glycine into the unprotected L-amino acids without affecting the D-enantiomer[153]] and of *Erwinia carotovora* immobilized in κ-carrageenan gel for the catalysed hydrolysis of the L-enantiomer of *N*-acetyl-DL-methioninamide.[154]

Absolute Configuration Studies. – While enzyme stereospecificity, as exploited in resolution studies (preceding section), has yielded absolute configurational assignments, classical chemical correlation studies continue to be used routinely. 4-Hydroxy-4-methylglutamic acids that occur naturally have hitherto been assigned the (2*S*,4*R*) and (2*S*,4*S*) configurations – wrongly, it now appears, through combined degradative and enzymic studies.[155] Whereas the response of the natural amino acids to L-amino acid oxidase confirms the (2*S*) assignment, oxidative deamination and decarboxylation using $Ca(OCl)_2$ give (*R*)-(−)- or (*S*)-(+)-citramalic acids of known absolute configuration. Absolute configuration was assigned to (2*S*,4*S*)-hydroxyglutamic acid through the same approach, the resulting L-malic acid being identified by optical rotation and response to L-malic acid dehydrogenase.[155]

5 Physical and Stereochemical Studies of Amino Acids

Crystal Structures of Amino Acids and Their Derivatives. – Some of the reports[83,156–159] briefly listed here combine *X*-ray crystallographic studies with

150 J. M. Roper and D. P. Bauer, *Synthesis*, 1983, 1041.
151 K. Folkers, T. Kubiak, and J. Stepinski, *Int. J. Pept. Protein Res.*, 1984, **24**, 197.
152 S. Bory, J. Dubois, M. Gaudry, A. Marwuet, L. Lacombe, and S. Weinstein, *J. Chem. Soc., Perkin Trans. 1*, 1984, 475.
153 Y. Suhara, S. Itoh, K. Yokose, R. Ninomiya, K. Watanabe, and H. B. Maruyama, *Can. J. Microbiol.*, 1984, **30**, 1301.
154 Y. Nishida, K. Nabe, S. Yamada, and I. Chibita, *Enzyme Microbiol. Technol.*, 1984, **6**, 85.
155 B. Bjerg, O. Olsen, and H. Soerensen, *Acta Chem. Scand., Ser. B*, 1983, **37**, 321.
156 I. Buchanan, M. Minelli, M. T. Ashby, T. J. King, J. H. Enemark, and C. D. Garner, *Inorg. Chem.*, 1984, **23**, 495.
157 D. L. Eng-Wilmot, A. Rahman, J. V. Mendenhall, S. L. Grayson, and D. Van der Helm, *J. Am. Chem. Soc.*, 1984, **106**, 1285.
158 C. P. Huber, P. R. Carey, S. C. Hsi, H. Lee, and A. C. Storr, *J. Am. Chem. Soc.*, 1984, **106**, 8263.
159 G. Valle, G. M. Bonora, and C. Toniolo, *Can. J. Chem.*, 1984, **62**, 2661.

other physical methods and are therefore mentioned again in later sections of this chapter. Further attention has been given (see Vol. 15, p. 20) to three-centre (bifurcated) hydrogen bonds in amino acid crystals involving the $\overset{+}{N}H_3$ group, through assessment of more than 50 structures.[160] A relatively high proportion of three-centre bonds is concluded to be due to a deficiency in the number of functional protons needed to satisfy the normal acceptor co-ordination of the carboxylate oxygens (two per oxygen) and the chloride ions (four per Cl).

Amino acid derivatives that have been subjected to crystal-structure determination include the following: *N*-carboxy-L-valine anhydride,[161] N^{α}-acetyl-L-arginine methyl ester hydrochloride,[162] 1-lithio-3,6-diethoxy-2,5-dimethyl-1,2-dihydropiperazine (the functionalized alanine diketopiperazine derivative of the type used in asymmetric synthesis of α-methyl-α-amino acids; *cf.* refs. 42, 44, 45, and 76),[163] *threo*- and *erythro*-3-fluorophenylalanine,[83] the L-cysteine methyl complex MoO_2(L-Cys-OMe)$_2$,[156] ferric neurosporin–6MeCN (containing the amino acid moiety N^{α}-acetyl-N^{δ}-hydroxy-N^{δ}-[(*R*)-3-hydroxybutyryl]-D-ornithine),[157] *N*-benzoylglycine *S*-ethylthioester and its *N*-(β-phenylpropionyl) analogue (evidence of N···S interaction),[158] and Fmoc-L-alanine and -α-amino-isobutyric acid.[159] In the last-mentioned study, a result emerges from the structural analysis that is relevant to a practical observation: that the unexpected removal of the Fmoc-protecting group by catalytic hydrogenolysis might be explained by the long C(sp^3)—O bond determined for the Fmoc moiety.[159]

Nuclear Magnetic Resonance Spectrometry. – N.m.r. studies of common amino acids themselves mainly concern the pushing back of frontiers in terms of n.m.r. rather than advancing structural knowledge of the amino acids. Solid-state ^{13}C n.m.r. with cross-polarization and magic-angle spinning of alanine and its derivatives reveals specific downfield shifts in methyl resonances due to van der Waals' interactions.[164] Other solid-state studies concern phenylalanine (partial collapse of dipolar and chemical-shift tensors for carbon atoms on and off the C_2 symmetry axis confirms that some sites in one of the crystal modifications permit high-frequency 180° ring flips, or, in other words, loosely packed crystals allow molecular reorientations to occur),[165] and ^2H n.m.r. has been applied to deuteriated α-glycine, to suggest a weak C—H···O hydrogen bond.[166]

Solution studies of more familiar types to regular readers of this *Specialist Periodical Report* deal with ^{13}C n.m.r. demonstration of the adoption by *N*-Boc-sarcosine esters of *cis* and *trans* isomers involving the urethane bond and with a

160 G. A. Jeffrey and J. Mitra, *J. Am. Chem. Soc.*, 1984, **106**, 5546.
161 H. Kanazawa, Y. Ohashi, and Y. Sasada, *Acta Crystallogr., Sect. C*, 1984, **40**, 1094.
162 M. Coll, X. Solans, M. Font-Altaba, and J. A. Subirana, *Int. J. Pept. Protein Res.*, 1984, **23**, 242.
163 D. Seebach, W. Bauer, J. Hansen, T. Laube, W. B. Schweizer, and J. D. Dunitz, *J. Chem. Soc., Chem. Commun.*, 1984, 853.
164 C. F. Brewer, *Eur. J. Biochem.*, 1984, **143**, 363.
165 J. Schaefer, E. O. Stejskal, R. A. McKay, and T. Dixon, *J. Magn. Reson.*, 1984, **57**, 85.
166 C. Mueller, W. Schajor, H. Zimmermann, and U. Haeberlen, *J. Magn. Reson.*, 1984, **56**, 235.

similar behaviour for *N*-glycylsarcosine (no such splitting of the CO carbon resonance is seen in leucine, alanine, and glycine analogues).[167] β-Deuteriated histidine methylamide and *N*-acetylhistidine and its ethyl ester and methylamide of known absolute configuration were used to assign the β-proton resonances for the normal (non-deuteriated) species;[168] lower- and higher-field components of the β-proton resonances are assigned to the pro-*R*- and pro-*S*-protons, respectively, in deuteriated polar solvents and the other way round in non-polar solvents. Rotamer populations about the C^{α}—C^{β} bond depend little on the state of ionization of the basic and acidic groups but greatly on solvent polarity (similar results were found for Phe, Trp, and Tyr derivatives). Stabilizing interactions between a fluorine atom and an NH_2 group are important in determining conformations of β-fluoro-α-amino acids, even overriding other apparently dominating effects, and have been seen in action with *erythro*- and *threo*-3-fluorophenylalanines.[83] A 180° dihedral angle about the C^{α}—C^{β} bond has been established for (3*R*)-^{2}H-D-phenylalanine-α-^{2}H.[121] On the more general question of precise values of vicinal coupling constants (as used to obtain the foregoing conformational information), values of ^{13}C component vicinal coupling constants have been calculated for the three minimum-energy staggered rotamers for the $C^{\alpha}HC^{\beta}H_2$ side chains of amino acids.[169]

Solution studies also lend themselves to some of the more state-of-the-art variants of n.m.r. spectrometric techniques, with longitudinal and transverse ^{1}H relaxation rates for L-histidine in water,[170] ^{2}H relaxation studies to reveal intra- and inter-molecular interactions for simple amino acids in aqueous solution,[171] and ^{13}C spin-lattice relaxation times for serine, threonine, phosphoserine, and phosphothreonine, to indicate that phosphorylation causes only minor motional changes for serine and threonine,[172] and for L-aspartic acid, L-alanine, phosphoserine, and 2-mercapto-L-succinic acid in the presence of paramagnetic metal ions (Cu^{2+} and Mn^{2+}), to indicate sites of co-ordination (Asp-Cu^{2+} involves the amino and β-carboxy groups; Asp-Mn^{2+} involves only the two carboxy groups).[173] ^{13}C spin exchange occurs rapidly in natural-abundance samples of L-isoleucine hydrochloride with magic-angle sample spinning; this important observation allows resonances from carbon atoms bonded together to be identified and to have their chemical shifts correlated.[174] It also allows the detection of carbon atoms that are near to each other in space, even when they are on separate residues.[174]

Conformational studies of the γ-amino acid statine and its analogues indicate a dihedral angle near 165° or 0° for the NH—C_4H bond and near 90° for the

[167] J. M. Matsoukas, *Spectrosc. Lett.*, 1984, **17**, 21.
[168] J. Kobayashi, T. Hifashijima, and T. Miyazawa, *Int. J. Pept. Protein Res.*, 1984, **24**, 40.
[169] M. C. Reddy, B. P. N. Reddy, K. R. Sridharan, and J. Ramakrishna, *Org. Magn. Reson.*, 1984, **22**, 464.
[170] C. Rossi, L. Pogliani, F. Laschi, and N. Niccolai, *J. Chem. Soc., Faraday Trans. 1*, 1983, **79**, 2955.
[171] Y. Van Haverbeke, R. N. Muller, and L. Van der Elst, *J. Phys. Chem.*, 1984, **88**, 4978.
[172] L. Pogliani, N. Niccolai, and C. Rossi, *Spectrosc. Lett.*, 1984, **17**, 159.
[173] S. Khazaeli and R. E. Viola, *J. Inorg. Biochem.*, 1984, **22**, 33.
[174] M. H. Frey and S. J. Opella, *J. Am. Chem. Soc.*, 1984, **106**, 4942.

C_4H—C_3H bond.[175] Four possible conformations are compatible with the n.m.r. data for *N*-Boc-statine methyl ester in solution (although molecular-mechanics calculations strongly favour one of these over the others).

N.m.r. studies involving nuclei with higher atomic numbers continue to make progress, ^{14}N and ^{2}H n.q.r. spectra of cytosine complexes of *N*-formylglycine, *N*-benzoylglycine, and *N*-phthaloylglutamic acid giving electric-field gradient parameters to all ^{14}N and ^{2}H atoms in the molecules.[176] Amongst these heavier nuclei, ^{19}F has long held a place for routine n.m.r. work, applied in this context to *p*-fluorobenzoylamino acids.[177] ^{95}Mo n.m.r. contributes to a multi-pronged study of the cysteine complex MoO_2(L-Cys-OMe)$_2$.[156]

Circular Dichroism. – Thorough studies of N^{α}-acetylamino acid methylamides aimed at interpretations of c.d. data in terms of conformation[178] and the influence of solvent polarity indicate complex equilibria for the aliphatic protein amino acids L-alanine and L-leucine. While L-valine shows similar behaviour in many polar solvents, it behaves like the non-protein amino acid t-butylglycine in fluorinated alcohols in its adoption of a right-handed α-helical conformation.

Established configurational relationships for the two geometrical isomers of the *N*-oxide of *N*-benzyl-L-proline can be transferred to the *N*-oxides of *N*-benzyl-*N*-methyl-L-amino acids through the signs of characteristic Cotton effects.[179]

C.d. in conjunction with other physical methods has contributed to the structure determination of ferric neurosporin, a minor siderophore-like compound containing N^{δ}-hydroxy-D-ornithine; the Λ-*cis* absolute configuration about the central ferric ion and the assignment of the D-configuration to the amino acid moiety followed from empirical correlations of Cotton-effect behaviour.[157]

Mass Spectrometry. – All the papers reviewed here deal with the newer instrumental possibilities, from which the analysis of amino acids and peptides in particular has benefited dramatically, especially the soft-ionization techniques compatible with relatively involatile samples. Even so, some of these 'newer methods' have now become routine in the amino acid field.

Fast-atom-bombardment spectra of samples as small as 10^{-7} g isolated from thin-layer chromatograms were unambiguous criteria for identification,[180] and for *N*-Boc-amino acids there are characteristic features, including a McLafferty-type rearrangement.[181] F.a.b. spectra show more fragmentation, compared with field-desorption mass spectra, for protonated arginine, for which the site of

175 D. H. Rich, Y. Terada, and M. Kawai, *Int. J. Pept. Protein Res.*, 1983, **22**, 325.
176 E. A. Keiter, Y. Hiyama, and T. L. Brown, *J. Mol. Struct.*, 1983, **111**, 1.
177 M. P. Spratt, Y. Meng, and H. C. Dorn, *Anal. Chem.*, 1985, **57**, 76.
178 P. Malon, P. Pancoska, M. Budesinsky, J. Hlavacek, J. Pospisek, and K. Blaha, *Collect. Czech. Chem. Commun.*, 1983, **48**, 2844.
179 I. Z. Siemion, K. Marks, and A. Sucharda-Sobczyk, *Bull. Pol. Acad. Sci., Chem.*, 1983, **31**, 1.
180 G. D. Tantsyrev, M. I. Povolotskaya, and V. A. Saraev, *Bioorg. Khim.*, 1984, **10**, 848.
181 G. V. Garner, D. B. Gordon, L. W. Tetler, and R. D. Sedgwick, *Org. Mass Spectrom.*, 1983, **18**, 486.

ammonia loss after ionization is the guanidinium grouping.[182] Both positive- and negative-ion modes were adopted in some of these studies[181,182] and in chemical-ionization spectrometry of the aromatic protein amino acids.[183] Fast heavy-ion-induced desorption of valine and leucine represents a study of damage cross-sections of biologically important molecules for its own sake, rather than advocating the virtues of a technique that is impossibly expensive for routine laboratory work (at least as of 1985).[184]

Secondary-ion mass spectra of amino acids embedded in a glycerol solution of camphorsulphonic acid as matrix have been reported.[185]

Creation of a supersonic molecular beam of tryptophan molecules by combined thermospray and seeded molecular-beam techniques, then photo-ionization, employs a number of commercially available mass-spectrometer accessories.[186]

Infrared and Raman Spectrometry. – Routine i.r. spectrometry has been used in establishing crystal types deposited from supersaturated DL-amino acid solutions[145] (other routine uses are excluded from this section). Intermolecular association (between N and S atoms) has been established for *N*-acylglycine ethylthiolesters[158] by Fourier-transform i.r. studies, and similar self-association occurs with peptide oxazolin-5(4*H*)-ones based on α-aminoisobutyric acid[187] and for phospho-L-serine (hydrogen bonds with the carboxylate group as acceptor).[188] Far-i.r. spectra (30–650 cm^{-1}) of the 22 naturally occurring amino acids (*sic*) accumulated by the Fourier-transform technique[189] have been reported, with as yet little progress in interpretation of the data. The same complexity arises with a similar study, this time[190] based on one amino acid derivative, *S*-nitroso-L-cysteine. Data for isotopically substituted ($S^{15}NO$ and N^2H_3) analogues assisted the full assignment of absorption features by use also of the available potential-energy distribution data; of a total of 96 fundamentals occurring above 300 cm^{-1}, 65 were classified as group vibrations by the potential-energy criterion.

An enterprising study[191] reveals the trend from zwitterionic to 'free-amino–free-carboxy' tautomer with rising temperature for *N*-dialkylglycines and glycine itself.

182 J. J. Zwinselman, N. M. M. Nibbering, J. Van der Greef, and M. C. Ten Noever de Brauw, *Org. Mass Spectrom.*, 1983, **18**, 525.
183 T. Hayashi, H. Naruse, Y. Iida, and S. Daishima, *Shitsuryo Bunseki*, 1983, **31**, 205 (*Chem. Abstr.*, 1984, **100**, 210 366).
184 M. Salehpour, P. Haakansson, and B. Sundqvist, *Nucl. Instrum. Methods Phys. Res., Sect. B*, 1984, **230**, 752.
185 E. De Pauw, G. Pelzer, V. D. Dao, and J. Marien, *Biochem. Biophys. Res. Commun.*, 1984, **123**, 27.
186 T. R. Rizzo, Y. D. Park, and D. H. Levy, *J. Am. Chem. Soc.*, 1985, **107**, 277.
187 C. Toniolo, G. M. Bonora, M. Crisma, E. Benedetti, A. Bavoso, B. DiBlasio, V. Pavone, and C. Pedone, *Int. J. Pept. Protein Res.*, 1983, **22**, 603.
188 R. A. Dluhy, D. G. Cameron, and H. H. Mautsch, *Biochim. Biophys. Acta*, 1984, **792**, 182.
189 S. K. Husain, J. B. Hasted, D. Rosen, E. Nicol, and J. R. Birch, *Infrared Phys.*, 1984, **24**, 201.
190 D. M. Byler, H. Susi, and W. V. Gerasimowicz, *Appl. Spectrosc.*, 1984, **38**, 200.
191 M. A. Peterson and C. P. Nash, *J. Phys. Chem.*, 1985, **89**, 522.

Tyrosine and iodotyrosine Raman resonance intensity studies reveal exaltation of certain lines on ionization of the phenolic OH group, suggesting that an ionized residue in the presence of un-ionized tyrosines in a peptide might be located.[192]

Other Physical Studies. – Most of the studies surveyed in this section involve purely physico-chemical measurements: partial molar volumes of α-amino acids in water,[193] viscosity studies,[194–196] including indications of amino acid–detergent interactions[194] through volume changes for solutions of β-alanine, histidine, or glutamic acid,[197] volume and adiabatic compressibility of amino acids in water–urea mixtures,[198] hydration numbers determined from ultrasonic velocity and density data,[199] monolayer formation from long-chain *N*-acyl-L-amino acids,[200] dissociation constants,[201–203] differential-scanning calorimetric evidence for solid-state phase transitions,[204,205] including correction of earlier structural assignments deduced from crystal structures,[205] and electron transfer between metal atoms in cobalt(III)–ruthenium(III) complexes, *e.g.* $[(NH_3)_4Ru(SO_4)$-py-CO-X-$OCo(NH_3)_5]$, where X is an amino acid residue.[206]

Other physical studies range through electron microscopy [helical aggregates of chiral *N*-(2-hydroxydodecyl)amino acids[207]], e.s.r. studies of amavadin, bis-[*N*-hydroxy-*N*-(1-carboxyethyl)alaninato]oxovanadium(IV),[208] luminescence and excitation luminescence spectra of powdered DL-tryptophan, L-tyrosine, and DL-phenylalanine,[209] stoicheiometric complex formation between β-alanine or γ-aminobutyric acid and DNA, partly to displace phosphate counter-ions and to alter the DNA–water interactions,[210] and a wholly biological study, the transport

192 M. H. Baron, C. De Loze, T. Mejean, M. J. Coulange, P. Y. Turpin, and L. Chinsky, *J. Chim. Phys. Phys.-Chim. Biol.*, 1983, **80**, 729.
193 M. V. R. Rao, M. Atreyi, and M. R. Rajeswari, *J. Phys. Chem.*, 1984, **88**, 3129; *J. Chem. Soc., Faraday Trans. 1*, 1984, **80**, 2027.
194 T. Ogawa, K. Mizutani, and M. Yasuda, *Bull. Chem. Soc. Jpn.*, 1984, **57**, 2064.
195 S. J. Kim, Y. J. Oh, K. S. Choi, and Y. K. Shin, *Bull. Korean Chem. Soc.*, 1983, **4**, 284.
196 M. M. Bhattacharyya and M. Sengupta, *Z. Phys. Chem. (Leipzig)*, 1984, **265**, 109.
197 A. Vallejo, C. Abad, M. Trueba, and J. M. Macarulla, *An. Quim., Ser. C*, 1984, **80**, 164.
198 T. Ogawa, M. Yasuda, and K. Mizutani, *Bull. Chem. Soc. Jpn.*, 1984, **57**, 662.
199 J. K. Sinha and S. C. Srivastava, *Indian J. Phys., Sect. B*, 1984, **58**, 88.
200 K. Takahashi, F. Tanaka, and K. Motomura, *Bull. Chem. Soc. Jpn.*, 1984, **57**, 944.
201 A. Kuusk and S. Faingol'd, *Eest N.S.V. Tead. Akad. Toim., Keem.*, 1984, **33**, 28 (*Chem. Abstr.*, 1984, **100**, 175 242).
202 R. S. Saxena and S. K. Dhawan, *Acta Cienc. Indica, Ser. Chem.*, 1983, **9**, 151.
203 B. Rodriguez Rios, J. Fuentes Diaz, and R. Sierra Rodriguez, *An. Quim., Ser. B*, 1984, **80**, 54.
204 A. Gruenenberg, D. Bougeard, and B. Schrader, *Thermochim. Acta*, 1984, **77**, 59.
205 M. Matsumoto and K. S. Kunihisa, *Chem. Lett.*, 1984, 1279.
206 S. S. Isied and A. Vassilian, *J. Am. Chem. Soc.*, 1984, **106**, 1726, 1732.
207 H. Hikada, M. Murata, and T. Onai, *J. Chem. Soc., Chem. Commun.*, 1984, 562.
208 P. Krauss, E. Bayer, and H. Kneifel, *Z. Naturforsch., B*, 1984, **39**, 829.
209 E. I. Timoshkin, *Deposited Doc.*, 1984, VINITI 2907-83 (*Chem. Abstr.*, 1984, **101**, 152 288).
210 V. M. Aslanyan, S. G. Arutyunyan, and Yu. S. Babayan, *Biofizika*, 1984, **29**, 564; V. M. Aslanyan and S. G. Arutyunyan, *Biofizika*, 1984, **29**, 148; A. A. Akhrem, V. M. Aslanyan, S. G. Arutyunyan, and D. Yu. Lando, *Dokl. Akad. Nauk B.S.S.R.*, 1984, **28**, 272.

of imino and non-α-amino acids across the brush-border membrane of guinea-pig small intestine.[211]

Molecular-orbital Calculations. – A number of glycine-based studies have appeared, some dealing with aspects not easily amenable to experimentation. Conformational analysis of glycine aldehyde[212] (potential-energy curve for N—C—C=O torsion is very similar to that calculated for glycine methyl ester[213]) has been reported, also for glycine itself[214] for aqueous solutions, including both zwitterionic and neutral tautomers. Aqueous hydration of the glycine zwitterion at 25 °C has been simulated by Monte Carlo methods,[215] and calculations have indicated two pathways, one concerted, the other two-step, for the reaction of glycine with ammonia in the presence of Mg^{2+} ions.[216]

Calculations along similar lines yield data for serine–water interactions,[217] free energy of transfer of amino acids from water to apolar solvents (leading to the definition of the apolar surface being of common amino acids),[218] hydrogen-bond energies for the acidic and basic protein amino acids,[219] and the aromatic and heteroaromatic protein amino acids[220] with bases and base pairs of nucleic acids and with conformational-energy features for *N*-acetylalanine methylamide.[221]

6 Chemical Studies of Amino Acids

Racemization. – Results of mechanistic interest continue to accumulate under this heading, as well as applications of data for dating fossils.

A salutary example with a message for those preparing samples for careful determination of enantiomer ratios is the observed mechanochemical racemization of L-leucine in a ball-mill in the presence of such diluents as SiO_2 with dilute hydrochloric acid.[222] Salts of optically active amino acids with mineral acids or sulphonic acids underwent racemization on heating in acetic acid at 80–100 °C

211 G. M. Hanozet, B. Gordana, P. Parenti, and A. Geuvritore, *J. Membr. Biol.*, 1984, **81**, 233; B. G. Munck, *Biochim. Biophys. Acta*, 1984, **770**, 35.
212 L. Van den Enden, C. Van Alsenoy, J. N. Scarsdale, V. J. Klimkowski, and L. Schaefer, *Theochem.*, 1983, **14**, 407.
213 V. J. Klimkowski, L. Schaefer, L. Van den Enden, C. Van Alsenoy, and W. Caminati, *Theochem.*, 1983, **14**, 169.
214 R. Bonaccorsi, P. Palla, and J. Tomasi, *J. Am. Chem. Soc.*, 1984, **106**, 1945.
215 M. Mezei, P. K. Mehrotra, and D. L. Beveridge, *J. Biomol. Struct. Dyn.*, 1984, **2**, 1.
216 T. Oie, G. H. Loew, S. K. Burt, and R. D. MacElroy, *J. Am. Chem. Soc.*, 1984, **106**, 8007.
217 X. Shi, G. Ye, and X. Ni, *Fenzi Kexue Yu, Huaxue Yanjiu*, 1984, **4**, 405.
218 C. Froemnel, *J. Theor. Biol.*, 1984, **111**, 247.
219 N. V. Kumar and G. Govil, *Biopolymers*, 1984, **23**, 1995.
220 N. V. Kumar and G. Govil, *Biopolymers*, 1984, **23**, 2009.
221 L. Schaefer, V. J. Klimkowski, F. A. Momany, H. Chuman, and C. Van Elsenoy, *Biopolymers*, 1984, **23**, 2335.
222 A. Ikekawa and S. Hayakawa, *Bull. Chem. Soc. Jpn.*, 1984, **57**, 889.

during 1 hour in the presence of 0.1 molar equivalents of an aldehyde.[223] Some success was achieved[223] in an attempted asymmetric transformation involving the preferential crystallization of L-alanine *p*-chlorobenzenesulphonate (16%); the D-enantiomer underwent racemization in the liquid phase.

Dating studies continue to be supported by more detailed knowledge of the factors determining racemization rates of amino acids in their free state and bound within proteins and peptides. Fossil bones from Upper Paleolithic times, from Languedoc caves, are correctly dated on the basis of D:L ratios for their alanine and aspartic acid content only if rather higher racemization rates are assumed (based on ^{14}C dates) than those used hitherto.[224] Isoleucine isolated from fossil shells is highly epimerized before it is released from its protein sources;[225] epimerization rates for isoleucine in proteins are dependent on the position in the sequence, being fastest at the N-terminus. Similar studies have been undertaken for free and peptide-bound serine and aspartic acid, to attempt to account for the contributions of different phases through which these amino acids would have passed during ageing.[226]

General Reactions of Amino Acids. – Functional-group transformations mainly involving the $-NH_2$ and $-CO{\cdot}O-$ moieties of an α-amino acid are collected in this section, with the following section emphasizing the role of the side chain and transformations of functional groups present in it.

Reviews have appeared covering the electrochemistry of amino acids,[227] α-methoxylation of α-amino acids and β-lactams by electro-oxidation,[228] and reactions of oxazolinones derived from α-amino acids.[229]

Reactions at the amino group include condensation with 3,5-dinitro-1-(4-nitrophenyl)-4-pyridone to give conveniently protected derivatives[230] and cleavage of a phthaloyl group in an efficient two-stage, one-flask operation using $NaBH_4$ in 2-propanol followed by acetic acid.[231] *N*-Alkylation of *N*-Boc-amino acids with an alkyl iodide after treatment with KH and 18-crown-6 proceeds in good yield, and products with excellent optical purity can be obtained.[232] *N*-Alkylation of amino acids through condensation with an aldehyde in the

[223] C. Hongo, R. Yoshioka, M. Tohyama, S. Yamada, and I. Chibata, *Bull. Chem. Soc. Jpn.*, 1983, **56**, 3744.
[224] R. Lafont, G. Perinet, F. Bazile, and M. Icole, *C.R. Acad. Sci., Ser. 2*, 1984, **299**, 447.
[225] R. M. Mitterer and N. Kriausakul, *Org. Geochem.*, 1984, **7**, 91.
[226] S. M. Steinberg, P. M. Masters, and J. L. Bada, *Bioorg. Chem.*, 1984, **12**, 349.
[227] I. Vodrazka, I. Stibor, and M. Janda, *Chem. Listy*, 1984, **78**, 803.
[228] T. Shono and Y. Matsumura, *Kagaku (Kyoto)*, 1984, **39**, 114.
[229] W. Steglich, Proceedings of the 29th IUPAC Congress, ed. H. Grünwald, Pergamon Press, Oxford, 1983, p. 211.
[230] E. Matsumura, H. Kobayashi, T. Nishikawa, M. Agira, Y. Tohda, and T. Kawashima, *Bull. Chem. Soc. Jpn.*, 1984, **57**, 1961.
[231] J. O. Osby, M. G. Martin, and B. Ganem, *Tetrahedron Lett.*, 1984, **25**, 2093.
[232] R. T. Shuman, E. L. Smithwick, D. L. Smiley, G. S. Brooke, and P. D. Gesellchen in 'Peptides: Structure and Function', Proceedings of the 8th American Peptide Symposium, ed. V. J. Hruby and D. H. Rich, Pierce Chemical Company, Rockford, Illinois, U.S.A., 1983, p. 143.

presence of $NaBH_3CN$ in MeOH gives excellent yields[233] (for a two-stage equivalent of this process, using formaldehyde, see ref. 108). Amino acids react with formaldehyde and secondary phosphines to give *N*-phosphinomethyl derivatives $R^1PPhCH_2NHCHR^2CO_2H$.[234] *N*-Diphenylphosphinylamino acids, advocated for use in peptide synthesis, are obtained by reaction of the corresponding esters with $Ph_2P(O)Cl$.[235]

Reactions at the carboxy group to bring about conversion into other carbonyl derivatives include formation of symmetrical anhydrides and aryl esters from *N*-protected amino acids, using di-t-butyl pyrocarbonate,[236] and the formation of symmetrical anhydrides of various N-alkoxycarbonyl-L-valines through condensation with 0.5 equivalents of *N*-ethyl-*N*′-(3-dimethylaminopropyl) carbodiimide hydrochloride in CH_2Cl_2.[237] Excess of reagent leads to the 2-alkoxyoxazolin-5(4H)-one (except for 2,2,2-trichloroethyloxycarbonylvaline) as well as the anhydride.[237] There are more conventional esterifications of the amino acids themselves achieved by adding toluene-*p*-sulphonic acid to a suspension of the amino acid in methanol or ethanol and refluxing the mixture for 24 h;[238] methyl toluene-*p*-sulphonate in refluxing methanol was used in the preparation of amino acid methyl esters.[239] N-Protected amino acids can be converted into methylthiomethyl esters using Bu^tBr and Me_2SO under mild conditions.[240] Fmoc-amino acid trichlorophenyl esters have been prepared from the N-protected amino acid and the phenol, using DCCI,[241] and *N*-Boc- or -Z-amino acid amides from the N-protected amino acids, conc. aqueous NH_4OH, and isopropyl chloroformate in THF containing *N*-methylmorpholine.[242]

Reduction of the carboxy group proceeds in high yield, *via* N-protected amino acid esters, when $NaBH_4$ in Bu^tOH with slow addition of MeOH is used.[243] Decarboxylation of N-protected amino acid *N*-hydroxypyridine-2-thione esters through photolysis in the presence of Bu^tSH (aspartic and glutamic acids also undergo reductive loss of their side-chain carboxy groups) extends the already broad applicability demonstrated for this reaction.[244]

Thermal decomposition of amino acids has been followed by thermovoltaic detection, thermogravimetry, and differential scanning calorimetry.[245] While the

233 Y. Ohfune, N. Higuchi, M. Saito, M. Hashimoto, and T. Tanaka, Proceedings of the 21st Peptide Chemistry Conference, p. 89 (*Chem. Abstr.*, 1984, **101**, 131 048).
234 K. Kellner, W. Hanke, and A. Tzschach, *Z. Chem.*, 1984, **24**, 193.
235 R. Ramage, D. Hopton, M. J. Parrott, G. W. Kenner, and G. A. Moore, *J. Chem. Soc., Perkin Trans. 1*, 1984, 1357.
236 V. F. Pozdnev and M. Yu. Chernaya, *Khim. Prir. Soedin.*, 1984, 357.
237 A. Paquet, F. M. F. Chen, and N. L. Benoiton, *Can. J. Chem.*, 1984, **62**, 1335.
238 M. Bodanszky, *Int. J. Pept. Protein Res.*, 1984, **23**, 111.
239 K. Ueda, M. Waki, and N. Izumiya, *Mem. Fac. Sci., Kyushu Univ., Ser. C*, 1984, **14**, 307.
240 A. Dossena, G. Palla, R. Marchelli, and T. Lodi, *Int. J. Pept. Protein Res.*, 1984, **23**, 198.
241 K. M. Sivanandaiah and S. Gurusiddappa, *Indian J. Chem., Sect. B*, 1984, **23**, 372.
242 B. Rzeszotarska, M. Makowski, and Z. Kubica, *Org. Prep. Proced. Int.*, 1984, **16**, 136.
243 K. Soai, H. Oyamada, and M. Takase, Proceedings of the 21st Peptide Chemistry Conference, p. 85 (*Chem. Abstr.*, 1984, **101**, 152 278).
244 D. H. R. Barton, Y. Herve, P. Potier, and J. Thierry, *J. Chem. Soc., Chem. Commun.*, 1984, 1298.
245 S. Contarini and W. W. Wendlandt, *Thermochim. Acta*, 1983, **70**, 283.

study indicated the thermovoltaic method to be the more reproducible of the techniques, no attention was given to the nature of the degradation products.

Decarboxylative transamination is a term coined to describe the consequences of condensation of an amino acid with carbonyl compounds in the presence of a dipolarophile.[246–248] The initial condensation creates a 1,3-dipole $R^1\overset{+}{C}HN\overset{-}{C}HR^2$ and its mesomeric equivalent from an aldehyde R^1CHO and an amino acid $\overset{+}{N}H_3CHR^2CO_2^-$ through decarboxylation; 1,3-dipolar cycloaddition then ensues if there is a suitable opportunity. A spectacular example is the formation of the pyrrolidine (13) through the reaction of ninhydrin, alanine, and *N*-phenylmaleimide.[248]

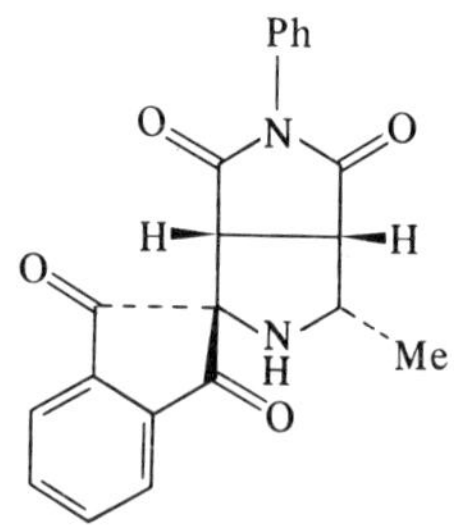

(13)

Addition of L-alanine benzyl ester to $PhCOCH{=}CHCO_2Et$ proceeds regioselectively and diastereoselectively, to give *N*-[(1*S*)-ethoxycarbonyl-3-phenylpropyl]-L-alanine after catalytic hydrogenation.[249] Other reactions involving the amino group of an α-amino acid in more unfamiliar processes are carbamoylation by *N*-nitroso-*N*-butylurea in aqueous solution (for 3 weeks),[250] polymerization of L-phenylalanine by triphenyl phosphite in a matrix support,[251] and grafting of L-amino acids onto silica pretreated with a 3-chloropropyl- or 3-aminopropyltrichloro- or -triethoxy-silane.[252] Phenyl isothiocyanate is well known for its addition to an amino group, but more vigorous reaction with an *N*-acylamino acid gives corresponding anilides,[253] while *N*-dimedonylamino acids give products of phenylthiocarbamoylation of the cyclohexenonyl moiety.[254] Lawesson's

246 R. Grigg and H. Q. N. Gunaratne, *Tetrahedron Lett.*, 1983, **24**, 4457.
247 R. Grigg, M. F. Aly, V. Sridharan, and S. Thianpatanagul, *J. Chem. Soc., Chem. Commun.*, 1984, 182.
248 R. Grigg and S. Thianpatanagul, *J. Chem. Soc., Chem. Commun.*, 1984, 180.
249 H. Urbach and R. Henning, *Tetrahedron Lett.*, 1984, **25**, 1143.
250 A. Suzuki, M. Fukui, S. Nakayasu, R. Takitani, and K. Tada, *Kyoritsu Yakka Daigaku Kenkyu Nenpo*, 1983, 1 (*Chem. Abstr.*, 1984, **101**, 111 368).
251 R. L. Bernard and S. P. Sawan, *Polym. Prepr. (Am. Chem. Soc., Div. Polym. Chem.)*, 1983, **24**, 178.
252 V. A. Malinovskii, S. M. Staroverov, and G. V. Lisichkin, *Vestn. Mosk. Univ., Ser. 2, Khim.*, 1984, **25**, 80.
253 R. Ashare, R. N. Ram, and A. K. Mukerjee, *Indian J. Chem., Sect. B*, 1984, **23**, 759 (*Chem. Abstr.*, 1984, **101**, 192 425).
254 M. Gomez Guillen and J. P. Garcia Martin, *An. Quim., Ser. C*, 1983, **79**, 109.

reagent yields *N*-thioacyl derivatives by mild treatment of corresponding acylamino acids, from which 1,2,4-triazines are easily formed by reaction with hydrazine.[255] Another reaction involving both amino and carboxy groups, 5-amino-2-phenyloxazolium ion formation from an *N*-benzoyl *N*-alkylamino nitrile as an unexpected product during the attempted hydrolysis by 60% $HClO_4$,[256] his precedents (Vol. 5, p. 304). Curtius rearrangement of *N*-acylamino acid hydrazides followed by addition to an alcohol gives diacylated *gem*-diaminoalkyl compounds used in retro-inverso peptide synthesis, but with troublesome side reactions in the isocyanate–alcohol reactions that have been given attention.[257]

Regeneration of amino acids from N-protected derivatives is discussed in another chapter in the context of peptide synthesis as far as common protecting groups are concerned. 2,4,6-Trinitrophenyl derivatives, however, do not fall in this category; their treatment with aqueous hydrazine at 30 °C is mild, so that fully active proteins can be recovered from trinitrophenylated materials.[258]

A crop of papers, no fewer than usual for a cull of a year's literature, deals with routine oxidation-kinetics studies of amino acids by standard oxidants.[259] Mechanistic interest is the common factor in a variety of reports of stereoselective hydrolysis of *N*-acyl- or -alkoxycarbonyl-DL-amino acid *p*-nitrophenyl esters in chiral (L-histidine-containing) micelles[260] and similar media.[261–263] In all such studies, the possibility of alternative pathways has to be recognized, and the hydrolysis of *N*-acetyl phenyl-alaninyl- or -valyl-imidazolides is much faster than for simple models, thus implicating an oxazolinone intermediate.[264] The role of the *N*-substituent, crucial in determining the ease of oxazolinone forma-

[255] T. P. Andersen, A. B. A. G. Ghattas, and S. O. Lawesson, *Tetrahedron*, 1983, **39**, 3419.
[256] R. P. Iyer, M. S. Sonaseth, S. P. Kulkarni, R. Gopalan, K. R. Ratnam, and A. V. Prabhu, *Indian J. Chem., Sect. B*, 1984, **23**, 289 (*Chem. Abstr.*, 1984, **101**, 230971).
[257] M. Chorev, S. A. MacDonald, and M. Goodman, *J. Org. Chem.*, 1984, **49**, 821.
[258] S. Takahashi, T. Yamamura, M. Kamo, and K. Satake, *Chem. Lett.*, 1984, 127.
[259] D. S. Mahadevappa, S. Ananda, M. B. M. Gowda, and K. S. Rangappa, *J. Indian Chem. Soc.*, 1984, **61**, 323; D. S. Mahadevappa, S. Ananda, A. S. A. Murthy, and K. S. Rangappa, *Indian J. Chem., Sect. A*, 1984, **23**, 17; D. S. Mahadevappa, S. Ananda, A. S. A. Murthy, and K. S. Rangappa, *React. Kinet. Catal. Lett.*, 1983, **23**, 181; M. S. Ramachandran and T. S. Vivekanandam, *J. Chem. Soc., Perkin Trans. 2*, 1984, 1341; M. S. Ramachandran, T. S. Vivekanandam, and R. P. M. M. Raj, *ibid.*, p. 1345; M. K. Reddy, C. S. Reddy, and E. V. Sundaram, *Indian J. Chem., Sect. A*, 1984, **23**, 197; P. A. Gidde, M. B. Hogale, M. H. Jagdale, and A. Y. Nimbalkar, *J. Indian Chem. Soc.*, 1984, **61**, 366; S. C. Ameta, P. N. Pande, H. L. Gupta, and H. C. Chowdhry, *Cienc. Cult. (Sao Paolo)*, 1983, **35**, 1885; R. N. Mehrotra, R. C. Kapoor, and S. K. Vajpai, *J. Chem. Soc., Dalton Trans.*, 1984, 999; K. C. Gupta and K. K. Gupta, *Natl. Acad. Sci. Lett. (India)*, 1983, **6**, 53; I. Ahmad, *Arab Gulf J. Sci. Res.*, 1983, **1**, 121; U. C. Verma and B. S. Yadav, *J. Indian Chem. Soc.*, 1984, **61**, 58; A. Lal and M. C. Agrawal, *Indian J. Chem., Sect. A*, 1984, **23**, 411.
[260] Y. Matsumoto and R. Ueoka, *Bull. Chem. Soc. Jpn.*, 1983, **56**, 3370.
[261] R. Ueoka, Y. Matsumoto, T. Nagamatsu, and S. Hirohata, *Chem. Lett.*, 1984, 583.
[262] Y. Kimura, M. Nango, Y. Ihara, and N. Kuroki, *Chem. Lett.*, 1984, 429.
[263] S. Sasaki, N. Hayashida, Y. Nakano, and K. Ohkubo, *J. Mol. Catal.*, 1984, **26**, 7; Y. Kimura, A. Tanaka, M. Nango, N. Kuroki, and Y. Ihara, *J. Polym. Sci., Polym. Chem. Ed.*, 1984, **22**, 407; Y. Ihara, N. Kunikiyo, T. Kunimasa, Y. Kimura, M. Nango, and N. Kuroki, *J. Chem. Soc., Perkin Trans. 2*, 1983, 1741.
[264] R. L. Kogan and T. H. Fife, *J. Org. Chem.*, 1984, **49**, 5229.

tion, has been identified as the major factor in determining the rates of alkaline hydrolysis of *N*-substituted alanine, glycine, and β-alanine *p*-nitrophenyl esters.[265] Hydrolysis, aminolysis, and alcoholysis of Boc-glycine-activated esters have been studied, particularly the competition between hydrolysis and aminolysis.[266] Butylaminolysis of *cis*-2-hydroxycyclopentyl benzoylglycinate (a simple model for a peptidyl-tRNA) occurs some 300 times faster than the corresponding reaction with cyclopentyl benzoylglycinate, thus implicating anchimeric assistance by the hydroxy group.[267] Hydrolysis of *N*-acetylphenylalaninyl adenylate anhydride and the corresponding free amine has been followed spectrophotometrically for 10^{-5}M solutions.[268]

Other mechanistic studies deal with ozonation of amino acids,[269] decomposition of *N*-bromoalanine in aqueous solutions,[270] and protonation rates of proline in various media.[271]

Specific Reactions of Amino Acids. – The topics discussed deal mainly with the reactions undergone or facilitated by the side chains of common amino acids.

Among common aliphatic amino acids, 1-aminocyclopropanecarboxylic acid has become familiar as the archetypal alkene producer in plants; increasingly more papers are acknowledged here through representative citations[272] covering ethylene production. Formation of 1-butene in pea (*Pisum sativum*) epicotyls and in a cell-free pea epicotyl enzyme preparation is only feasible from (1*R*,2*S*)-1-amino-2-ethylcyclopropanecarboxylic acid[273] and not from other stereoisomers.

Chlorination of alanine by HOCl in dilute aqueous solutions gives acetaldehyde and acetonitrile, the latter arising from the reaction of $ClNH_2$ with the aldehyde.[274]

The mixture of DL-isoleucine and DL-alloisoleucine formed by chemical synthesis can be separated because the 1,5-naphthalenedisulphonic acid salt of DL-isoleucine is less soluble.[275] Moreover, epimerization of the alloisoleucine through heating in acetic acid in the presence of salicylaldehyde followed by separation allows up to 95% recovery of DL-isoleucine from such mixtures.[275]

265 P. Ascenzi, G. Sleiter, and E. Antonini, *Gazz. Chim. Ital.*, 1983, **113**, 859.
266 S. K. Girin and Yu. P. Shvachkin, *Zh. Obshch. Khim.*, 1983, **53**, 2779.
267 M. Julia and H. Mestdagh, *Tetrahedron*, 1984, **40**, 327.
268 J. C. Lacey, N. Senaratne, and D. W. Mullins, *Origins Life*, 1984, **15**, 45.
269 W. A. Pryor, D. H. Giamalva, and D. F. Church, *J. Am. Chem. Soc.*, 1984, **106**, 7094.
270 W. D. Stanbro and M. J. Lenkevich, *Int. J. Chem. Kinet.*, 1983, **15**, 1321.
271 A. M. Slifkin and S. M. Ali, *J. Mol. Liq.*, 1984, **29**, 75.
272 M. Knee, *J. Exp. Bot.*, 1984, **35**, 1799; C. Vinkler and A. Apelbaum, *F.E.B.S. Lett.*, 1984, **167**, 64; S. Satoh and Y. Esashi, *Plant Cell. Physiol.*, 1984, **25**, 1277; M. Guy and H. Kende in 'Ethylene; an International Symposium', ed. Y. Fuchs and E. Chalutz, Nijhoff, The Hague, Netherlands, 1984, p. 89; N. Amrhein, U. Dorzok, C. Kionka, U. Kondziolka, H. Skorupka, and S. Tophof, *ibid.*, p. 11.
273 T. A. McKeon and S. F. Yang, *Planta*, 1984, **160**, 84.
274 C. Le Cloirec, J. Poncin, and G. Martin, *C.R. Acad. Sci., Ser. 2*, 1984, **298**, 559.
275 C. Hongo, R. Yoshiola, M. Tohyama, S. Yamada, and I. Chibata, *Bull. Chem. Soc. Jpn.*, 1984, **57**, 1328.

General reactions applied to specific amino acids include the following: the synthesis of fructosylglycine from glycine and D-glucose heated in MeOH for 3 h,[276] the Maillard reaction between glucose and lysine, to give mono- and di-fructosyl-lysine,[277] and decarboxylation processes (L-glutamic acid to γ-amino-butyric acid and proline to δ-aminovaleric acid by *Clostridium sordelli*,[278] phenylglycine to benzaldehyde using the coenzyme methoxatin in the presence of cetyltrimethylammonium bromide,[279] and thermal decarboxylation of γ-carboxyglutamic acid[280]). The last-mentioned paper[280] contains a useful broad coverage of the chemistry of γ-carboxyglutamic acid, including specific exchange of the γ-proton for ^{3}H, and the first synthesis of pyro-γ-carboxyglutamic acid (a reversible reaction with possible applicability in the estimation of the γ-carboxy-glutamic acid content of proteins). Glutamic acid has been found to undergo deamination using pyridoxal phosphate and copper(II) smectite (a swelling phyllosilicate) to give ammonia and 2-ketoglutaric acid.[281]

Biosynthetic studies are not reviewed systematically; representative papers cover the conversion of threonine into glycine and aminoacetone in rat liver mitochondria[282] and the demonstration of 1,2-hydride shift from C-3 of (2*RRSS*, 3*RSRS*)-[1-^{14}C, 3-^{3}H]phenylalanine to the carbon atom that ultimately becomes the hydroxymethyl group of tropic acid *in vivo*.[283]

Nucleophilic reactivity towards alkylating agents of valine methylamide has been studied in view of the finding that the α-amino group shows a pK'_a of 7.65, somewhat higher than the value for a terminal *N*-valyl peptide.[284] Nucleophilic reactivity of the imidazole nitrogen atoms of N^α-acetylhistidine and its methylamide was studied in the same laboratory.[285] Lysine side-chain reactions feature in uncatalysed N^ϵ-methylation and -formylation reactions and their inhibition by excess ascorbic acid,[286] in a one-pot synthesis of N^α-Z-N^ϵ-Boc-L-lysine from L-lysine by successive masking of the α- and ϵ-amino groups,[287] and in a synthesis of the nitrosocarbamates $RO_2CN(NO)(CH_2)_4CH(\overset{+}{N}H_3)CO_2^-$.[288] Side-chain protected arginine suitable for solid-phase synthesis carries the 4-methoxy-2,3,6-trimethylbenzenesulphonyl group, removable by trifluoroacetic acid.[289]

276 E. A. Karpova and V. K. Gorodetskii, *Vopr. Med. Khim.*, 1984, **30**, 128.
277 C. M. Lee, B. Sherr, and Y. N. Koh, *J. Agric. Food Chem.*, 1984, **32**, 379.
278 W. Huckenbeck and T. Daldrup, *Zentralbl. Bakteriol., Mikrobiol., Hyg., Ser. A*, 1984, **258**, 51.
279 S. Itoh, N. Kato, Y. Ohshiro, and T. Agawa, *Tetrahedron Lett.*, 1984, **25**, 4753.
280 P. A. Price, C. Nelson, and M. K. Williamson, *Anal. Biochem.*, 1984, **136**, 119.
281 M. M. Mortland, *J. Mol. Catal.*, 1984, **27**, 143.
282 M. I. Bird, P. B. Nunn, and L. A. J. Lord, *Biochim. Biophys. Acta*, 1984, **802**, 229.
283 E. Leete, *J. Am. Chem. Soc.*, 1984, **106**, 7271.
284 V. Poirier and C. J. Calleman, *Acta Chem. Scand., Ser. B*, 1983, **37**, 817.
285 C. J. Calleman and V. Poirier, *Acta Chem. Scand., Ser. B*, 1983, **37**, 809.
286 L. Trezl, I. Rusznak, E. Tyihak, T. Szarvas, and B. Szende, *Biochem. J.*, 1983, **214**, 289.
287 E. P. Heimer, C. T. Wang, T. J. Lambros, and A. M. Felix, *Org. Prep. Proced. Int.*, 1983, **15**, 379.
288 V. F. Gopko, G. M. Anoshina, and L. B. Radina, *Khim.-Farm. Zh.*, 1984, **18**, 301.
289 E. Atherton, R. C. Sheppard, and J. D. Wade, *J. Chem. Soc., Chem. Commun.*, 1983, 1060.

Tryptophan side-chain reactivity continues to attract attention (its photochemistry is featured in a later section), with Pictet–Spengler reactions with aldehydes giving β-carbolines.[290] Oxidative breakdown occurs under the conditions of iodination of tyrosine residues in peptides (N^{in}-formylation protects efficiently against this degradation).[291] N^{in}-Boc-L-Tryptophan formed with Boc-anhydride and 4-dimethylaminopyridine in MeCN[292] should also be a welcome new protected amino acid. Dye-sensitized photo-oxygenation of tryptophan in alkaline phosphate buffer gives the corresponding 4-hydroxypyrrolo-indole, which on air oxidation yields 5-hydroxy-*N*′-formylkynurenine.[293]

Ozonation degrades tryptophan into aspartic acid, kynurenine, and melanin through the intervention of hydroxy radicals, probably, since similar results are obtained through radiolysis and oxidation using Fenton's reagent.[294] Tryptophan radicals formed by electron transfer from azide radicals or $Br^{\bullet}$[295] can be 'repaired' by their treatment with an antioxidant.[296] Hydroxyl radicals formed by pulse radiolysis of aqueous solutions saturated with N_2O give 50% conversion of tyrosine into the radical resulting from addition *ortho* to the tyrosine hydroxy group and about 35% of the *meta* isomer.[297] Hydroxy radicals from $ADP–Fe^{II}–H_2O_2$ yield long-lived free radicals with proline and hydroxyproline, shown by e.p.r. studies to be nitroxyls (already known as the products of reaction with t-butyl hydroperoxide).[298] One-electron oxidation of cysteine by bromine or iodine radical anions is persuasively suggested to involve cysteinyl radical bromide or iodide complexes, respectively.[299] Oxidation of cysteine by iodine is much more complex than hitherto reported, involving I atoms, I^-, and I_3^-.[299]

Reactions based on the serine hydroxy group include that of the *O*-tosylated protected serine with a lithium diorganocuprate (see also ref. 79) to give a mixture of alkyl-substitution products and the protected dehydroalanine.[300] An *O*- → *N*-acetyl shift accompanies hydrogenolysis of *threo*-*O*-acetyl-β-phenyl-L-serine, leading to *N*-acetyl-L-phenylalanine in good yield.[301] β-Hydroxy-α-amino acid amides and Mitsunobo reagents give proline, pipecolic acid, and higher homologue derivatives, and lactams in some cases, together with ethers formed unexpectedly between two molecules of the amino acid derivative.[302]

[290] M. Jawdosiuk and J. M. Cook, *J. Org. Chem.*, 1984, **49**, 2699.
[291] G. Mourier, L. Moroder, and A. Previero, *Z. Naturforsch., B*, 1984, **39**, 101.
[292] H. Franzen, L. Grehn, and U. Ragnarsson, *J. Chem. Soc., Chem. Commun.*, 1984, 1699.
[293] M. Nakagawa, Y. Yokoyama, S. Kato, and T. Hino, *Heterocycles*, 1984, **22**, 59.
[294] S. V. Sikorskaya, A. V. Ignatenko, and S. N. Cherenkevich, *Zh. Prikl. Chim. (Leningrad)*, 1984, **57**, 2066.
[295] J. Butler, E. J. Land, A. J. Swallow, and W. Prutz, *Radiat. Phys. Chem.*, 1984, **23**, 265.
[296] B. M. Hoey and J. Butler, *Biochim. Biophys. Acta*, 1984, **791**, 212.
[297] S. Solar, W. Solar, and N. Getoff, *J. Phys. Chem.*, 1984, **88**, 2091.
[298] R. A. Floyd and I. Z. Nagy, *Biochim. Biophys. Acta*, 1984, **790**, 94.
[299] J. E. Packer, *J. Chem. Soc., Perkin Trans. 2*, 1984, 1015.
[300] J. A. Bajrowicz, A. El Hallaoui, R. Jacquier, C. Pigiere, and P. Viallefont, *Tetrahedron Lett.*, 1984, **25**, 2759.
[301] J. S. Tou and B. D. Vineyard, *J. Org. Chem.*, 1984, **49**, 1135.
[302] K. Nakajima, M. Morishita, and K. Okawa, Proceedings of the 21st Peptide Chemistry Conference, p. 77 (*Chem. Abstr.*, 1984, **101**, 192 411).

Reactions of sulphur-containing amino acids range from a down-to-earth electrochemical reduction for cleavage of L-cystine (over-reduction causes racemization)[303] to cyclization of *N*-acylpenicillamines with isopropyl chloroformate–NEt_3 to give D-β-thiolactones[304] and synthesis of (2*R*)- and (2*S*)-vinylglycinols CH_2=CHCH(NHR)CH_2OH from D- or L-methionine (R = Z or Boc),[305] employing standard sulphonium salt chemistry.

All the above narrative in this section has dealt with α-amino acids, on which most biological interest resides; but not all – the hydroxamic acid of γ-aminobutyric acid formed from the ethyl ester by reaction with NH_2OH possesses anticonvulsant and cardiovascular activity.[306]

Non-enzymic Models of Biochemical Processes Involving Amino Acids. – This section is not the only location for studies covered by this general title (amino acid–nucleotide interactions have been discussed in an earlier section[210] for example). Photochemical addition of amino acids to denatured calf thymus DNA,[307] homopolyribonucleotides,[308] polyuridylic acid,[309] or thymine[310] has been fully studied. Lysine and thymine react in basic solutions under irradiation at 254 nm to give 6-amino-2-(1-thyminyl)- and 2-amino-6-(1-thyminyl)-hexanoic acid.[310] This result is notable, side by side with the report[311] that tyrosine and tryptophan protect aqueous thymine from radiation damage.

Effects of Electromagnetic Radiation on Amino Acids. – Radiation-damage studies provide a linking theme from year to year for this section, and amino acids have been studied in this context for uranyl-sensitized photolysis in pressed KBr pellets,[312] for *X*-irradiated single crystals of L-asparagine hydrate (e.s.r.–e.n.d.o.r. study),[313] for radical yields of a mixture of amino acids (less than for the irradiated peptide of which the amino acids are constituents),[314] for γ-irradiated phenylalanine–glycylglycylphenylalanyl-leucine mixtures (three different adducts are formed),[315] and for ion-forming irradiation of amino acid–D-lactose mixtures (lyoluminescence as an index of the radical–oxygen and radical–radical reaction rates).[316]

303 R. Yang and B. Wu, *Shengwu Huaxue Yu Shengwu Wuli Jinzhan*, 1984, **56**, 67 (*Chem. Abstr.*, 1984, **101**, 131 064).
304 W. Reid and U. Reiher, *Chem.-Ztg.*, 1984, **108**, 152.
305 Y. Ohfune and N. Kurokawa, *Tetrahedron Lett.*, 1984, **25**, 1071.
306 H. Kehl, *Bol. Soc. Quim. Peru*, 1983, **49**, 131 (*Chem. Abstr.*, 1984, **101**, 111 369).
307 M. D. Shetlar, J. Christensen, and H. Horn, *Photochem. Photobiol.*, 1984, **39**, 125.
308 M. D. Shetlar, K. Horn, J. Carbone, D. Moy, E. Steady, and M. Watanabe, *Photochem. Photobiol.*, 1984, **39**, 135.
309 M. D. Shetlar, J. Christensen, E. Steady, and K. Horn, *Photochem. Photobiol.*, 1984, **39**, 141.
310 M. D. Shetlar, J. A. Taylor, and K. Horn, *Photochem. Photobiol.*, 1984, **40**, 299.
311 M. Li and G. Wang, *Fushe Yanjii Yu Fushe Goongye Xuebao*, 1984, **2**, 11.
312 A. K. Bansal, S. Goyal, and R. D. Dubey, *Acta Cienc. Indica Ser. Chim.*, 1983, **9**, 215.
313 G. C. Moulton and J. M. Coleman, *J. Chem. Phys.*, 1984, **80**, 4748.
314 C. Wiezorek, *Int. J. Radiat. Biol. Relat. Stud. Phys., Chem., Med.*, 1984, **45**, 93.
315 H. J. Kim, L. K. Mee, S. J. Adelstein, I. A. Traub, S. A. Carr, and V. N. Reinhold, *Radiat. Res.*, 1984, **100**, 30.
316 H. Kundu and B. Mitra, *Proc. Nucl. Phys. Solid State Phys. Symp. 1980*, 1983, **23**, 708 (*Chem. Abstr.*, 1984, **101**, 131 058).

Asymmetric *X*-ray decomposition of DL-alanine or DL-phenylalanine in ^{18}O-enriched H_2O has been monitored through detecting the ^{18}O-labelled products by nuclear reactions.[317] It transpires that *X*-irradiated D-enantiomers are more sensitive ^{18}O-traps than their L-counterparts, and the significance of this will surely be followed up.

Pulse radiolysis and flash photolysis bring about one-electron oxidation of dopa, to give dopasemiquinone, that disproportionates to the quinone *en route* to melanin.[318] Photo-oxidation of dopa sensitized by haematoporphyrin, using 2,2,6,6-tetramethyl-4-piperidone-1-oxyl ('Tempone') as a convenient probe for monitoring oxygen consumption, proceeds mainly by a singlet-oxygen mechanism.[319] γ-Irradiation of aqueous solutions of amino acids in the presence of 2-methyl-2-nitrosopropane as a spin trap gives stable spin adducts that were analysed by h.p.l.c.[320–322] This useful method allows the identification of short-lived radicals, some the result of hydrated electron reactions, others from attack by the hydroxyl radical.

4′-(1-Azi-2,2,2-trifluoroethyl)phenylalanine eliminates N_2 rapidly at wavelengths longer than 300 nm, yielding a highly reactive carbene capable of OH and CH insertion reactions.[323]

The usual high level of interest in tyrosine and tryptophan continues. Trifluoroacetamide has been observed[324] to quench tryptophan fluorescence. The more subtle details of redistribution of the absorbed energy take many forms; fluorescence polarization for tyrosine and tryptophan has been linked to the thermal coefficient of frictional resistance to rotation in these molecules.[325] Several papers at a NATO Advanced Study Institute relate to the 'tryptophan problem',[326] also concerned with the decay processes available to the irradiated molecule. Laser flash photolysis of indole at 265 nm in the presence of glycine, proline, or hydroxyproline shows variations in yields of hydrated electrons, triplet-state intermediates, and indole cation radicals.[327]

A radioprotective effect has been ascribed[328] to 2-mercaptopropionylglycine against 3Gy γ-radiation.

317 C. Wiezorek, *Int. J. Radiat. Biol. Relat. Stud. Phys., Chem., Med.*, 1984, **46**, 233.
318 M. R. Chedekel, E. J. Land, A. Thompson, and T. G. Truscott, *J. Chem. Soc., Chem. Commun.*, 1984, 1170.
319 K. Reszka and R. C. Sealy, *Photochem. Photobiol.*, 1984, **39**, 293.
320 K. Makino, F. Moriya, and H. Hatano, *Radiat. Phys. Chem.*, 1984, **23**, 217.
321 N. Iguchi, F. Moriya, K. Makino, S. Rokushika, and H. Hatano, *Can. J. Chem.*, 1984, **62**, 1722.
322 F. Moriya, K. Makino, N. Iguchi, N. Suzuki, S. Rokushika, and H. Hatano, *J. Phys. Chem.*, 1984, **88**, 2373.
323 M. Nassal, *J. Am. Chem. Soc.*, 1984, **106**, 7540.
324 P. Midoux, P. Wahl, J. C. Auchet, and M. Monsigny, *Biochim. Biophys. Acta*, 1984, **801**, 16.
325 G. Weber, S. Scarlata, and M. Rholam, *Biochemistry*, 1984, **23**, 6785, 6789.
326 A. G. Szabo, *NATO Adv. Sci. Inst. Ser., Ser. A*, 1983, 621; R. Lopez-Delgado, *ibid.*, p. 615; D. M. Jameson, *ibid.*, p. 623; G. S. Beddard, *ibid.*, p. 629; I. Saito, H. Sugiyama, A. Yamamoto, S. Muramatsu, and T. Matsuura, *J. Am. Chem. Soc.*, 1984, **106**, 4286.
327 C. M. Previtali, *Photochem. Photobiol.*, 1984, **40**, 689.
328 P. U. Devi and R. Gupta, *Radiobiol. Radiother.*, 1984, **25**, 585.

7 Analytical Methods

Gas–Liquid Chromatography. – The year's literature amounts to consolidation of existing methods, with *N*-heptafluorobutyroyl isobutyl esters as the most favoured derivatization objective (for analysis of D-alanine in bacterial cell walls,[329] for glycine,[330] and more generally,[331,332] especially[332] for micro-scale sampling of biological fluids). *N*-Trifluoroacetyl butyl esters[330] and pentafluoropropionyl methyl esters[333] have also been used, the latter for quantitative analysis of enantiomers of lanthionines, cystathionines, β-methyl-lanthionines, and *S*-alkyl cysteines using 'Chirasil-Val' as the chiral stationary phase.[333] Less familiar derivatization routines based on *N*-ethoxycarbonyl isopropyl esters and trimethylsilyl analogues,[334] reaction of phosgene with *N*-methylamino acids (for enantiomeric analysis using XE-60 rendered chiral with L-valyl-(*R*)-phenylethylamide as stationary phase),[335] and comparison of oxazolidinones formed between amino acids and 1,3-dichlorotetrafluoroacetone with perfluorinated derivatives for analysis at femtomole levels, have been described.[336]

Ion-exchange Chromatography. – Routine amino acid analyser regimes are not covered in this review, although indications of developments are to be found in triple-column procedures with fluorimetric quantitation,[337] appraisal of sources of error in 'high-performance' amino acid analysers,[338] and modern automated techniques for the analysis of 1- and 3-methylhistidines, tyrosine, phenylalanine, tryptophan, lysine, histidine, and arginine in urine.[339] 3-Methylhistidine features in a related study[340] (see also ref. 331 for g.l.c. analysis of 1- and 3-methylhistidines), and cation-exchange analysis of lysine–glutamic acid mixtures[341] and of threonine, alanine, proline, and aspartic acid[342] has been reported. In the last-mentioned study, use is made of the different stability constants of complexes formed between amino acids and copper(II) or zinc(II) counter-ions incorporated in the stationary phase.[342]

329 A. Tunlid and G. Odham, *Biomed. Mass Spectrom.*, 1984, **11**, 428.

330 J. Jiang, S. Wang, J. Pan, Z. Xu, and G. Wang, *He Dianzixue Yu Tance Jishu*, 1984, 4, 19.

331 F. Marcucci, L. Colombo, and E. Mussini, *J. Chromatogr.*, 1984, **336**, 356; S. L. Mackenzie and K. R. Holme, *J. Chromatogr.*, 1984, **299**, 387.

332 D. Labadarios, G. C. Shepherd, I. M. Moodie, and E. Botha, *S. Afr. J. Sci.*, 1984, **80**, 240.

333 E. Kuesters, H. Allgaier, G. Jung, and E. Bayer, *Chromatographia*, 1984, **18**, 287.

334 H. J. Chaves das Neves and A. M. Pestana de Vasconcelos, *Rev. Port. Quim.*, 1983, **25**, 184.

335 W. A. Koenig, E. Steinbach, and K. Ernst, *J. Chromatogr.*, 1984, **301**, 129.

336 P. Husek and V. Felt, *J. Chromatogr.*, 1984, **305**, 442.

337 T. N. Ferraro and T. A. Hare, *Anal. Biochem.*, 1984, **143**, 82.

338 D. E. C. Cole and L. Libadia, *Clin. Chem. (Winston-Salem, N.C.)*, 1984, **30**, 331.

339 R. C. Feldhoff, D. J. Ledden, M. C. Steffen, J. M. Steffen, and X. J. Musacchia, *J. Chromatogr.*, 1984, **311**, 267.

340 G. Zuric, S. Stanomirovic, and J. Savic, *J. Chromatogr.*, 1984, **311**, 69.

341 V. D. Verenko, E. D. Nestorovskaya, V. E. Kabal'skii, and A. N. Burya, *Khim. Tekhnol. (Kiev)*, 1984, 21.

342 J. Maslowska and E. Gasinska, *Chem. Anal. (Warsaw)*, 1984, **29**, 163.

Post-column ninhydrin colour formation has been compared with *o*-phthaldialdehyde fluorimetry for amino acid analysis.[343] Ninhydrin reduced with $TiCl_3$ compares favourably with the usual ninhydrin–hydrindantin cocktail in terms of reproducible colour yield.[344]

Thin-layer Chromatography. – A useful reversed-phase technique for amino acids using MeCN–0.4% trifluoroacetic acid as the mobile phase has been introduced.[345] The crosslinking amino acids desmosine, isodesmosine, merodesmosine, and lysinonorleucine are separated from each other and from all other amino acids present in elastin hydrolysates[346] by routine t.l.c. methods.

Amino acid derivatives feature prominently in the recent literature, dansyl derivatives lending themselves well to the identification of hydroxylysine through double-labelling techniques in collagen hydrolysates.[347] Reversed-phase[348] and polyamide t.l.c.[349] have been shown to offer useful advantages in the analysis of dansylamino acids, and the use of silica gel with the solvent mixtures that constitute the overpressured layer-chromatography technique has been shown to be suitable for identification of phenylthiohydantoins of the common amino acids.[350] Sarcosine has been derivatized using NBD-Cl to permit its analysis by t.l.c.[389]

High-performance Liquid Chromatography. – This has become the major separation technique for most areas of amino acid analysis, though the relative size of this section and of the preceding sections distorts the relationship since some ion-exchange results, for example, are mentioned here rather than in the 'Ion-exchange Chromatography' section earlier in this section.

Papers referring to post-column derivatization methods are covered first; of these,[351–363,17,102] several[351,352] refer to introduction of a fluorophore, through reaction of the eluate with *o*-phthaldialdehyde and a thiol, others employ electrochemical detection,[353–355] and another[358] describes the creation of fluorescent products through post-column treatment with an L-amino acid oxidase and peroxidase immobilized on aminopropylated glass beads (for the analysis of tyrosine, phenylalanine, tryptophan, and methionine). Brief details of the analytical objectives of these studies are: trimethyl-lysine,[351] collagen hydrolysates (NaOCl–*o*-phthaldialdehyde as the reagent system),[352] cysteine, homocysteine, and glutathione,[353] biogenic amines and their precursor amino

343 J. G. Vaughn, 'Clinical Liquid Chromatography', ed. P. M. Kabra and L. J. Marton, Chemical Rubber Company, Boca Raton, Florida, U.S.A., 1984, Vol. 2, p. 1.
344 L. B. James, *J. Chromatogr.*, 1984, **284**, 97.
345 J. C. Touchstone, E. J. Levin, and S. G. Lee, *J. Liq. Chromatogr.*, 1984, **7**, 2719.
346 S. Keller, A. K. Ghosh, A. K. Ghosh, G. M. Turina, and I. Mandl, *J. Chromatogr.*, 1984, **305**, 461.
347 J. Kelley and L. Chrin, *J. Chromatogr.*, 1984, **311**, 400.
348 N. Grinberg and S. Weinstein, *J. Chromatogr.*, 1984, **303**, 251.
349 Z. Wang, X. Tang, Z. Wang, Y. Wei, S. Dong, and M. Fan, *Zhongwa Yixue Jian Yan Zazhi*, 1984, **7**, 144.
350 S. Fater and E. Mincsovics, *J. Chromatogr.*, 1984, **298**, 534.
351 A. T. Davis, S. T. Ingalls, and C. L. Hoppel, *J. Chromatogr.*, 1984, **306**, 79.
352 H. Konomi, T. Hayashi, Y. Takehana, and M. Arima, *J. Chromatogr.*, 1984, **311**, 375.
353 E. G. Demaster, F. N. Shirota, B. Redfern, D. J. W. Goon, and H. T. Nagasawa, *J. Chromatogr.*, 1984, **308**, 83.

acids,[354] phenylalanine and tyrosine in serum,[355] tryptophan[356–358] together with tyrosine,[357] dityrosine in wool hydrolysates,[359] 3-methylhistidine in urine,[360] pipecolic acid in human plasma,[361] ε-(γ-glutamyl)lysine in protein hydrolysates,[17] and (4′-amino-3′-hydroxyphenyl)alanine and (7′-hydroxybenzothiazol-4′-yl)alanine and other degradation products in phaeomelanin hydrolysates.[102] Several of these studies employ cation-exchanger phases;[352,353,360] one[362] concentrates on reversed-phase applicability, and the potential of ternary gradient systems is fully explored.[363]

Derivatives of amino acids are formed from sample mixtures prior to h.p.l.c.[364–379] (pre-column treatment). Dansylamino acids[364–367] feature in several papers, and two papers[368] continue the recent interest in dabsylamino acids. Chemiluminescence generated by post-column reaction of dansylamino acids in eluates with bis-(2,4,6-trichlorophenyl)oxalate and H_2O_2 (see also Vol. 16, p. 42) offers sensitivity that is not available in other detection methods. The separation of glutamic acid, glutamine, and γ-aminobutyric acid[366] and entantiomeric analysis of dansylated amino acids using the chiral mobile-phase technique [copper(II)acetate-*N*,*N*-di-n-propyl-L-alanine[365,367] or copper(II) aspartame[367]] are notable applications of h.p.l.c. methods. Pre-column treatment with *N*-(9-acridinyl)maleimide yields a fluorescent derivative with cystine that is conveniently separated and quantitated by h.p.l.c.[369] 9-Fluorenylmethyl chloroformate reacts with amino acids under mild conditions (within 30 s) to give stable fluorescent derivatives that can be extracted into pentane and separated by h.p.l.c.[370] *o*-Phthaldialdehyde–thiol derivatization remains popular for pre-column introduction of a fluorophore.[376,377] This crop of papers includes a novel approach to the preparation of diastereoisomeric derivatives that allow enantiomeric analysis using sensitive fluorescence assay; instead of a simple thiol

[354] K. Oka, K. Kojima, A. Togari, T. Nagatsu, and B. Kiss, *J. Chromatogr.*, 1984, **308**, 43.
[355] W. T. Kok, U. A. T. Brinkman, and R. W. Frei, *J. Pharm. Biomed. Anal.*, 1983, **1**, 369.
[356] P. Rocchini, M. Bizzarri, M. Pompei, D. Ciani, M. Panicucchi, and S. Gallo, *Rass. Chim.*, 1984, **36**, 15 (*Chem. Abstr.*, 1985, **102**, 58 640).
[357] S. M. Lasley, I. A. Michaelson, R. D. Greenland, and P. M. McGinnis, *J. Chromatogr.*, 1984, **305**, 27.
[358] N. Kiba and M. Kaneko, *J. Chromatogr.*, 1984, **303**, 396.
[359] M. S. Otterburn and P. E. Gargan, *J. Chromatogr.*, 1984, **303**, 429.
[360] J. C. Robert and P. Serog, *Clin. Chim. Acta*, 1984, **142**, 161.
[361] H. Nishio and T. Segawa, *Clin. Chim. Acta*, 1984, **143**, 57.
[362] W. S. Hancock, *Chem. N.Z.*, 1983, **47**, 145.
[363] A. Henshall, M. J. Pickering, and D. Soto, *Spectra 2000*, 1984, **12**, 29 (*Chem. Abstr.*, 1985, **102**, 75 081).
[364] K. Miyaguchi, K. Honda, and K. Imai, *J. Chromatogr.*, 1984, **303**, 173; *ibid.*, 1984, **316**, 501.
[365] S. Weinstein and S. Weiner, *J. Chromatogr.*, 1984, **303**, 244.
[366] S. L. Paliya, J. Albert, and T. W. Reid, *J. Liq. Chromatogr.*, 1984, **7**, 2261.
[367] S. Lam, H. Azumaya, and A. Karmen, *J. Chromatogr.*, 1984, **302**, 21.
[368] J. Y. Chang, *J. Chromatogr.*, 1984, **295**, 193; J. C. Lin and S. Y. L. Shian, *J. Chin. Biochem. Soc.*, 1983, **12**, 47.
[369] S. Matsui, K. Kitabakate, H. Takahashi, and H. Meguro, *J. Inst. Brew.*, 1984, **90**, 20.
[370] S. Einarsson, B. Josefsson, and S. Lagerkvist, *J. Chromatogr.*, 1983, **282**, 609.
[371] A. S. Bhown, T. W. Cornelius, and J. C. Bennett, *L.C., Liq. Chromatogr., H.P.L.C. Mag.*, 1983, **1**, 50.

(2-mercaptoethanol[372–375] as a general rule) *N*-acetyl-L-cysteine[376] or its Boc analogue[377] is used as the derivatization reagent with *o*-phthaldialdehyde. Other points of interest from these papers include a rapid analysis (16 min),[371] multiple-step gradient procedures,[374] and accurate determination of D-: L-aspartic acid ratio (5 picomoles of D-aspartic acid in the presence of 500 picomoles of the L-enantiometer[376]). In other derivatization procedures, 1-fluoro-2,4-dinitrophenyl-5-L-alanineamide (referred to in instrumentation manufacturers' advertisements as 'Marfey's reagent') enables the separation and determination of enantiometer ratios for amino acids by reversed-phase h.p.l.c.,[378] and the use of *N*-benzyloxycarbonylamino acids amidated with glycyl *p*-nitrophenyl-L-alanine methyl ester achieves the same objective.[379]

The conversion of amino acid mixtures into *N*-phenylthiocarbamoyl derivatives through pre-column treatment with phenyl isothiocyanate is advocated repeatedly in the recent literature[380] (see also Vol. 16, p. 41), 1 picomole sensitivity being claimed. The process is based on the coupling step of the Edman degradation, although, since the conditions appear to be the same as those yielding phenylthiohydantoins in Edman's hands,[381] the certainty of analysing phenylthiocarbamoyl amino acids rather than mixtures of these with their isomeric phenylthiohydantoins seems in doubt.

Phenylthiohydantoins continue to generate analytical interest, h.p.l.c. on octadecylsilane–Ultrasphere[382] and reversed-phase h.p.l.c. on phenylthiohydantoins of *N*-methyl amino acids[383] featuring in a number of careful studies.[382–384] Structurally related *p*-dimethylaminoazophenylthiohydantoins also feature in recent papers.[385,386] Use of a chiral derivatizing agent, 2,3,4,6-tetra-*O*-acetyl-β-D-glucopyranosyl isothiocyanate, for generating diastereoisomeric mixtures of the resulting thiohydantoins with partly racemic amino acids has been compared for h.p.l.c. enantiomeric analysis with the use of simpler analogues but with a chiral mobile phase [copper(II) *N*-tosyl-D-phenylglycine, *cf.* refs. 365 and 367).[387]

372 J. D. H. Cooper, G. Ogden, J. McIntosh, and D. C. Parnell, *Anal. Biochem.*, 1984, **142**, 98.
373 B. N. Jones and J. P. Gilligan, *Am. Biotechnol. Lab.*, 1983, **46**, 48.
374 G. A. Qureshi, L. Fohlin, and J. Bergstroem, *J. Chromatogr.*, 1984, **297**, 91.
375 C. Cloete, *J. Liq. Chromatogr.*, 1984, **7**, 1979.
376 D. W. Aswad, *Anal. Biochem.*, 1984, **137**, 405.
377 R. H. Buck and K. Krimmen, *J. Chromatogr.*, 1984, **315**, 279.
378 P. Marfey, *Carlsberg Res. Commun.*, 1984, **49**, 591.
379 T. Yamada, M. Shimamura, T. Miyazawa, and S. Kuwata, Proceedings of the 21st Peptide Chemistry Conference, p. 31.
380 B. A. Bidlingmeyer, S. A. Cohen, and T. L. Tarvin, *J. Chromatogr.*, 1984, **336**, 93; S. A. Cohen, T. L. Tarvin, and B. A. Bidlingmeyer, *Am. Lab. (Fairfield, Conn.)*, 1984, **16**, 48, 50, 56; S. A. Cohen, *BioTechniques*, 1984, **2**, 273.
381 P. Edman, *Acta Chem. Scand.*, 1950, **25**, 585.
382 A. S. Bhown and J. C. Bennett, *Anal. Biochem.*, 1984, **137**, 256.
383 A. S. Bhown and J. C. Bennett, *J. Chromatogr.*, 1984, **314**, 467.
384 R. L. Cunio, R. Simpson, L. Correia, and C. T. Wehr, *J. Chromatogr.*, 1984, **336**, 105.
385 C. Y. Yang and S. J. Wakil, *Anal. Biochem.*, 1984, **137**, 54.
386 A. Lehmann and B. Wittmann-Liebold, *F.E.B.S. Lett.*, 1984, **176**, 360.
387 K. Nimura, A. Toyama, and T. Kinoshita, *J. Chromatogr.*, 1984, **316**, 547.

The foregoing paragraphs underline the growing preoccupation with assessments of both identity and enantiomeric purity of amino acid samples, and an example of a covalently bonded chiral stationary phase for the purpose, Lichrosorb-NH_2, to which an alkylaminocarbonyl-L-valine is condensed, has been evaluated.[388]

Fluorescence Methods. – Reference has been made in preceding sections to a number of common fluorogenic reagents used in conjunction with a chromatographic procedure. Imino acids do not react with *o*-phthaldialdehyde, and they can be estimated in an amino acid mixture after all amino acids have been derivatized with this reagent by treatment with 4-chloro-7-nitrobenzo-2-oxa-1,3-diazole (giving derivatives with $\lambda_{emission}$ 524–540 nm).[16] The procedure has been used to assess the presence of *cis*- and *trans*-3- and -4-hydroxyprolines in collagen hydrolysates[16] and also the estimation of sarcosine (using t.l.c. for separation).[389]

Determination of Specific Amino Acids. – Colorimetric methods have been used for assays of arginine (thymol with NaOBr yields a yellow product, λ_{max} 440 nm),[390] hydroxyproline (Chloramine-T with *p*-dimethylaminobenzaldehyde gives a red product, λ_{max} 560 nm),[391] and γ-carboxyglutamic acid (red product, λ_{max} 530, with 4-diazobenzenesulphonic acid, not yielded by other amino acids).[392]

Enzymatic methods also continue to provide reliable assay procedures because of their specificity; recent examples are arginase for the conversion of arginine in serum into urea,[393] asparaginase on an ammonia-gas sensor for the estimation of asparagine,[394] and a radioenzymatic method for *S*-adenosyl-L-methionine based on doubly labelled melatonin formation.[395] A spectacular example of things to come, perhaps, is the use of a column of immobilized exopeptidase as an inlet to a thermospray mass-spectrometry ion source, for the analysis of amino acids released sequentially from the C-terminus of a peptide.[396]

[388] N. Oi and H. Kitahara, *J. Chromatogr.*, 1984, **285**, 198.
[389] G. Bellon, A. M. Lundy, A. Malgras, and J. P. Borel, *J. Chromatogr.*, 1984, **311**, 405.
[390] C. S. P. Sastry and M. K. Tummuru, *Food Chem.*, 1984, **15**, 257.
[391] A. Knorr, *Dtsch. Gesundheitswes.*, 1984, **39**, 1713.
[392] N. D. Danielson, Y. P. Wu, D. K. Morgan, and J. L. Glajch, *Anal. Chem.*, 1985, **57**, 185.
[393] I. Roberts and I. M. Smith, *Ann. Clin. Biochem.*, 1984, **21**, 515.
[394] D. P. Nikolelis, *Anal. Chim. Acta*, 1984, **161**, 343.
[395] P. Guilidori and G. Stramentinoli, *Anal. Biochem.*, 1984, **137**, 217.
[396] D. Pilosof, H. Y. Kim, M. L. Vestal, and D. F. Dyckes, *Biomed. Mass Spectrom.*, 1984, **11**, 403.

2
Peptide Synthesis

BY I. J. GALPIN
Appendices compiled by
C. M. GALPIN

1 Introduction*

During the past year there has once again been a large increase in the number of published papers concerned with the synthesis of peptides and new methods for use in peptide synthesis. Polyamide- and polystyrene-based supports are both now widely used for solid-phase work, and many peptides are still produced by solution chemistry. Indeed, solution methods still probably provide the best means of preparing modified peptides, such as the 'retro-inverso' peptides or peptides containing thio-amide linkages. Increasingly, we see modification of the peptide backbone, giving rise to compounds with additional enzymic stability, or indeed enzyme inhibitors.

The characterization of synthetic peptides prior to biological evaluation is an important area, and h.p.l.c. is now probably the most used chromatographic technique. Fast-atom-bombardment mass spectroscopy is now widely used for peptide characterization;[1-4] both positive- and negative-ion modes may be useful, and under ideal conditions with the appropriate instrument m/z values of up to 5000–6000 may be measured.[2]

The proceedings of the 18th European Peptide Symposium held in Djuronaset[5] and of the 22nd Symposium on Peptide Chemistry held in Fukuoka, Japan,[6] have been published. As in previous volumes of this series no attempt has been made to cover the content of these meetings.

Volume 6 in the series entitled 'The Peptides, Analysis, Synthesis, Biology' has been published.[7] This volume, which is the first under the new editorial

* For unusual abbreviations employed in this chapter see the introduction to Appendix II, p. 96.

[1] K. B. Tomer, F. W. Crow, M. L. Gross, and K. D. Kopple, *Anal. Chem.*, 1984, **56**, 880.
[2] A. M. Buko, L. R. Phillips, and B. A. Fraser, *Biomed. Mass Spectrom.*, 1983, **10**, 408.
[3] J. M. Gilliam, P. W. Landis, and J. L. Occolowitz, *Anal. Chem.*, 1984, **56**, 2285.
[4] M. E. Rose, M. C. Prescott, A. H. Wilby, and I. J. Galpin, *Biomed. Mass Spectrom.*, 1984, **11**, 10.
[5] Proceedings of the Eighteenth European Peptide Symposium, Djuronaset, Sweden, ed. U. Ragnarson, Almqvist and Wiksell International, Stockholm, 1985.
[6] Proceedings of the 22nd Symposium on Peptide Chemistry, Fukuoka, Japan, 1984, ed. Ph. D. N. Izumiya, Peptide Institute, Protein Research Foundation, Osaka, 1984.
[7] 'The Peptides, Analysis, Synthesis, Biology', ed. S. Udenfriend and J. Meienhoffer, Academic Press, New York, 1984, Vol. 6 (Opioid Peptides: Biology, Chemistry and Genetics).

combination of Udenfriend and Meienhoffer, deals with opioid peptides covering biology, chemistry, and genetic aspects of these important compounds. A new text[8] dealing with the chemistry and biochemistry of the amino acids has appeared, as has a new series of peptide and protein reviews.[9] Specialist texts dealing with metalloproteins have been published; the first volume[10] deals with metal proteins with redox roles and the second volume[11] with those proteins that do not have a redox role.

As in previous years, this chapter covers virtually all of the published literature in this area, and because of the large number of references involved many papers have only been mentioned in the Appendices.

2 Methods

Protective Groups. – *Established Methods of Amino-group Protection.* New methods continue to be developed for the introduction of the t-butyloxycarbonyl (Boc) group. t-Butyl-2-pyridyl carbonate has been developed[12] for the introduction of the Boc group and is prepared by the route outlined in Scheme 1. The stable, crystalline mixed carbonate (1) routinely gives 80–90% yield of the Boc-amino acid when the parent amino acid is reacted with it in the presence of triethylamine, using aqueous DMF as the solvent. A full account of the use of t-butylamino carbonate (2) has now been published.[13] This compound, which may be prepared from hydroxylamine and Boc-anhydride, reacts between one-and-a-half and two-and-a-half times as rapidly as Boc-anhydride and also has the ability to acylate in acidic solution.

The use of 3-alkoxycarbonyl-2-oxazolones and their homopolymers for the introduction of urethane-protecting groups has been reported.[14] The oxazolones

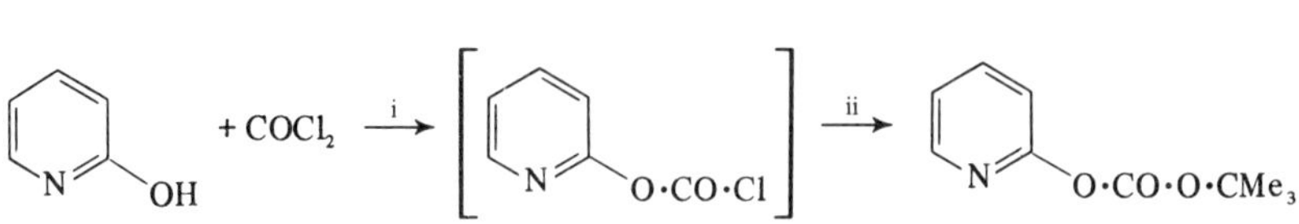

(1)

Reagents: i, pyridine/CH_2Cl_2, 0 °C, $\frac{1}{2}$ h; ii, Bu^tOH, pyridine/CH_2Cl_2/RT, 5 h

Scheme 1

$Me_3C·O·CO·ONH_2$

(2)

[8] 'Chemistry and Biochemistry of the Amino Acids', ed. G. C. Barrett, Chapman and Hall, London, 1984.

[9] 'Peptide and Protein Review', ed. M. T. W. Hearn, Marcel Dekker A.G. Verlag, Basel, Switzerland, 1984, Vols. 1–3.

[10] 'Metalloproteins', ed. P. M. Harrison, Macmillan Press, Basingstoke, 1984, Vol. 1.

[11] 'Metalloproteins', ed. P. M. Harrison, Macmillan Press, Basingstoke, 1984, Vol. 2.

[12] S. Kim and J. I. Lee, *Chem. Lett.*, 1984, 237.

[13] R. B. Harris and I. B. Wilson, *Int. J. Pept. Protein Res.*, 1984, **23**, 55.

[14] T. Kunieda, T. Higuchi, Y. Abe, and M. Hirobe, *Chem. Pharm. Bull.*, 1984, **32**, 2174.

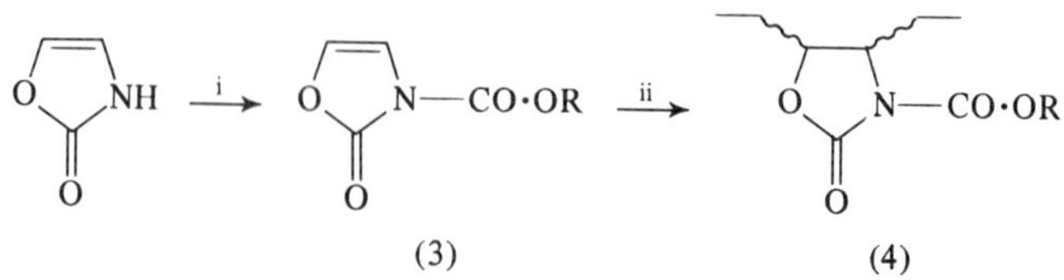

Reagents: i, (*a*) $CO{\cdot}Cl_2$ or $Cl{\cdot}CO{\cdot}CCl_3$, (*b*) ROH (R = Bu^t, CH_2Ph, or *p*-$CH_2{\cdot}C_6H_4{\cdot}OMe$); ii, benzoyl peroxide

Scheme 2

(3) or their homopolymers (4) are prepared according to the route outlined in Scheme 2. Most of the oxazolones of type (3) are easily prepared and may be polymerized following the addition of benzoylperoxide to give the homopolymer (4). However, the 3-t-butoxycarbonyl-oxazolone (Boc-Ox) could not be polymerized. The use of Boc-Ox and the related benzyloxycarbonyl derivative (Cbz-Ox) was demonstrated. In the case of Boc-Ox the Boc group was introduced in the presence of triethylamine and dimethylaminopyridine, whereas the Cbz-Ox gave a satisfactory yield of the benzyloxycarbonyl derivative in the presence of triethylamine without the addition of dimethylaminopyridine. These reagents are claimed to be selective in that they are able to acylate the α-amino function of serine and threonine, without reaction with the hydroxyl side chain.

Occasionally *N*-formyl peptides are required, and a simple method whereby these may be produced from Boc peptides has been developed.[15] The Boc peptide is deprotected with 98% formic acid, then after removal of excess formic acid ethoxycarbonylethoxydihydroquinoline (EEDQ) is added. After 2–4 hours at room temperature the formylation is complete and the product may be isolated by precipitation with hexane.

An improved method for the preparation of Nps amino acids has been reported,[16] as the Nps chloride which is usually used in the preparation of these derivatives is relatively unstable. In this modification 2,2-dinitrophenyl disulphide is treated with a stock solution of chlorine for 10 minutes to prepare the Nps chloride *in situ*. The Nps chloride is then reacted with the amino acid in aqueous NaOH, and whilst the solution is still alkaline extraction with ether is carried out to remove any unreacted Nps chloride and other side products. The current modification emphasizes that the extraction procedure is crucial if Nps amino acids of high quality are to be produced.

New Methods of Amino-group Protection. Several new urethane-based protecting groups have been studied. The 1,3-dibromo-2-methyl-2-propyloxycarbonyl (DBTBoc) group has been evaluated as an acid-stable, solvolytically removable amino-protecting group.[17] The protecting group (5) is stable to TFA, HCl in

[15] G. Lajoie and J.-L. Kraus, *Peptides*, 1984, **5**, 653.

[16] R. Katakai, H. Shida, and T. Takada, *Biopolymers*, 1984, **23**, 1397.

[17] L. A. Carpino, N. W. Rice, E. M. E. Mansour, and S. A. Triolo, *J. Org. Chem.*, 1984, **49**, 836.

nitromethane, and HCl in ethyl acetate; however, on warming overnight in methanol or ethanol, cleavage takes place to liberate the free α-amino function. The group has a half-life of 4.8 hours at 37 °C when treated with methanol: water (95:5), and it was noted that the mono-bromo compound has a half-life of 11 hours under similar conditions. A series of 2-(4-pyridyl)ethoxycarbonyl-(4-Pyoc) amino acids has been prepared.[18] These derivatives are prepared by reaction of the nitrophenyl mixed carbonate (6) with the amino acid sodium salt, using acetonitrile/water as the solvent. The protecting group has similar properties to the 2-Pyoc group,[19] being stable to acidic and basic conditions and being considerably hydrophilic in nature, thus facilitating synthesis in aqueous media. The group, which is removed by treatment with morpholine after quaternization with methyl iodide in acetonitrile, has been applied to the synthesis of serine glycopeptides.

Allyloxycarbonyl amino acids have been synthesized by reaction of allyl-chloroformate with the corresponding amino acid between pH 10 and 11, although the use of a pH stat is recommended.[20] The allyloxycarbonyl group may be removed by hydrogenolysis in the presence of a catalyst; however, side products that do not undergo cleavage may be produced. A very mild method of cleavage involves treatment with tetrakis(triphenylphosphane)palladium(0) with 5,5-dimethyl-1,3-cyclohexane dione (dimidone) as the allyl acceptor. Using dimidone in an eight-fold excess, removal may be affected in 30 min at room temperature. The protecting group is unaffected by treatment with trifluoroacetic acid, and therefore t-butyl-based protecting groups may be used in combination with the allyloxycarbonyl group.

The use of isopropyl- or cyclohexyl-3-ethoxy-4-nitro-2-oxo-3-pyrroline (7) as amino-protecting groups has been suggested.[21] These protecting groups facilitate solubility in partially aqueous media and are easily removed by treatment with ammonium hydroxide at room temperature.

The 3,5-dinitro-1-(4-nitrophenyl) or pyridone derivatives of amino acids provide a method of totally blocking the amino group of an amino acid.[22] The group is introduced by a reaction of an amino acid with the compound (8) as indicated in Scheme 3. Removal of the protection requires treatment with an

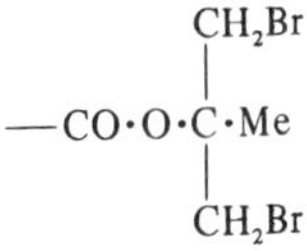

(5)

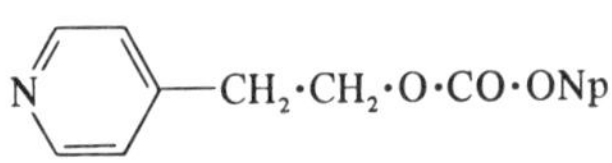

(6)

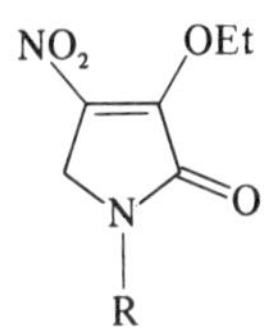

(7) R = Me_2CH or cyclohexyl

[18] H. Kunz and S. Birnbach, *Tetrahedron Lett.*, 1984, **25**, 3567.

[19] H. Kunz and R. Barthels, *Angew. Chem., Int. Ed. Engl.*, 1983, **22**, 783.

[20] H. Kunz and C. Unverzagt, *Angew. Chem.*, 1984, **96**, 426.

[21] P. L. Southwick, G. K. Chin, M. A. Koshute, J. R. Miller, K. E. Niemela, C. A. Siegel, R. T. Nolte, and W. E. Brown, *J. Org. Chem.*, 1984, **49**, 1130.

[22] E. Matsumura, H. Kobayashi, T. Nishikawa, M. Ariga, Y. Tohda, and T. Kawashima, *Bull. Chem. Soc. Jpn.*, 1984, **57**, 1961.

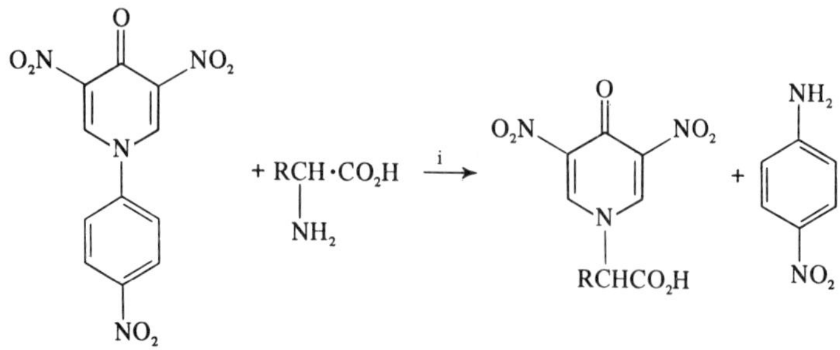

(8)

Reagents: i, aqueous pyridine, 2–24 h

Scheme 3

amine that is more nucleophilic than the α-amino function of an amino acid; thus deblocking may readily be achieved by treatment with hexylamine in pyridine for 1–2 hours. The reaction involves amine interchange, and under these conditions no racemization was observed.

A detailed study of the use of the diphenylphosphinyl-protecting group has appeared.[23] Diphenylphosphinyl (Dpp) protection is normally introduced by reacting an amino acid methyl or benzyl ester salt with diphenylphosphinyl chloride in the presence of *N*-methyl morpholine. The resulting Dpp amino acid ester may then be hydrolysed by alkaline treatment, or by hydrogenolysis in the case of benzyl esters. The resulting Dpp amino acid is obtained in yields of between 65 and 85%. Cleavage under acidic conditions was monitored by use of ^{31}P n.m.r.; the cleavage proceeds by *N*-protonation followed by nucleophilic attack. The reaction proceeds *via* a trigonal-bipyramidal intermediate, the nitrogen geometry taking up the form of a flattened tetrahedron and the lone pair on nitrogen almost becoming in plane with the NPO plane. This is in comparison with the amide hydrolysis in which the lone pair on nitrogen is orthogonal to the NCO plane. These stereoelectronic factors lead to a rate difference of 10^5 between cleavage of diphenylphosphinamide and benzamide. Other methods of introducing the Dpp group were explored; Dpp azide and Dpp chlorophenyl ester were not as effective, and the nitrophenyl ester did not react at all. It was found that benzyloxycarbonyl and Dpp amino acids behaved in a similar manner during coupling, showing little racemization, although in the Dpp case no cyclic intermediates in the form similar to an oxazolone were observed. The group was found to be slightly more labile to acid than a Boc group, and under ideal conditions selective cleavage of Dpp in the presence of a t-butyl ester could be achieved using 15% TFA in deuteriochloroform. Ideally, deprotection was carried out using 2 equivalents of *p*-toluene sulphonic acid in methanol over 1–6 hours or 6 equivalents of HCl in methanol over $2\frac{1}{2}$ hours. When phenyl esters were used, *p*-toluene sulphonic acid in isopropanol or DMF was employed in order to avoid the danger of transesterification.

[23] R. Ramage, D. Hopton, M. J. Parrott, G. W. Kenner, and G. A. Moore, *J. Chem. Soc., Perkin Trans. 1*, 1984, 1357.

$$(Pr^{i}O)_2P(=O)—$$

(9)

$$Ph_2CH·O·P(=O)(OPh)_2$$

(10)

The di-isopropyl phosphoryl group (9) has also been explored,[24] and it was found that this group had considerable acid stability.

Carboxy-group Protection. Bodanszky has drawn attention[25] to the fact that the original method of Curtius developed 100 years ago[26] is still frequently satisfactory for the preparation of alkyl esters of amino acids. Generally, the acid-catalysed reaction of methyl or ethyl alcohol with the amino acid gives a good yield of the amino acid ester; both HCl and *p*-toluene sulphonic acid may be used as catalysts. Bodanszky's paper goes on to discuss alkyl sulphites and alkyl tosylates and indicates that they are not necessarily the best reagents to choose for the preparation of amino acid alkyl esters.

The compound diphenylmethyl diphenylphosphate (10)[27] is an activated ester that may be used in the preparation of the *O*-diphenylmethyl derivative of alcohols and carboxylic acids. The compound (10) is prepared by reaction of diphenylmethyl chloride with the silver salt of diphenylphosphoric acid; alternatively it may be prepared by reaction of diphenyldiazomethane with diphenylphosphoric acid. This reagent generally reacts faster with hydroxyl groups than with carboxyl groups and therefore allows selective reaction. However, it may be used to prepare diphenylmethyl esters when there is no competing hydroxyl group in the side chain. In the presence of an acid catalyst (*p*-toluene sulphonic acid or trifluoroacetic acid) the rate of reaction with both hydroxyl and carboxyl groups is enhanced, and under these conditions no selectivity is observed and reaction with both functional groups occurs.

An example of the type of complex procedure outlined by Bodanszky as being unnecessary[25] is provided by the route shown in Scheme 4.[28] In this reaction sequence the ethyl acetoacetate is reacted with an amino acid to give an intermediate enamine (11), which is subsequently alkylated and ultimately hydrolysed to give the protected amino acid. The method is claimed to be an improvement on established procedures; however, it clearly involves more stages than the direct acid-catalysed esterification that has been satisfactory on many occasions.

Methyl thiomethyl esters have recently been proposed as a new carboxy-protecting group.[29] The esters are prepared by reaction of an N-protected amino acid with t-butyl bromide in DMSO in the presence of sodium bicarbonate. These esters may be cleaved by HCl in anhydrous ether, although an improved

[24] Y.-F. Zhao, S.-K. Xi, G.-J. Ji, A.-T. Song, H.-G. Tang, and Y.-F. Tian, *Acta Chim. Sin.*, 1984, **42**, 358.
[25] M. Bodanszky, *Int. J. Pept. Protein Res.*, 1984, **23**, 111.
[26] Th. Curtius, *Ber. Dtsch. Chem. Ges.*, 1883, **16**, 753.
[27] M. Kolovos and C. Froussios, *Tetrahedron Lett.*, 1984, **25**, 3909.
[28] A. M. Kotodziejczyk and M. Slebioda, *Synthesis*, 1984, 865.
[29] A. Dossena, G. Palla, R. Marchelli, and T. Lodi, *Int. J. Pept. Protein Res.*, 1984, **23**, 198.

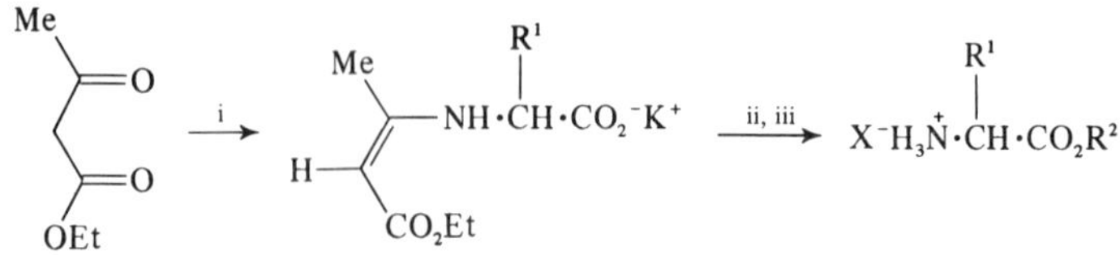

Reagents: i, amino acid/DMSO; ii, alkylation/R^2X (R^2 = Me, Et, or Bzl, X = Cl, Br, or Tos·O); iii, hydrolysis

Scheme 4

cleavage which allows selective removal may be brought about by treatment with hydrogen peroxide in the presence of ammonium molybdate followed by alkaline hydrolysis. It is claimed that no racemization of the amino acid is observed during removal of the methyl thiomethyl ester.

2-(2-Pyridyl)ethyl esters (12) and 2-(4-pyridyl)ethyl esters (13) have been utilized for the protection of carboxy groups.[30,31] The 2-pyridyl esters (12) can be introduced by a reaction of an N-protected amino acid with the corresponding alcohol in the presence of DCCI and HOBt[30] or by reaction with the alcohol in the presence of DCCI and dimethylamino pyridine.[31] Deprotection is achieved by quaternization with methyl iodide followed by treatment with morpholine. A possible advantage lies in the use of the 4-pyridyl methyl esters (13) as they are methylated using methyl iodide at 20 °C in acetonitrile in 3 hours, whereas the corresponding 2-pyridyl esters require 4 days for complete alkylation. One major disadvantage of this type of protecting group is that they cannot be used on peptides that contain easily alkylated centres, such as the sulphur-containing amino acids; therefore their use is rather restricted.

Allyl esters[32] have also been used and may be useful in such areas as the preparation of *O*-glycopeptides. The allyl esters are selectively removable under very mild conditions in the presence of benzyloxycarbonyl or t-butyloxycarbonyl; removal utilizes tetrakis(triphenylphosphane)palladium(0) to bring about isomerization of the allyl ester group. In the presence of a 10% excess of morpholine the allyl residue is transferred to morpholine, giving a virtually quantitative cleavage.

2-(Diphenylphosphino)ethyl esters (14) have also been evaluated as carboxy-protecting groups.[33] These esters are prepared by reaction of the corresponding

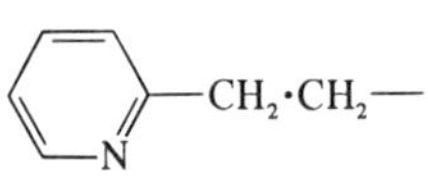

(12)

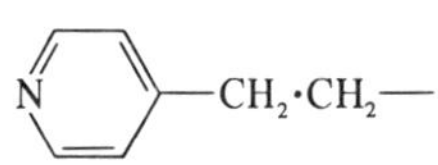

(13)

$Ph_2P \cdot CH_2 \cdot CH_2$—

(14)

[30] H. Kessler, G. Becker, H. Kogler, and M. Wolff, *Tetrahedron Lett.*, 1984, **25**, 3971.
[31] H. Kunz and M. Kneip, *Angew. Chem., Int. Ed. Engl.*, 1984, **23**, 716.
[32] H. Kunz and H. Waldmann, *Angew. Chem., Int. Ed. Engl.*, 1984, **23**, 71.
[33] D. Chantreux, J.-P. Gamet, R. Jacquier, and J. Verducci, *Tetrahedron*, 1984, **40**, 3087.

alcohol with an N-protected amino acid, using DCCI and dimethylaminopyridine to form the ester linkage. The esters are relatively stable and may be deprotected by quaternization with methyl iodide followed by β-elimination brought about by treatment with fluoride ion or potassium carbonate. Clearly this protecting group also suffers from the problems associated with an alkylation step being part of the removal procedure.

An interesting anionic protecting group for use in peptide synthesis has been investigated.[34,35] The 4-sulphobenzyl esters (15) are introduced by reaction of the caesium or DCHA salts of an N-protected amino acid with the corresponding bromomethyl sulphonic acid. The 4-sulphobenzyl esters may be removed by hydrogenolysis using a palladium catalyst or by alkaline saponification. They may also easily be converted to amides or hydrazides. They are stable to HBr in acetic acid and trifluoroacetic acid and may therefore be used in conjunction with a wide range of amino-protecting groups. Their considerable hydrophilicity facilitates purification by ion exchange, and their utility has been demonstrated by a step-wise synthesis of enkephalin using a variety of coupling methods.

Side reactions involving the side-chain carboxy group of aspartic and glutamic acids have again been reported. It was noted that aspartimide formation was particularly prevalent when β-benzylaspartic acid was adjacent to glycine; however, the reaction seemed particularly facile in DMF or methanol, whereas in THF it did not appear to be a problem.[36]

An interesting study of peptides of type (16) has been carried out to examine the formation of internal pyroglutamic acid residues.[37] On treatment of the peptide in DMF with carbonyldi-imidazole it was found that an internal pyroglutamic acid could be formed by cyclization at the amide nitrogen (a) whereas cyclization to the 6-membered imide by reaction with the NH (b) could also occur. High yields of such products could be obtained, and the course of reaction was determined by steric factors depending on the bulk of the R^1 and R^2 groups.

Side-chain Protection. The development of new protecting groups for the thiol function of cysteine has received considerable attention. Arylvinyl sulphones of type (17) have been studied by two groups of workers.[38,39] The protecting

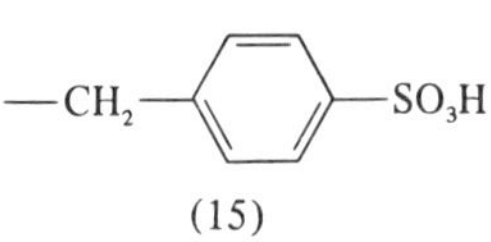

(15)

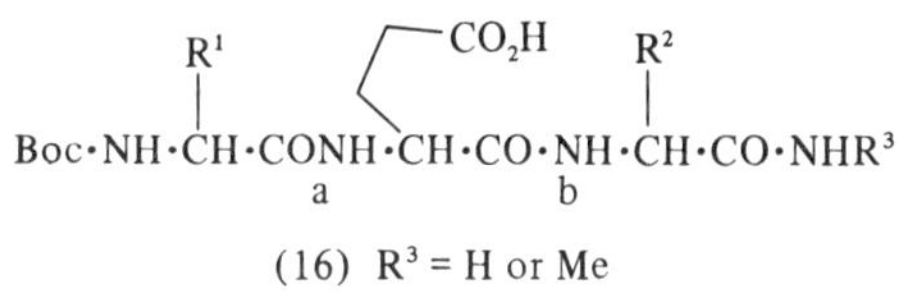

(16) R^3 = H or Me

[34] R. Bindewald, A. Hubbuch, W. Danho, E. E. Buellesbach, J. Foehles, and H. Zahn, *Int. J. Pept. Protein Res.*, 1984, **23**, 368.
[35] R. Bindewald, W. Danho, E. Buellesbach, M. Bodanszky, and H. Zahn, *Int. J. Pept. Protein Res.*, 1984, **23**, 376.
[36] E. A. Hagen, T. Loennechen, L. K. Sydnes, and A. J. Aasen, *Int. J. Pept. Protein Res.*, 1984, **23**, 642.
[37] S. A. Khan and B. W. Erickson, *J. Am. Chem. Soc.*, 1984, **106**, 798.
[38] L. Horner and H. Lindel, *Liebigs Ann. Chem.*, 1985, 22.
[39] Y. Kuroki and R. Lett, *Tetrahedron Lett.*, 1984, **25**, 197.

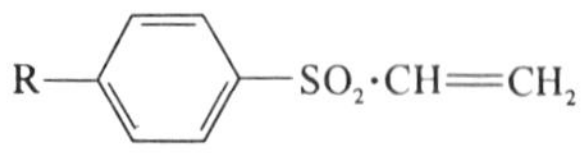

(17) R = H or CO_2Et (ref. 38) or H or Me (ref. 39)

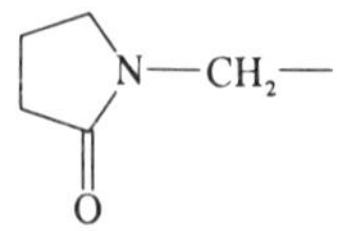

(18)

$PhI(OAc)_2$

(19)

group is introduced by Michael addition of the thiol to the arylvinyl sulphone and may be removed under basic conditions by the use of potassium t-butoxide or sodium methoxide to bring about β elimination.

The (2-oxo-1-pyrolidinyl)methyl group (18) has been evaluated for thiol protection.[40] The group, which bears some similarity to the acetamidomethyl group, is introduced by an acid-catalysed reaction between cysteine and hydroxymethyl pyrolidinone over 30 min. The protected amino acid may easily be converted to its Boc derivative or to its methyl ester, and these compounds have been used in the synthesis of model cysteine peptides. Removal of the group may be achieved by treatment with iodine in methanol and is complete in about 40 min. The protecting group showed stability over a wide range of pH, being stable between a pH of 0.5 and 13.

Use of the phosphinothioyl-protecting group for cysteine has now been evaluated,[41] and it has been found that removal of this group may be readily achieved using potassium fluoride in the presence of an equimolar quantity of 18-crown-6. An instantaneous removal was observed in aprotic solvents; however, in methanol the reaction was slow owing to solvation of the fluoride ion. However, mixtures of acetonitrile or dichloromethane and methanol seem to be satisfactory and up to 50% methanol may be tolerated. Selective removal of the α-amino phosphinothioyl-protecting group from MPT-Cys(MPT)-OH may readily be achieved using 0.4M HCl in methanol.

A study of the use of diacetoxyphenyliodine for the direct oxidative conversion of cysteine derivatives to cystine compounds has been carried out.[42] This compound (19) has been reacted with Cys(Trt), Cys(Dpm), and Cys(Acm) derivatives, giving the cystine peptide directly. The reaction may be carried out in the presence of acid-labile t-butyl-based groups or of diphenylmethyl esters.

Oxidative deblocking of *S*-4-methoxybenzyl cysteinyl peptides has been studied using homogeneous electron transfer employing the tris-4-bromophenylammoniumyl radical cation.[43] The methoxybenzyl group may be removed in the presence of Boc, Z, and t-butyl and benzyl esters. The ammoniumyl radical cation may be generated by indirect electrolysis at the anode in acetonitrile/2,6-lutidine or by use of the hexachlorostilbnate salt. As the methoxybenzyl group is unaffected by oxidative deblocking of trityl or diphenylmethyl groups with iodine, this method of removal allows ordered disulphide-bond formation by the use of a combination of thiol-protecting groups.

[40] O. S. Papsuevich, G. I. Aukone, S. Y. Miksta, and U. O. Kalei, *J. Gen. Chem. U.S.S.R.*, 1982, **52**, 404.
[41] M. Ueki and K. Shinozaki, *Bull. Chem. Soc. Jpn.*, 1984, **57**, 2156.
[42] M. G. Kolovos and P. Moutevelis-Minakakis, *Tetrahedron Lett.*, 1984, **25**, 4153.
[43] M. Platen and E. Steckhan, *Liebigs Ann. Chem.*, 1984, 1563.

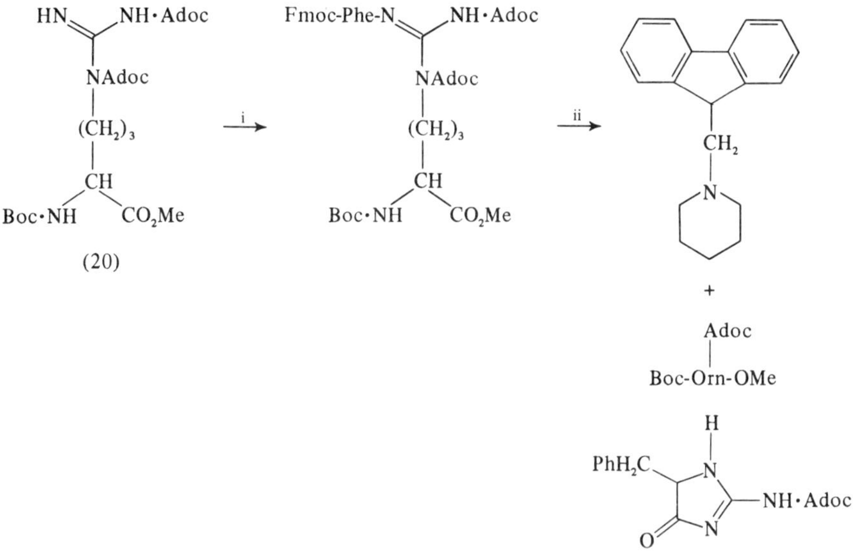

Reagents: i, $(Fmoc\text{-}Phe)_2O$, 1.5 equiv., 1 h, T_R; ii, piperidine/DMF (1/1), 10 min, T_R

Scheme 5

In a synthesis of the C-terminal fragment of human leukocyte interferon α-F, partial conversion of arginine to ornithine was observed when the bis-adamantyloxycarbonyl was used for protection of the guanidine function of arginine.[44] The formation of ornithine, which has been known for some time, was investigated using *N*-α-Boc-bis-Adoc-arginine-methyl ester (20). The compound was subjected to the reaction sequence shown in Scheme 5, and from this it can be seen that under these conditions ornithine derivatives may be produced that are similar to those employed during solid-phase synthesis employing Fmoc for N-α-protection and piperidine for deblocking. It is clear, therefore, that care should be exercised when employing urethane-protected arginine derivatives; however, it is interesting to note that the Boc group has been successfully used in a solid-phase synthesis employing Fmoc-Arg(Boc)-OH.[45]

In a recent synthesis[46] of an opiate receptor mimetic peptide, ornithine residues at positions 1, 7, 24, and 30 were incorporated as their phthaloyl derivatives with a view to hydrazinolysis and subsequent guanidination with *O*-methylisourea. During the guanidination, incomplete reaction was observed, and eventually only 13.6% of the required arginine peptide was produced. This finding reinforces results of other workers which suggest that when several ornithine residues are to be converted to arginine within the same molecule complete reaction is difficult to achieve.

[44] H. Rink, P. Sieber, and F. Raschdorf, *Tetrahedron Lett.*, 1984, **25**, 621.
[45] R. Colombo, *Int. J. Pept. Protein Res.*, 1982, **19**, 71.
[46] W. Kullmann, *J. Med. Chem.*, 1984, **27**, 106.

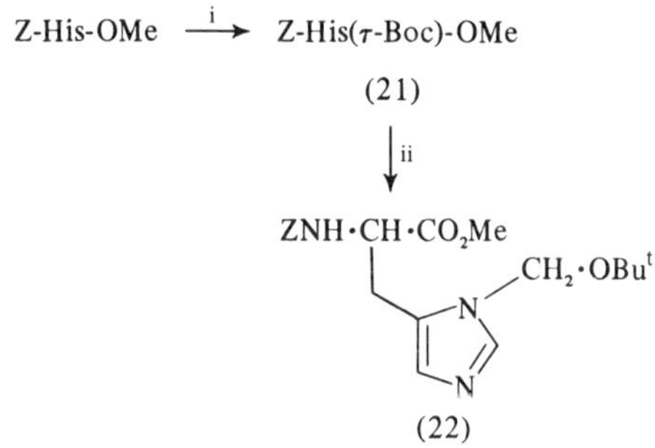

Reagents: i, Boc_2O/MeOH; ii, $Me_3C{\cdot}O{\cdot}CH_2Cl$

Scheme 6

The t-butoxymethyl-protecting group has been developed as an acid-labile protection for the π-nitrogen of histidine.[47] The protecting group (Bum) is stable to base and to hydrogenolysis but is ultimately removed by mild acidolysis. The method of introduction, which is shown in Scheme 6, involves preparation of the intermediate *N*-τ-Boc derivative (21), which is ultimately converted to the π-Bum derivative (22). Compound (22) may then be subjected to alkaline hydrolysis and hydrogenolysis to give the π-Bum-protected histidine. Reprotection of the *N*-α-amino group with Fmoc chloride allows preparation of a derivative that is useful in solid-phase synthesis.

Protection of the indole nitrogen of tryptophan has been shown to be advantageous on several occasions. For example, when N^{in}-formyl tryptophan was used,[48] oxidative destruction of tryptophan during iodination of tyrosine was prevented. The formyl group was subsequently removed by mild alkaline treatment.

Bis-Boc-tryptophan methyl ester was prepared by treatment of *N*-α-Boc-Trp-OMe with Boc_2O in the presence of dimethylaminopyridine.[49] The methyl ester could be readily hydrolysed by treatment with sodium hydroxide, and both Boc groups could be removed on treatment with TFA. N-α-Deprotection could be achieved selectively by using 2.7M HCl in dioxan over 3 hours at room temperature. Although the N^{in}-Boc group deactivates the aromatic ring to electrophilic attack, preliminary experiments have shown that Boc-Trp(Boc)-OMe is partially destroyed on hydrogenolysis over palladium.

Mesitylene sulphonyl tryptophan has also been prepared.[50] This compound was prepared by a reaction of Z(OMe)-Trp-OBzl with mesitylene sulphonyl chloride in the presence of pulverized sodium hydroxide, using a catalytic amount of cetyl trimethyl ammonium chloride. The free zwitterion was prepared by alkaline hydrolysis of this compound followed by TFA treatment to remove

[47] R. Colombo, F. Colombo, and J. H. Jones, *J. Chem. Soc., Chem. Commun.*, 1984, 292.
[48] G. Mourier, L. Moroder, and A. Previero, *Z. Naturforsch., Teil B*, 1984, **39**, 101.
[49] H. Franzen, L. Grehn, and U. Ragnarsson, *J. Chem. Soc., Chem. Commun.*, 1984, 1699.
[50] N. Fujii, S. Futaki, K. Yasumura, and H. Yajima, *Chem. Pharm. Bull.*, 1984, **32**, 2660.

the Z(OMe) group. This group is stable to dilute sodium hydroxide, hydrazine hydrate, TFA, 4M HCl in dioxan, and 25% HBr in acetic acid. It is not cleaved by HF but is removed by treatment with 1M trifluoromethane sulphonic acid in trifluoroacetic acid. Thus, the protecting group may be removed at the end of a synthesis along with other acid-labile groups, employing 3,5-dimethylanisole and 2% ethane dithiol as a scavenger. The utility of this protecting group was demonstrated in a synthesis of a heptapeptide from CCK.

There is growing interest in phosphonopeptides, and to this end bis-(2-methylsulphonylethyl)phosphochloridate (23) has been developed as a new phosphorylating agent.[51] The group is very stable to acid but is rapidly removed under mildly alkaline conditions. The group was used for the conversion of tyrosine to tyrosine phosphate and may be quite useful for the synthesis of peptides containing this residue. The synthesis of peptides containing phosphoserine[52] has also been examined by the use of dibenzylphosphonoserine during the synthesis. This dibenzyl-protected phosphate was used in combination with acid-labile t-butyl-based protection. Debenzylation could be achieved by the usual methods to give the free phosphoserine peptide. The sulphation of tyrosine is also an important reaction, and to this end the introduction of sulphate esters using pyridinium acetyl sulphate (24) has been examined.[53] Serine, threonine, tyrosine, and hydroxyproline have all been investigated by the use of this reagent. The reagent (24) is prepared by initial reaction of acetic anhydride and pyridine, followed by addition of concentrated sulphuric acid. No oxidation of the methionine or sulphation of tryptophan was observed.

$$\mathrm{Cl{\cdot}P(=O)(OCH_2CH_2{\cdot}SO_2Me)_2} \qquad (23)$$

$$\mathrm{Me{\cdot}CO{\cdot}O{\cdot}SO_3^{-}PyH^{+}} \qquad (24)$$

General Deprotection. – The use of trifluoromethane-sulphonic acid in the presence of thioanisole and TFA for the removal of benzyl-based protecting groups has increased.[54–56] In the synthesis of hylambatin,[54] deprotection using this combination was employed and satisfactory deprotection of Arg(MTS) was also achieved. *S*-Alkylation of methionine was found to be suppressed more effectively by using 3,5-dimethylanisole in the presence of 2% ethane dithiol than by using anisole plus 2% ethane dithiol, although on some occasions ethane

[51] A. Beld, C. A. A. Claesen, E. S. Roersma, W. J. M. Schippers, L. M. Keizer, and G. I. Tesser, *Recl. Trav. Chim. Pays-Bas*, 1984, **103**, 196.

[52] P. F. Alewood, J. W. Perich, and R. B. Johns, *Tetrahedron Lett.*, 1984, **25**, 987.

[53] B. Penke, F. Hajnal, J. Lonovics, G. Holzinger, T. Kadar, G. Teleggy, and J. Rivier, *J. Med. Chem.*, 1984, **27**, 845.

[54] K. Okamoto, K. Yasumura, K. Fujitani, S. Katakura, K. Akaji, H. Yajima, Y. Nakata, A. Inoue, and T. Segawa, *Chem. Pharm. Bull.*, 1984, **32**, 430.

[55] S. Kiyama, N. Fujii, H. Yajima, M. Moriga, and A. Takagi, *Int. J. Pept. Protein Res.*, 1984, **23**, 174.

[56] N. Fujii, M. Nomizu, K. Akaji, K. Watanabe, M. Shimokura, S. Katakura, H. Yajima, F. Shono, M. Tsuda, A. Yoshitake, and H. Imura, *Chem. Pharm. Bull.*, 1984, **32**, 4797.

dithiol alone[55] appears to be satisfactory. A good example of the use of this deprotection medium is provided by the synthesis of human corticotrophin-releasing factor;[56] in this work the final deprotection of the 41-residue peptide was carried out by this method and a good yield of the product was obtained.

It has been noted that the removal of *N*-benzyloxycarbonyl and benzyl esters may be accelerated by the presence of 4-methylthiophenol.[57] This compound acts as an acceptor of benzyl groups, replacing thioanisole/*m*-cresol mixtures; alkylation of tyrosine is also prevented. The use of this deprotection regime was illustrated in a synthesis of [Gly]-oxytocin; however, the limitations of the method are still to be determined.

The use of trifluoroacetic acid for Boc-group removal in the presence of benzyl-based protection may be made more selective by dilution with acetic acid.[58] Unfortunately the TFA/acetic acid mixture may cause acetylation of OH groups and the formation of t-butyl esters. Also the method is thus only useful if both hydroxy and carboxy groups are blocked. TFA in the presence of phenol and *p*-cresol(4:3:3) provided improved selectivity and suppressed any alkylation of side chains. Boc removal from a hexapeptide was achieved in three minutes, and it was noted that cleavage of benzyloxycarbonyl from a similar peptide was 6000 times slower.

In a synthesis of the wasp venom, Mastopyran,[59] the 4-methoxy-2,3,6-trimethylbenzene sulphonyl (MTR) was used. In this case accelerated cleavage was achieved using a mixture of methane sulphonic acid/TFA thioanisole in the presence of dimethyl sulphide.

The cleavage of t-butyloxycarbonyl and 4-methoxybenzyloxycarbonyl groups by treatment with *p*-toluenesulphonic acid in acetonitrile has been investigated.[60] The deprotection was carried out as indicated in Scheme 7, the urethane-protected amino acid (25) being treated with tosic acid in acetonitrile to give the *p*-toluenesulphonate salt of the amine (26) and the intermediate (27). This intermediate is formed by reaction of acetonitrile with the carbonium ion produced on cleavage of the urethane. Ultimately hydrolysis of the activated intermediate (27) gives a simple amide (28) as a by-product.

Cleavage of amino acid benzyl esters using triethylsilyl halides has been investigated.[61] The cleavage, which is carried out using palladium(II) chloride as a catalyst, generally involves a rather high temperature and will probably not ultimately be very useful in peptide synthesis.

Cleavage of the dibutyl-1-phenyl-4-oxo-2,5-cyclohexadienyl (PChd) group has been investigated using cathodic reduction.[62] The cathodic reduction carried out in a divided cell at a controlled potential allows selective cleavage of the PChd group when benzyloxycarbonyl groups are also present. The use of sodium

[57] M. Bodanszky and A. Bodanszky, *Int. J. Pept. Protein Res.*, 1984, **23**, 287.
[58] M. Bodanszky and A. Bodanszky, *Int. J. Pept. Protein Res.*, 1984, **23**, 565.
[59] K. Saito, T. Higashijima, T. Miyazawa, M. Wakimasu, and M. Fujino, *Chem. Pharm. Bull.*, 1984, **32**, 2187.
[60] H. Yamada, H. Tobiki, N. Tanno, H. Suzuki, K. Jimpo, S. Ueda, and T. Nakagome, *Bull. Chem. Soc. Jpn.*, 1984, **57**, 3333.
[61] Y. Watanabe, Y. Maki, and K. Kikuchi, *Chem. Ind.*, 1984, 272.
[62] M. H. Khalifa and A. Rieker, *Tetrahedron Lett.*, 1984, **25**, 1027.

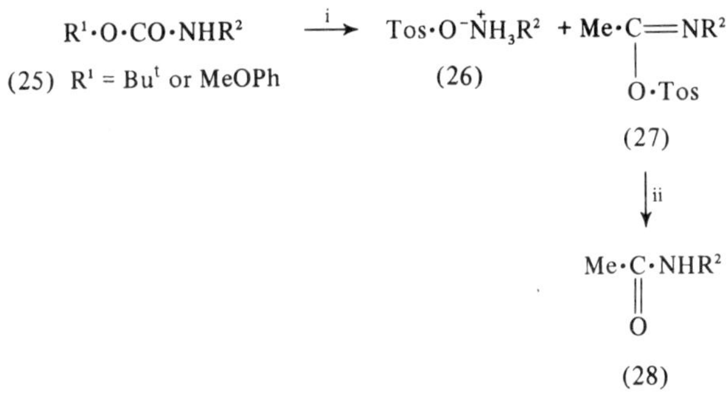

Reagents: i, Tos·OH/Me·CN; ii, H_2O

Scheme 7

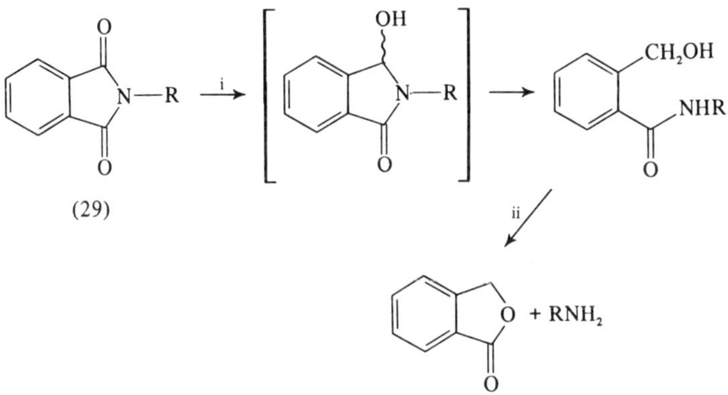

Reagents: i, $NaBH_4/Me_2CHOH/H_2O$; ii, AcOH

Scheme 8

in liquid-ammonia reduction in peptide chemistry has been reviewed, and one clear conclusion is that the optimal amount of sodium should always be determined in order to avoid side reactions.[63]

A two-stage, one-pot procedure for the removal of the phthaloyl protecting group is outlined in Scheme 8.[64] Using a 6:1 mixture of isopropanol and water the reduction of the phthaloyl amino acid (29) is easily achieved in high yield. The cyclization in the second stage is best carried out using aqueous acetic acid (pH 5) and is usually achieved in two hours at 80 °C. There is some danger of racemization under these conditions; however, in several cases no loss of optical purity was observed.

[63] I. Schoen, *Chem. Rev.*, 1984, **84**, 287.

[64] J. O. Osby, M. G. Martin, and B. Ganem, *Tetrahedron Lett.*, 1984, **25**, 2093.

Formation of the Peptide Bond. – A large number of new coupling techniques have been evaluated. Frequently the physical conditions under which a coupling is carried out are not varied, although coupling under high pressure has been examined.[65] In this work benzyloxycarbonyl amino acid hydroxysuccinamide esters were coupled to a variety of amino components. It was found that when pressures of up to 10 kbar were used the yield increased considerably; in one case the reaction at a pressure of 1 bar gave a yield of 10.5% whereas at 10 kbar the yield was 82.2%. The reactions were carried out in a Teflon capsule at room temperature for seven days, dry dichloromethane being used as the solvent. Reactions in which the amino acids had bulky substituents were dramatically improved, although often the yields were still quite low.

An interesting paper relating to the solubility of protected peptides in organic solvents has been published.[66] The predictions are based on the use of Chuand–Fassman-type parameters, which allow prediction of solubility in the dipolar aprotic solvents that are normally employed in peptide synthesis. Prediction of β-structure could be correlated with high insolubility, whereas α-helical structures generally led to high solubility. Also, hydrophobic sequences were found to be very insoluble unless a tertiary amide was present. The use of 2,4-dimethoxybenzyl or 2,4,6-trimethoxybenzyl for the protection of intermediate amide bonds in a peptide was advocated, as the formation of these tertiary amides improves solubility considerably and the protecting groups may be removed at the end of the synthesis by treatment with HF. Such a prediction of solubility allows a reasoned approach to fragment condensations, and a strategy for the synthesis of large peptides may be developed based on the sequence under investigation.

The use of tertiary amines in couplings may give rise for concern, and therefore couplings in the absence of tertiary bases have been examined.[67] It was found that if a Boc group was cleaved with formic acid the resulting formate salt could be converted to a tetrazolate salt and then acylated with an active ester without the presence of additional base. It was found that formate salts alone could give some formylation but that this could be suppressed by the addition of hydroxybenzotriazole, although even then formylation was still detectable. Addition of 2,4-dinitrophenol gave no formylation. The use of tetrazole in formic acid as a cleavage agent gave encouraging results, but a more detailed investigation would be required before this procedure could be generally adopted.

The hexafluorophosphate salt (30) has been synthesized[68] and its use as a one-pot coupling reagent investigated. The reagent allowed coupling in 15 min in the presence of *N*-methyl morpholine or triethylamine, giving yields of not less than 87%. Little racemization was observed, and this was attributed to the intermediacy of a hydroxybenzotriazole active ester. Such active esters were

[65] T. Yamada, Y. Manabe, T. Miyazawa, S. Kuwata, and A. Sera, *J. Chem. Soc., Chem. Commun.*, 1984, 1500.
[66] M. Narita, K. Ishikawa, J.-Y. Chen, and Y. Kim, *Int. J. Pept. Protein Res.*, 1984, **24**, 580.
[67] M. Bodanszky and A. Bodanszky, *Int. J. Pept. Protein Res.*, 1984, **24**, 563.
[68] V. Dourtoglou, B. Gross, V. Lambropoulou, and C. Zioudrou, *Synthesis*, 1984, 572.

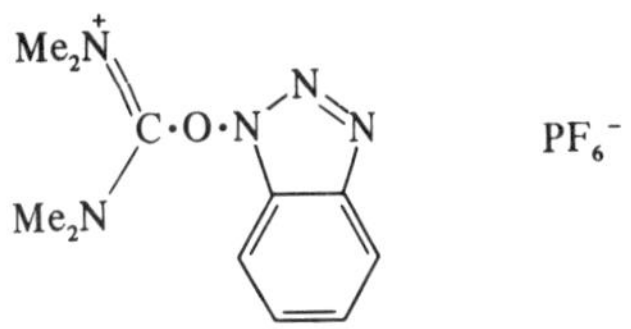

(30)

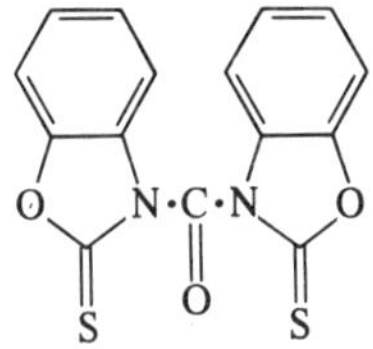

(31)

also implicated in other work,[69] which used di-isopropylcarbodi-imide/HOBt for the formation of the active ester.

The compound (31) has also been used as a condensing agent in peptide synthesis[70] and is synthesized by a reaction of 2-benzoxazolethiol with trichloromethylchloroformate in benzene. The reagent (31) is added to a solution of an N-protected amino acid in the presence of triethylamine, and after several hours of activation the amino component is added. When dichloromethane was used as a solvent the reaction could be carried out at room temperature and little racemization was observed.

The benzo- and naphtho-sultones (32) indicated in Scheme 9 have been used for peptide-bond formation.[71] Initial reaction with an N-protected amino acid

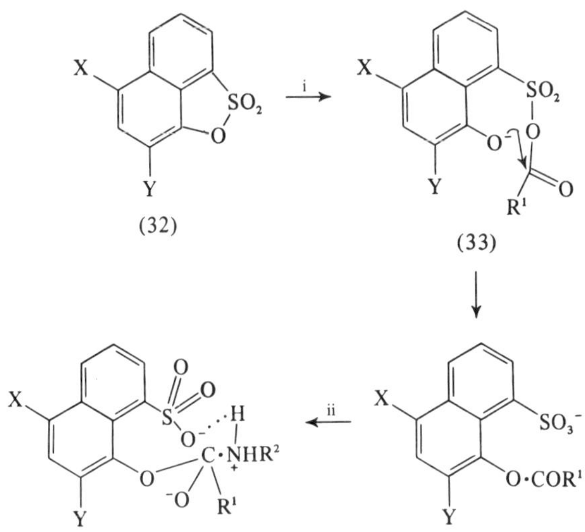

Reagents: i, $R^1CO_2^-$ for X = Y = H; ii, NH_2R^2 for X = NO_2 and Y = H or for X = Y = NO_2

Scheme 9

[69] A. Orlowska, E. Holodowicz, and S. Drabarek, *Pol. J. Chem.*, 1982, **56**, 1067.
[70] M. Ueda, N. Kawaharasaki, and Y. Imai, *Bull. Chem. Soc. Jpn.*, 1984, **57**, 85.
[71] F. Acher and M. Wakselman, *J. Org. Chem.*, 1984, **49**, 4133.

salt gives the mixed sulphonic carboxylic anhydride (33), which then rearranges to the activated ester (34). It is postulated that peptide-bond formation takes place on addition of the amino component after the formation of an intermediate of type (35). Some racemization was observed, and this is attributed to the formation of an oxazolone as a competing side reaction.

Acyl transfer was also observed when 3-(5-nitro-2-oxo-1,2-dihydro-1-pyridyl)-1,2-benzisothiazol-1,1-dioxide (36) was used as a coupling reagent.[72] The compound (36) is prepared by reaction of 3-chloro-1,2-benzisothiazol-1,1-dioxide with 5-nitro-2-pyridone. In coupling reactions it is reacted with an N-protected amino acid in the presence of triethylamine, the reaction following the course shown in Scheme 10. The initial intermediate (37) permits acyl transfer through a six-membered cyclic transition state leading to the activated ester (38); this activated ester then undergoes reaction with an amino component in the normal way. The optical rotation of synthetic peptides was measured and showed good agreement with literature values, thus confirming that little racemization takes place when this mode of activation is used.

Diethyl 2-(3-oxo-2,3-dihydro-1,2-benzisosulphonazolyl)phosphonate (DEPB) (39) has been used as a coupling agent for the preparation of amides, esters, and thioesters.[73] DEPB, when reacted with a carboxylic acid, gives the intermediate (40) as indicated in Scheme 11. This intermediate breaks down to yield the compound (41), once again through a cyclic six-membered transition state, and ultimately acylation may take place through either the intermediate (41) or (42).

Although many N-protected amino acid hydroxybenzotriazole esters have been difficult to isolate, it appears that *N*-trityl amino acid hydroxybenzotriazole

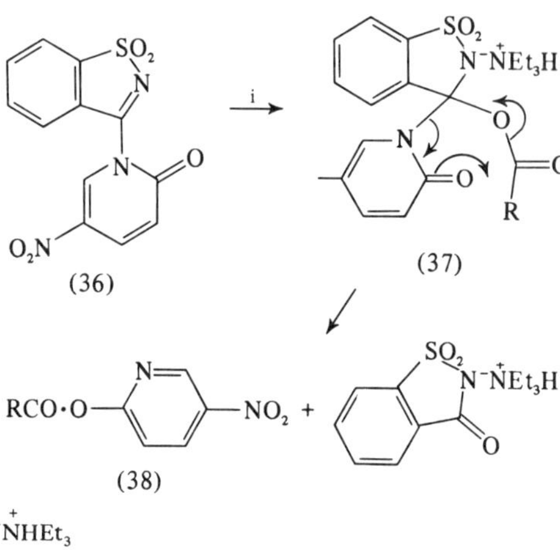

Reagents: i, $RCO_2^-\overset{+}{N}HEt_3$

Scheme 10

[72] A. Ahmed, N. Taniguchi, H. Fukuda, H. Kinoshita, K. Inomata, and H. Kotake, *Bull. Chem. Soc. Jpn.*, 1984, **57**, 781.

[73] M. Miyake, M. Kirisawa, and N. Tokutake, *Chem. Lett.*, 1985, 123.

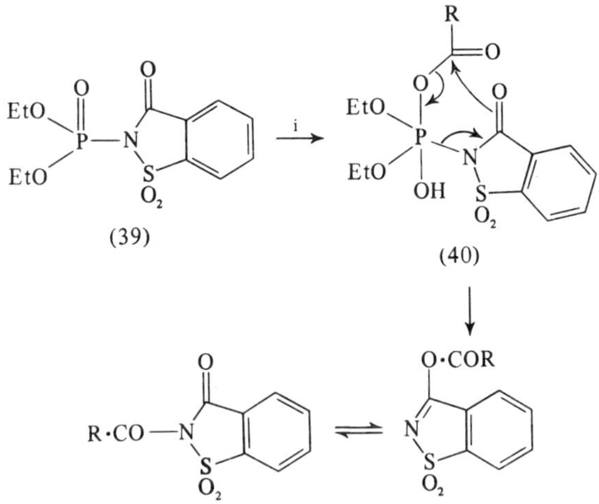

(39) (40) (42) (41)

Reagents: i, RCO_2H

Scheme 11

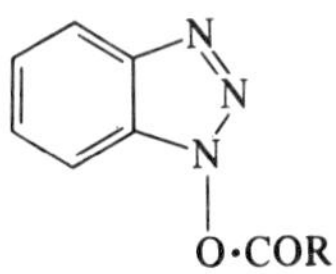

(43)

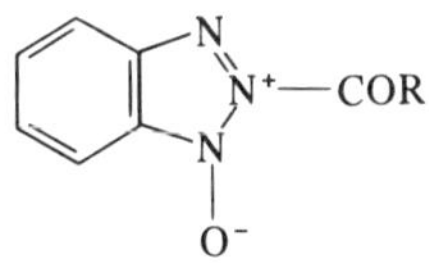

(44)

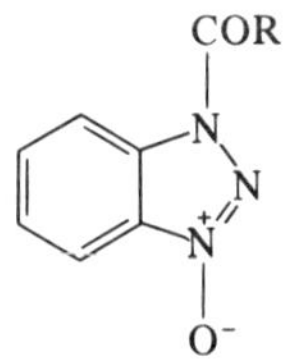

(45)

esters are relatively stable and in many cases may be isolated and used later in synthesis.[74–76] The esters are comparatively stable to hydrolysis and show few side reactions during acylation; this is attributed to the considerable steric bulk of the N-protecting group. The initially formed active ester generated from DCCI and HOBt has the structure (43), but this may rapidly isomerize to the form (44); a further form (45) may also be generated on prolonged storage. The use of mono- or tri-chloroacetic acid has been advocated for the removal of trityl protection during synthesis with hydroxybenzotriazole active esters.[75] Other acid reagents such as *p*-toluenesulphonic acid in isopropanol or trifluoroacetic acid may also be used,[76] but it appears that the use of the first two acids described is preferable. Attention was drawn to the fact that the absorbance of

[74] K. Barlos, D. Papaioannou, and D. Theodoropoulos, *Int. J. Pept. Protein Res.*, 1984, **23**, 300.

[75] J. Matsoukas, T. Tsegenidis, P. Cordopatis, and D. Theodoropoulos, *Tetrahedron*, 1984, **40**, 1869.

[76] K. Barlos, D. Papaioannou, and C. Sanida, *Liebigs Ann. Chem.*, 1984, 1308.

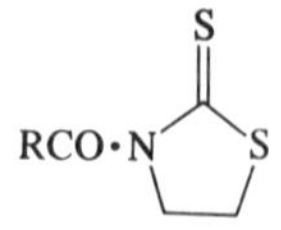

(46)

the triphenylmethylcarbinol generated during cleavage allows monitoring of the extent of the cleavage.

Peptide-bond formation using *N*-acyl-1,3-thiazolidine-2 thiones (46)[77] may also be monitored easily by the disappearance of the characteristic yellow colour of these compounds. Coupling may be carried out in a variety of organic solvents, although THF was preferred, and in the Young test racemization comparable with that observed with the azide method was observed. The reagent is claimed to be highly chemoselective, allowing acylation of the α-amino function of arginine, histidine, or serine, with no side reactions.

The *N*-hydroxy-5-norbornene-2,3-dicarboximide active esters[78] were also claimed to be chemoselective, allowing synthesis of seryl or threonyl peptides without protection of the side-chain hydroxy group. The kinetics of acylation using nitrophenyl,[79] pentafluorophenyl,[79] thiol esters,[80] and a range of active esters[81] have all been examined.

The pentafluorophenyl active esters[82] are claimed to be particularly reactive, and the brevity of the acylation time required prevents diketopiperazine formation in cases where this might be facile. Generally, the pentafluorophenyl active esters couple within minutes to give products with a high state of purity. The use of Fmoc-amino acid trichlorophenyl esters has also been advocated,[83] as these esters, in the presence of HOBt, react very rapidly, allowing them to be used in place of Fmoc symmetrical anhydrides in solid-phase peptide synthesis.

The use of active esters in salt couplings may be facilitated by addition of crown ethers;[84] this increases the nucleophilicity of the zwitterionic component, allowing more rapid reaction with the active ester.

Use of polymer-bound 4-hydroxy-3-nitrobenzophenone active esters (47) in peptide synthesis has been reported.[85] The resin-bound active ester is prepared from the corresponding resin-bound phenol by treatment with DCCI and the N-protected amino acid. Generally, a 40% excess of the polymeric reagent is used, 10–15 min being required for the formation of dipeptides; formation of

[77] Y. Nagao, T. Miyasaka, K. Seno, E. Fujita, D. Shibata, and E. Doi, *J. Chem. Soc., Perkin Trans. 1*, 1984, 2439.
[78] Y. P. Shvachkin, S. K. Girin, K.-D. Kaufmann, and R. Doelling, *J. Gen. Chem. U.S.S.R.*, 1983, **52**, 1471.
[79] S. K. Girin and Y. P. Shvachkin, *J. Gen. Chem. U.S.S.R.*, 1983, **53**, 2505.
[80] C. Shibuya, M. Murakami, and M. Shibukawa, *Nippon Nogei Kagaku Kaishi*, 1984, **58**, 675.
[81] S. K. Girin and Y. P. Shvachkin, *J. Gen. Chem. U.S.S.R.*, 1983, **53**, 2138.
[82] T. Szirtes, L. Kisfaludy, E. Palosi, and L. Szporny, *J. Med. Chem.*, 1984, **27**, 741.
[83] K. M. Sivanandaiah and S. Gurusiddippa, *Indian J. Chem., Sect. B*, 1984, **23**, 372.
[84] S. A. Andronati and A. A. Mazurov, *Bioorg. Khim.*, 1984, **10**, 1445.
[85] B. J. Cohen, H. Karoly-Hafeli, and A. Patchornik, *J. Org. Chem.*, 1984, **49**, 922.

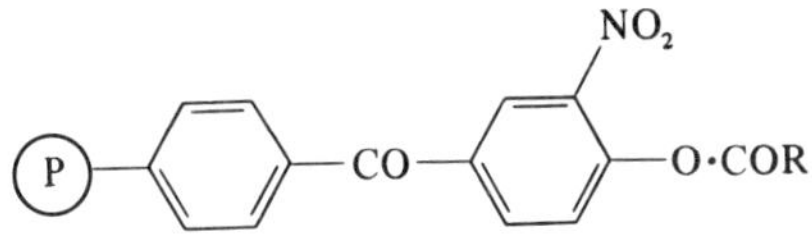

(47)

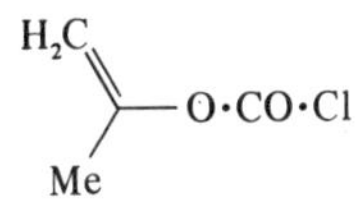

(48)

larger peptides requires longer. No significant racemization was detected, and the polymeric phenol is recyclable.

The preparation of pure symmetrical anhydrides of N-protected amino acids is often difficult, and for this reason it has been studied in detail.[86] Pure symmetrical anhydrides of a variety of urethane-protected amino acids were prepared by reaction of the N-protected amino acid with 0.5 equivalents of *N*-ethyl-*N*-dimethylaminopropylcarbodi-imide hydrochloride in dichloromethane. The product could be purified by washing with acid and base, as the symmetrical anhydride is moderately resistant to hydrolysis; side products were thus removed by an aqueous washing procedure. If an excess of carbodi-imide was used, then generally a high yield of the 2-alkoxy-5-(4H)-oxazolone was produced. When a Fmoc-amino acid is used, however, the oxazolone rapidly undergoes β-elimination on storage or on treatment with sodium bicarbonate.

Various mixed anhydrides have been studied, including the mixed anhydride formed by reaction of an N-protected amino acid with isopropenyl chloroformate (48).[87] The use of isopropenyl chloroformate in the formation of mixed anhydrides allows coupling at room temperature, and coupling yields were high, with urethane formation ranging between 3 and 10%. Triethylamine was used as the base during the formation of the mixed anhydride, 4 min being allowed for activation. Coupling to the amino component then required two hours at room temperature, making this particular mixed anhydride very useful in the REMA method.

It has been noted[88] that N-protected amino acids may be activated using Boc 'anhydride' in the presence of pyridine. This method of coupling was used mainly for the formation of esters, and in peptide-bond formation no racemization was observed.

The use of 3,3′-(chlorophosphoryl)-bis-(1,3-oxazolidine-2-one) has been investigated.[89] In coupling, the N-protected amino acid is treated with the reagent (49) in the presence of triethylamine, using dichloromethane as solvent; the amino component is then added, 5–6 hours being allowed for coupling at room temperature. It was noted that tyrosine peptides generally required no protection of the phenolic OH group and that couplings gave no appreciable side products.

1-Oxo-1-chlorophospholane (50) has been evaluated as a new coupling agent.[90] Mixed anhydrides are readily formed with N-protected amino acids in

[86] A. Paquet, F. M. F. Chen, and N. L. Benoiton, *Can. J. Chem.*, 1984, **62**, 1335.
[87] M. Jaouadi, C. Selve, J. R. Dormoy, and B. Castro, *Bull. Soc. Chim. Fr. II*, 1984, 409.
[88] V. F. Pozdnev, *Bioorg. Khim.*, 1984, **10**, 912.
[89] A. Omodei-Sale, G. Sindona, D. Sola and N. Uccella, *J. Chem. Res. (S)*, 1984, 50.
[90] R. Ramage, C. P. Ashton, D. Hopton, and M. J. Parrott, *Tetrahedron Lett.*, 1984, **25**, 4825.

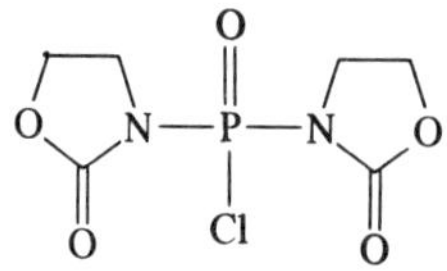

(49)

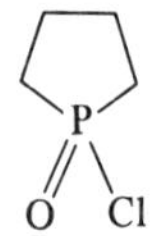

(50)

the presence of *N*-methyl morpholine, but the mixed anhydride is less reactive than that obtained from diphenylphosphinyl chloride. These mixed anhydrides are remarkably stable at 0 °C and give yields of between 80 and 90% on coupling, the cyclic phosphinic acid being easily removed during the work-up. Diphenyl phosphinic mixed anhydrides have also been used as acylating agents in solid-phase peptide synthesis.[91] The diphenyl phosphinic mixed anhydrides gave a rapid, efficient acylation, and neither wrong-way opening of the mixed anhydride nor disproportionation was observed. Use of the diphenyl phosphinic mixed anhydrides in a solid-phase context was exemplified by the synthesis of peptides related to bombesin.

Racemization. – The use of additives for the suppression of racemization has continued to be of interest, and dimethylaminopyridine has been compared with HOBt as an additive for use in DCCI couplings.[92] Whereas the addition of dimethylaminopyridine improved the yields slightly over the use of HOBt, racemization was generally equal to or slightly higher than that observed with HOBt. When dimethylaminopyridine is used as an additive, the choice of temperature and solvent may be critical, also the amount of dimethylaminopyridine that is added should be kept to an absolute minimum.

Liquid-crystal-forming organic molecules may also be used as additives in DCCI or mixed-anhydride couplings.[93] The effect on racemization was very variable, and in a n.m.r.-based test *p*-dimethoxyazoxybenzene, benzophenone, and stilbene gave no observable racemization.

Copper(II) chloride has also been used as an additive for the suppression of racemization.[94] In a DCCI-mediated coupling h.p.l.c. showed that with copper(II) chloride or copper(II) bromide used as the additive racemization was less than 0.1% whereas with HOBt as the additive it was 0.4%; when no additives were used 43% racemization was encountered. The mechanism of racemization of the serine dipeptide active ester (51) has been investigated.[95] In

Z-Gly-Ser(Bzl)-OPcp

(51)

[91] I. J. Galpin and A. E. Robinson, *Tetrahedron*, 1984, **40**, 627.

[92] J.-P. Gamet, R. Jacquier, and J. Verducci, *Tetrahedron*, 1984, **40**, 1995.

[93] H. Jeschkeit, M. Strube, J. Przybylski, and H. Miecznikowska, *J. Prakt. Chem.*, 1984, **326**, 638.

[94] T. Miyazawa, T. Otomatsu, T. Yamada, and S. Kuwata, *Tetrahedron Lett.*, 1984, **25**, 771.

[95] J. Kovacs, G. N. Jham, K. Y. Hui, E. M. Holleran, S. E. Kim, and T. Canavan, *Int. J. Pept. Protein Res.*, 1984, **24**, 161.

this work an isotope study using deuterium labelling of the α-CH of serine was employed to distinguish between an enolization mechanism and racemization proceeding through a 5-(4H)-oxazolone, using triethylamine as the base and the THF as solvent. The isotope study indicated that the racemization occurred through the intermediacy of an oxazolone, although racemization by α-proton abstraction may also account for a small degree of racemization. The authors claim that a unimolecular racemization of the oxazolone is not possible and that a solvent-catalysed bimolecular process should be invoked to explain racemization. Pentafluorophenyl active esters were found to minimize racemization,[95] and dipeptide active esters coupled much more rapidly than the corresponding protected amino acid.

Methods of detecting racemization are of considerable importance, and by the use of an n.m.r. method the effect of varying N-terminal substituents and side chains on the chemical shift of protons in model dipeptides has been studied.[96] The differences in chemical shift of the methyl ester proton in the two diastereomers may be maximized by emphasizing conformational changes that are maintained by hydrogen bonding. It was found that the changes in conformation had a greater effect than the electrical influence of residues in the N-terminal position; also, it was observed that the more bulky side chain of valine gave rise to large separations of the two methyl ester signals.

A chiral analysis of the reaction stages of the Edman degradation has been carried out with a view to determining the optical integrity of amino acids during Edman sequencing.[97] The chiral isothiocyanate (52) was used in the sequencing as it would form a diastereoisomer on reaction with the N-terminal amino acid; any diastereoisomers could then be separated using h.p.l.c. H.p.l.c. showed that the N-terminal residue is racemized during the cyclization and cleavage of the thiazoline; thus at present racemization during the analysis prevents a chiral Edman analysis. The racemization is thought to occur through the ready aromatization of the thiazolinone (53). Although the method is at present limited by this racemization, it is proposed to use a chiral isocyanate in place of the isothiocyanate, as this should lead to a reduction in aromatic character in the oxy analogue of the intermediate (53).

Repetitive Methods of Peptide Synthesis. – *Solid-phase Peptide Synthesis.* An excellent example of polystyrene-based solid-phase synthesis is provided by a

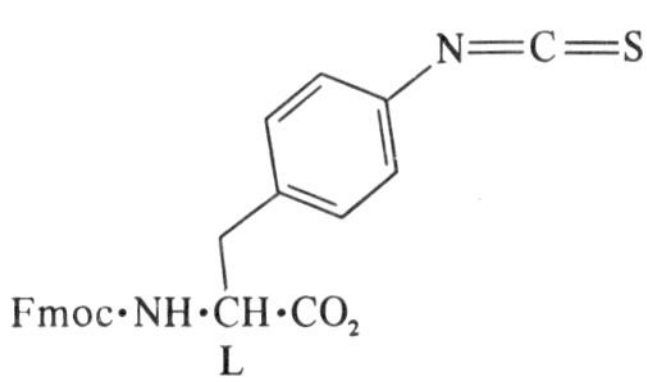

(52)

[96] J. S. Davies and E. Hakeem, *J. Chem. Soc., Perkin Trans. 2*, 1984, 1387.
[97] J. S. Davies and A. K. A. Mohammed, *J. Chem. Soc., Perkin Trans. 2*, 1984, 1723.

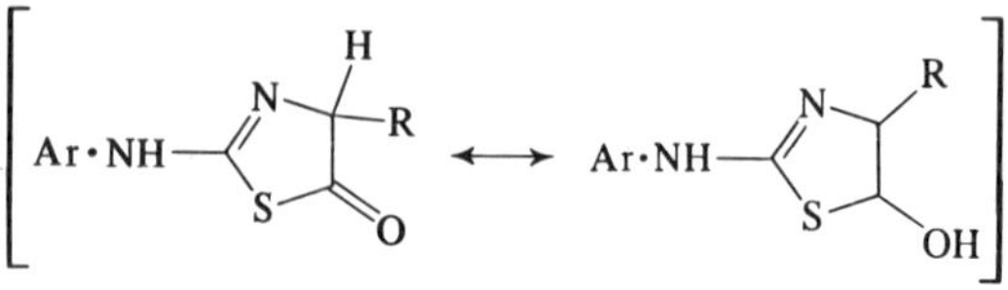

(53)

synthesis of mammalian glucagon.[98] In this work 4-(oxymethyl)phenylacetamidomethyl copolystyrene divinylbenzene (Pam) resin was used. This resin, which has improved acid stability, allowed a 75% yield after cleavage using a new HF procedure. The synthesis utilized β-cyclohexylaspartic acid, to minimize α–β rearrangement, and N^{in}-formyltryptophan. Chain extension was achieved using symmetrical anhydrides (four equivalents), active esters, or DCCI/HOBt. ^{14}C-Labelled glycine was incorporated at position 4 and tritiated leucine at position 25; thus by measuring the recovered radioactivity it was possible to monitor the loss of peptide during deprotection. The results showed a loss of 0.13% for each deprotection step.

The limitations of the step-wise solid-phase approach have prompted several workers to explore fragment condensations in solid-phase work.[99–101] In a synthesis of the (116–137) fragment of human leukocyte interferon[99] a traditional step-wise approach using a benzhydrylamine resin was used, except that two Boc-protected tetrapeptides were added during the synthesis. These had been prepared on a chloromethyl resin and had been released by hydrogenolysis. An efficient coupling was achieved using DCCI and two equivalents of HOBt, but carbonyl di-imidazole with HOBt was not satisfactory. The final product was released from the resin by treatment with HF and was easily purified. The ease of purification was attributed to the reduction in a number of stages owing to the incorporation of fragments. The synthesis of dynorphin(1–13) using fragment condensation has also been reported.[100] In this synthesis the (1–5) and (6–13) fragments were prepared on a high-loading phenolic polyacryloyl morpholine resin. The protected (1–5) fragment was removed by hydrazinolysis and after purification converted to the corresponding azide for use in an azide fragment condensation. The (6–13) fragment was assembled on the same resin, and, after coupling, the product was removed over 90 hours by transesterification with dimethylamino ethanol in DMF; the resulting dimethylamino ethyl ester was then hydrolysed at pH 9.7 by treatment with 1M sodium hydroxide in the presence of a catalytic amount of imidazole. Side-chain benzyl-based protection was removed by catalytic-transfer hydrogenolysis using ammonium formate and palladium.

In a synthesis of a number of apamin analogues[101] a fragment-condensation approach was again used. In this case the protected (1–12) sequence of apamin was prepared on the photosensitive α-(4-bromomethyl-3-nitrobenzamido)benzyl

[98] S. Mojsov and R. B. Merrifield, *Eur. J. Biochem.*, 1984, **145**, 601.
[99] A. Ljungqvist and K. Folkers, *Acta Chem. Scand., Ser. B*, 1984, **38**, 375.
[100] P. W. Small, R. Epton, G. Marr, and S. Twaij, *Int. J. Biol. Macromol.*, 1984, **6**, 189.
[101] F. Albericio, C. Granier, C. Labb-Jullie, M. Seagar, F. Couraud, and J. Van Rietschoten, *Tetrahedron*, 1984, **40**, 4313.

Ⓟ~CH_2·NH·CO·CH_2·Br

(54)

copolystyrene 1% DBV resin. Cleavage of the assembled (1–12) fragment was achieved by photolysis in trifluoroethanol/dichloromethane, giving an 89% yield. Purification of the (1–12) fragment was achieved by reprecipitation followed by gel filtration on Sephadex LH60 and semi-preparative h.p.l.c. on a reversed-phase column. The product, which was obtained in 55% overall yield, was coupled to three different (13–18) analogues that had been assembled on a benzhydrylamine resin. Following HF deprotection, yields of 77, 94, and 96% were achieved; after oxidation and purification, the position-13 analogues of apamin were biologically tested. A new acrylic-based polymer that allows attachment of the amino acid by reaction with the bromoacetamido resin (54) has been investigated.[102] This resin permits both synthesis with the minimum steric interference and coupling in polar media. Peptide cleavage may be achieved by treatment with 1M sodium hydroxide over 3 hours in isopropanol/water. Transesterification using triethylamine in methanol can also be used for deprotection; amide formation using ammonia in trifluoroethanol is also possible. The use of the polyacrylic resin was exemplified by a synthesis of the (1–14) fragment of angiotensinogen.

Synthesis using a macroporous, radiation-grafted Teflon–polystyrene resin was evaluated in the synthesis of a fragment of the β-chain of human haemoglobin.[103] With this polymer, deprotection was achieved using HBr in trifluoroacetic acid; this regenerates the bromomethyl polymer, which can then be reused. Increasing the proportion of Teflon to polystyrene ten-fold was found to have very little effect on the yield or purity of the products.

Sephadex LH20 has again been investigated as a solid-phase support.[104] In a synthesis of the (1–10) fragment of salmon calcitonin peptide assembly was achieved using symmetrical anhydrides or pentafluorophenyl active esters. The rate of coupling when using pentafluorophenyl active esters was found to be similar to that when using DCCI, although somewhat lower than that which may be achieved using symmetrical anhydrides; these esters, however, permit the use of unprotected serine and threonine. Cyclization through disulphide-bond formation was achieved with the peptide still attached to the resin. As well as the required cyclopeptide, higher oligomers were formed by intermolecular disulphide-bond formation.

Two other examples of the synthesis of cyclic peptides by solid-phase synthesis have been described.[105,106] In the preparation of a cyclic melanotropin

[102] F. Baleux, J. Daunis, R. Jacquier, and B. Calas, *Tetrahedron Lett.*, 1984, **25**, 5893.

[103] M. V. Sidorova, G. A. Zheltukhina, U. O. Kalei, G. I. Aukone, E. I. Filippovich, and R. P. Evstigneeva, *J. Gen. Chem. U.S.S.R.*, 1982, **52**, 1047.

[104] N. Y. Kozhevnikova, L. V. Krasnikov, and G. P. Vlasov, *J. Gen. Chem. U.S.S.R.*, 1983, **52**, 1456.

[105] M. Lebl, W. L. Cody, B. C. Wilkes, V. J. Hruby, A. M. De L. Castrucci, and M. E. Hadley, *Int. J. Pept. Protein Res.*, 1984, **24**, 472.

[106] M. Lebl and V. J. Hruby, *Tetrahedron Lett.*, 1984, **25**, 2067.

analogue[105] synthesis was carried out on a *p*-methylbenzhydrylamine resin; carbaoxytocin analogues[106] were also prepared, this time using the straightforward benzhydrylamine resin. In the former case 45% TFA in dichloromethane containing 2% anisole was used for deprotection, and in the second synthesis deprotection was carried out using HF in the presence of 10% anisole, plus 5% dithioethane.

An alternative approach to the segment-coupling method is illustrated in a synthesis of two omission analogues of human-growth-hormone-releasing factor.[107] In this work the (Gly^{15}-S)GRF(1–15), Tfa-GRF(20–44), Tfa-GRF (18–44), and Tfa-GRF(16–44) were synthesized on a benzhydrylamine resin. After removal from the resin the latter three fragments were reacted with citraconic anhydride and the Tfa group was removed by treatment with hydrazine hydrate. The thio-Gly^{15} (1–15) fragment was then coupled with the three citroconylated fragments using silver nitrate/HONSu in aqueous DMF, giving the three deletion analogues of GRF.

Coupling of the first residue can still cause problems, and for this reason a new method involving the use of DMF dineopentylacetal (56) has been proposed.[108] The compound is used as indicated in Scheme 12 to prepare the intermediate (58), *via* compounds (55) and (57), which is then coupled to an aminomethyl polymer (59) to give the polymer-linked protected amino acid (60); this can then be utilized in further synthesis. It is claimed that the method is superior to the use of DCC dimethylaminopyridine as it prevents the formation of dimers and does not cause racemization.

A new base-labile anchoring group based on compound (61) has been developed.[109] The linkage agent (61) embodies similar properties to Fmoc-amino acids, being easily cleaved by 15% piperidine in DMF, and is orthogonal to the use of Boc, Ddz, Bpoc, and Nps. In this work the linkage agent was used in liquid-phase synthesis, although it could readily be adapted for use with conventional solid-phase resins. Attachment of a Boc-amino acid through this linkage agent takes place in two stages. In the first step the compound (61) is reacted with DCCI and trichlorophenol to form the corresponding active ester. This active ester is then reacted with a Boc-amino acid in the presence of DCCI and dimethylaminopyridine. Reaction with aminopolyethylene glycol monomethyl ether gives the polymer-bound amino acid linked through unit (61). The use of the new linkage agent was demonstrated in the synthesis of some simple peptides.

The linkage agents (62)–(64) have been developed for use in peptide and nucleotide synthesis.[110] The hydroxy group of these compounds may be esterified by an amino acid or nucleotide, and an attachment to the polymer may be brought about by activation of the carboxy group. The compound CASET(2) (64) is the linkage agent of choice for peptides as these are poorer leaving groups than phosphodiesters. Cleavage is generally brought about by treatment with 0.1M sodium hydroxide or barium hydroxide.

[107] J. Blake, M. Westphal, and C. H. Li, *Int. J. Pept. Protein Res.*, 1984, **24**, 498.
[108] F. Albericio and G. Barany, *Int. J. Pept. Protein Res.*, 1984, **23**, 342.
[109] M. Mutter and D. Bellof, *Helv. Chim. Acta*, 1984, **67**, 2009.
[110] R. Schwyzer, E. Felder, and P. Failli, *Helv. Chim. Acta*, 1984, **67**, 1316.

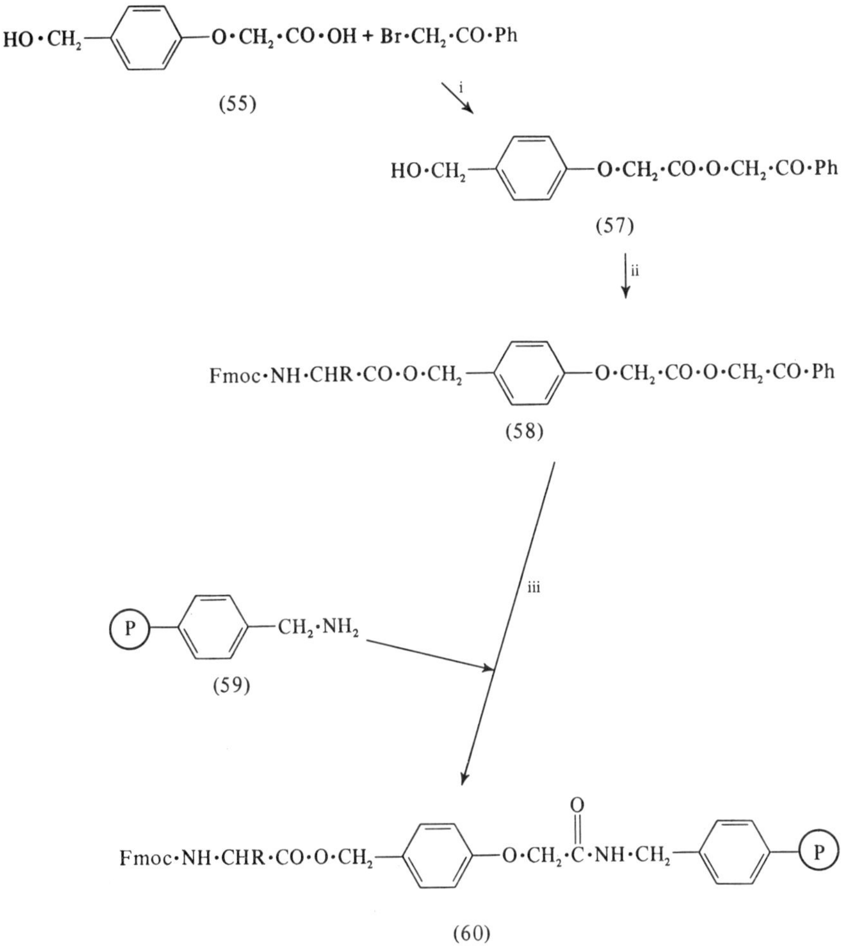

Reagents: i, $HNEt_2$/EtOAc; ii, Fmoc·NH·CHR·CO·OH/$Me_2N·CH(OCH_2CMe_3)_2$ (56)/CH_2Cl_2; iii, (*a*) Zn/HOAc, (*b*) DCCI/HOBt/DMF

Scheme 12

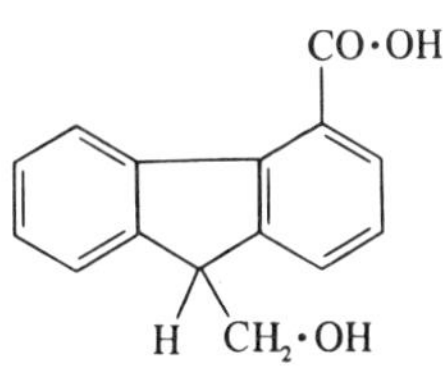

(61)

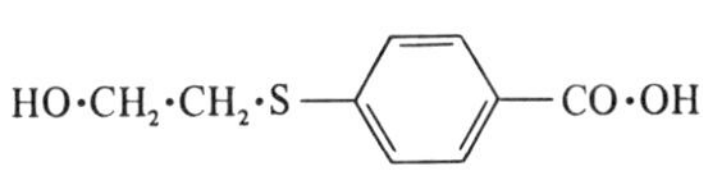

(62) H·CAMET·OH

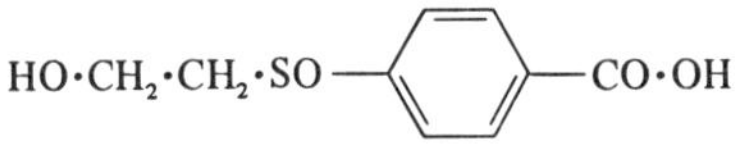

(63) H·CASET(1)·OH

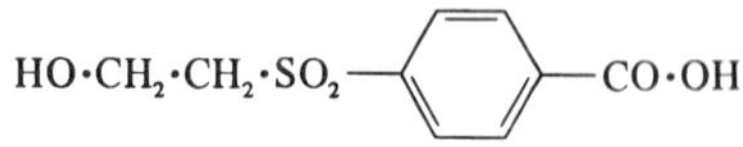

(64) H·CASET(2)·OH

(65)

A complex matrix containing polyethylene glycol grafted on to functionalized polystyrene has been proposed as the basis for a new multi-detachable approach.[111] The peptide may be synthesized on a matrix based on the system shown in (65). It is proposed that the groups A^1 and A^2 should be orthogonal, and generally the linkage A^1 is an acidic or photolabile linkage. The advantage of this system lies in the fact that the peptide is solubilized by attachment to the polyethylene glycol chain whilst still being anchored to the polystyrene backbone.

Although the use of symmetrical anhydrides still appears to be the most popular method of acylation in solid-phase synthesis, other methods have been explored. Non-symmetrical anhydrides formed from diphenyl phosphinyl chloride appear to be stable enough for use in solid-phase synthesis,[91] and their regioselective opening avoids the possibility of chain termination by reaction at phosphorous. Fmoc-amino acid trichlorophenyl esters have also been used in solid-phase synthesis,[83] as in combination with HOBt they provide a rapid efficient method of acylation.

A new method for the cleavage of 2-[4-(hydroxymethyl)phenylacetoxy]-propionyl resin linkages has been developed.[112] Tetramethyl guanidine or 1,8-diazabicyclo[5,4,0]undec-7-ene may be used to cleave this resin linkage, liberating protected peptide fragments; these may then be used in subsequent fragment condensations.

A new general approach to the quantitation of synthetic efficiency in solid-phase synthesis has been described.[113] The tetrapeptide Leu-Ala-Val-Gly was synthesized by solid-phase peptide synthesis at increasing distances from the backbone of a 1% crosslinked polystyrene resin. The oxymethylphenyl acetyl group was inserted between the test peptide, and peptide chains were used as spacers to the resin backbone. The deletion peptides Leu-Ala-Val and Leu-Gly-Val were estimated, allowing an evaluation of the efficiency of synthesis of the test peptide without interference from the spacer. It was found that the distance from the support and the peptide loading had little effect on the synthetic efficiency up to a spacer length of 60 residues with a peptide-to-resin weight ratio of 4:1.

[111] H. Hellermann, H.-W. Lucas, J. Maul, V. N. R. Pillai, and M. Mutter, *Makromol. Chem.*, 1983, **184**, 2603.

[112] D. B. Whitney, J. P. Tam, and R. B. Merrifield, *Tetrahedron*, 1984, **40**, 4237.

[113] V. K. Sarin, S. B. H. Kent, A. R. Mitchell, and R. B. Merrifield, *J. Am. Chem. Soc.*, 1984, **106**, 7845.

The use of gel-phase ^{13}C n.m.r. for the monitoring of solid-phase synthesis has been investigated.[114] The nature of the starting polymer and growing chain may be characterized by the method, and side products may be identified, using relaxation times to give an indication of the mobility of peptide chains.

An apparatus for the simultaneous, manual solid-phase synthesis of several peptide analogues has been devised.[115] The apparatus uses four vessels contained within one apparatus and is claimed to be particularly economical in the use of both time and space, the rate of synthesis being limited by the filtration time.

Liquid-phase Approach. The fragment-condensation approach has become more important in the liquid-phase-type synthesis, and in a synthesis of a hybrid insulin B-chain a nitrobenzoyl ester was used for attachment of the peptide to the polyethylene glycol. Photolysis allowed removal of the protected peptide from the polyethylene glycol, and thus it could be purified before fragment condensation.[116]

The efficiency of fragment condensation using a liquid-phase approach has been examined,[117] and the influence of peptide-chain length and the degree of polymer crosslinking were investigated. As with the solid-phase work described above, it was found that peptide-chain length did not have a significant effect on coupling efficiency; however, the loading of the peptide did significantly affect the coupling yields, and with styrene co-divinylbenzene (2%) as a support it was found that there are some sterically inaccessible sites.[117]

Repetitive Excess Mixed-anhydride Method. The REMA method has been used in the synthesis of several enkephalin analogues.[118] Also, syntheses of VIP and associated fragments of the hormone have been carried out.[119,120] In the REMA method terminal asparagine amide causes problems because of its poor solubility; however, this problem may be circumvented by the use of asparagine amide β-t-butyl ester. This can be carried through the synthesis as a t-butyl ester, then at the end of the synthesis on treatment with TFA the free-acid side chain is liberated and can be converted to the corresponding amide by treatment with ammonia following activation with isobutyl chloroformate. The efficiency of this rapid method was again demonstrated, but the problems that may arise from poor solubility are clear.

Enzyme-mediated Synthesis and Semi-synthesis. – Papain, chymotrypsin, and thermolysin are the most widely used enzymes in enzyme-mediated synthesis. All three enzymes were used in a fragment-condensation synthesis of the (6–11)

114 E. Giralt, J. Rizo, and E. Pedroso, *Tetrahedron*, 1984, **40**, 4141.
115 J. J. Gorman, *Anal. Biochem.*, 1984, **136**, 397.
116 B. Hemmasi, W. Stueber, and E. Bayer, *Hoppe-Seyler's Z. Physiol. Chem.*, 1984, **365**, 485.
117 S. Isokawa, N. Kobayashi, R. Nagano, and M. Narita, *Makromol. Chem.*, 1984, **185**, 2065.
118 A. M. Van Den Braken-Van and L. Maat, *Recl. Trav. Chim. Pays-Bas*, 1984, **103**, 110.
119 W. M. M. Schaaper and D. Voskamp, *Recl. Trav. Chim. Pays-Bas*, 1984, **103**, 17.
120 W. M. M. Schaaper and H. C. Beyerman, *Peptides*, 1984, **5**, 167.

fragment of Eledoisin;[121] 2 + 4 and 3 + 3 fragment condensations were used in aqueous, organic solvent systems, and both carbon tetrachloride and ethyl acetate were used. Yields in the 60–70% region were generally achieved, and ultimately the Boc hexapeptide amide was produced. In an investigation of papain-catalysed peptide synthesis[122] it was found that acetyl amino acid alkyl esters could be used as the carboxy component, with the amino component being protected as an amide or t-butyl ester. When an excess of carboxy component was used at pH 8–9.5, yields of between 50 and 94% were attainable. In this work, up to 40% methanol or butane-1,4-diol could be added to improve solubility. It was found that ionic strength had little effect on the efficiency of the synthesis but that pH was a critical factor. Also, the addition of EDTA is recommended to convert the papain into an active form; dithiothreitol, mercaptoethanol, or cysteine should also be present in the reaction medium.

Papain-catalysed formation of phenyl hydrazides has also been investigated.[123] This method was used to prepare a number of N^{α}-protected amino acid phenyl hydrazides; no side-chain protection was required, and with amino acids containing a second carboxylic acid group the phenyl hydrazide of the α-carboxylic acid was the only product. It was noted that a particularly low yield was obtained when basic amino acids were used, owing to the high solubility of the product hydrazide.

Biphasic aqueous/organic systems for use with chymotrypsin have been studied[124] by examining the synthesis of the model dipeptide Ac-Trp-Leu-NH_2. Both free and immobilized chymotrypsins were used, and particularly high yields were claimed with 0.1M acetate/phosphate/borate at pH 7 (1 ml) in ethyl acetate (100 ml).

The specificity of the nucleophile in chymotrypsin-catalysed synthesis has been investigated,[125,126] and it was found[125] that chymotrypsin-catalysed synthesis was much less efficient when there were charges or polar groups in the amino component. It was also found[126] that when coupling Z-Tyr-OMe to leucine amide the D-enantiomer was a much poorer nucleophile than the L-enantiomer. In fact, an eight-fold increase in the concentration of the D-isomer was required in order to bring about the same rate of coupling at pH 9.3 in carbonate buffer. In the general procedure a thick suspension was produced after coupling for 15 min, and after 30 min no acyl component remained. The reaction medium could then be filtered and new reagents added to allow further synthesis. It was also found that thermolysin showed more amino-component specificity than chymotrypsin, and it has been suggested[127] that nucleophile specificity reflects P1′ specificity in the corresponding hydrolytic process.

[121] P. Kuhl, G. Doering, K. Neubert, and H.-D. Jakubke, *Monatsh. Chem.*, 1984, **115**, 423.
[122] Yu. V. Mitin, N. P. Zapevalova, and E. Yu. Gorbunova, *Int. J. Pept. Protein Res.*, 1984, **23**, 528.
[123] V. Cerovsky and K. Jost, *Collect. Czech. Chem. Commun.*, 1984, **49**, 2557.
[124] Y. L. Khmel'Nitski, F. K. Dien, A. N. Semenov, K. Martinek, B. Veruovic, and V. Kubanek, *Tetrahedron*, 1984, **40**, 4425.
[125] D. D. Petkov and I. Stoineva, *Biochem. Biophys. Res. Commun.*, 1984, **118**, 317.
[126] D. D. Petkov and I. B. Stoineva, *Tetrahedron Lett.*, 1984, **25**, 3751.
[127] L. Riechmann and V. Kasche, *Biochem. Biophys. Res. Commun.*, 1984, **120**, 686.

A detailed kinetic investigation of chymotrypsin- and papain-catalysed syntheses of leucine- and methionine-enkephalin has been carried out.[128] In this work Boc-amino acid ethyl esters were the acyl donors and amino acid phenyl hydrazides were the acyl acceptors or amino component. The kinetic analysis was compatible with a 'ping-pong' mechanism, and it was shown that the mechanism of synthesis was the inverse of that of hydrolysis. At the end of the synthesis ferric chloride or *N*-bromosuccinimide was used to remove the phenyl hydrazide protection.

A number of other applications of enzymic synthesis have been reported,[129–133] although these will not be discussed in detail. The enzymic synthesis with chymotrypsin using a sulphobenzyl-protecting group for carboxy groups is of interest,[133] as is a review[132] that covers the use of enzymic synthesis using a solid-phase support.

The enzyme chymotrypsin in its zymogen form has been modified by treatment with 2,4-bis-(*O*-methoxypolyethylene glycol)-6-chloro-*S*-triazine.[134] Following this modification activation was achieved by treatment with trypsin; the resulting modified enzyme was soluble in benzene and retained its enzymic activity. The use of the benzene-soluble modified chymotrypsin was demonstrated by the synthesis of tyrosine and phenylalanine peptides.

Several semi-synthetic analogues of ribonuclease have been prepared.[135–137] On two occasions semi-synthetic analogues have been prepared by non-covalent interaction of (1–118) and (111–124) fragments;[135,136] the synthetic fragments contained asparagine at position 121[135] or 3-(3-pyrazolyl)alanine, *N*-π-, or *N*-τ-methylhistidine at position 119.[136] The (1–120) fragment was prepared by pepsin cleavage,[135] then carboxypeptidase A was used to remove the two terminal residues, giving a (1–118) fragment. Synthetic (111–124) fragments were then prepared and combined with the (1–118) fragment in non-covalent manner.

Ribonuclease analogues containing variations in the (1–15) region have also been prepared.[137] In this work analogue (1–15) fragments were prepared by DCCI/HOBt coupling of protected native (1–10) material, with resin-bound (11–15) analogues that had been prepared on a Merrifield resin. After cleavage and deprotection the (1–15) fragments were combined with *S*-protein and the activities of the semi-synthetic complex were studied.

Tetraguanidinated bovine trypsin–kallikrein inhibitors (Kunitz) have been prepared by sequential removal of amino acids from the amino terminus by

128 W. Kullmann, *Biochem. J.*, 1984, **220**, 405.
129 K. Breddam, F. Widmer, and J. T. Johansen, *Carlsberg Res. Commun.*, 1983, **48**, 231.
130 A. Koennecke, V. Schellenberger, H.-J. Hofmann, and H.-D. Jakubke, *Pharmazie*, 1984, **39**, 785.
131 Y. L. Khmel'Nitskii and K. Martinek, *Bioorg. Khim.*, 1984, **10**, 626.
132 N. A. Samoilova, Y. A. Davidovich, and S. V. Rogozhin, *Bioorg. Khim.*, 1984, **10**, 725.
133 P. Kuhl, J. Walpuski, and H.-D. Jakubke, *Pharmazie*, 1984, **39**, 280.
134 A. Matsushima, M. Okada, and Y. Inada, *FEBS Lett.*, 1984, **178**, 275.
135 M. S. Stern and M. S. Doscher, *FEBS Lett.*, 1984, **171**, 253.
136 J. Serdijn, W. Bloemhoff, K. E. T. Kerling, and E. Havinga, *Recl. Trav. Chim. Pays-Bas*, 1984, **103**, 50.
137 C. Di Bello, A. Lucchiari, O. Buso, and M. Tonellato, *Int. J. Pept. Protein Res.*, 1984, **23**, 61.

Edman degradation.[138] None of the four terminal residues was required for regeneration of activity on aerial oxidation of the reduced inhibitor, but residues 2, 3, and 4 were important to protein stability and affected both the conformation and the rate of reactivation on air oxidation. Removal of the positive charge of arginine-1 by diazotization gave an active deaminated compound with no salt bridge; however, on reduction and reoxidation this prevented the correct folding and inactive material was obtained.

3 Syntheses

Syntheses of several large peptides have been reported and include many synthesized by the solid-phase approach; in general these are reported in other places. An interesting solution synthesis is that of urogastrone,[139] which has been prepared by a fragment-condensation approach using maximal protection. The synthesis used benzyl-based side-chain protection with cysteine being protected at the acetamidomethyl derivative. The fragments were assembled using trichloroethoxycarbonyl hydrazide protection, which was removed after fragment assembly using zinc and acetic acid in DMF to give the free hydrazide; hydrazides were then employed in fragment condensation using the azide approach. On some occasions phenacyl esters were used and were cleaved using zinc/anthanilic acid/pyridine. In cases where solubility was a problem *N*-methylpyrrolidone or DMSO was used as solvent. Although the majority of fragments were combined by the azide method, a water-soluble carbodi-imide and HOBt were also employed. A step-wise fragment condensation was used, the build-up following the sequence (33–53), (27–53), and (13–53), and finally resulting in the assembly of the (1–53) peptide. The final deprotection was by treatment with HF in the presence of *p*-cresol, ethanedithiol, dimethyl sulphide, and excess methionine. Aerial oxidation at pH 7.6 followed by preparative h.p.l.c. gave a 10% yield of the final product, which was characterized by fast-atom-bomdardment mass spectrometry and had the required biological activity.

There has been a resurgence of interest in the synthesis of cyclic peptides, and frequently a solid-phase approach has allowed a rapid synthesis,[105,106,140,141] although frequently conventional solution methods have been used.[142,143] Diphenyl phosphoryl azide continues to be a popular method of cyclization, and it has been used effectively in the cyclization of a somatostatin analogue in which a methylenethio bond replaces an amide linkage.[141] Diphenyl phosphoryl

[138] L. Biondi, B. Filippi, F. Filira, M. Tolazzi, and R. Rocchi, *Int. J. Pept. Protein Res.*, 1984, **24**, 359.

[139] D. Hagiwara, M. Neya, Y. Miyazaki, K. Hemmi, and M. Hashimoto, *J. Chem. Soc., Chem. Commun.*, 1984, 1676.

[140] W. L. Cody, B. C. Wilkes, B. J. Muska, V. J. Hruby, A. M. De L. Castrucci, and M. E. Hadley, *J. Med. Chem.*, 1984, **27**, 1186.

[141] T. W. Gero, A. F. Spatola, I. Torres-Aleman, and A. V. Schally, *Biochem. Biophys. Res. Commun.*, 1984, **120**, 840.

[142] F. K. Mutulis, G. I. Chipens, O. E. Lando, and I. E. Mutule, *Int. J. Pept. Protein Res.*, 1984, **23**, 235.

[143] J. M. Berman and M. Goodman, *Int. J. Pept. Protein Res.*, 1984, **23**, 610.

azide has also been used in the preparation of cyclic enkephalin retro-inverso analogues.[143]

Several syntheses of retro-inverso peptides have been reported, mainly because of the fact that these peptides have enhanced resistance to enzymic hydrolysis.[144] In most work bis(trifluoroacetoxy)iodobenzene is used to prepare the diaminoalkyl component. Thus, for example, the Boc dipeptide carboxyamide (66) may be converted to the *gem*-diaminoalkyl dipeptide (67) as indicated in Scheme 13.[143] Retro-inverso analogues of the bradykinin potentiating peptide have been synthesized on a polyamide resin,[144] and as in most syntheses of this type half malonate esters were used. The diamino alkyl component can also be formed by Curtius rearrangement of *N*-acyl hydrazides;[145] the intermediate isocyanate may be trapped by alcohols, although side reactions upon the addition of alcohols depend on the nature of the original *N*-acyl hydrazide. These complications may be minimized by using a small excess of the alcohol over the isocyanate.

$$\underset{(66)}{\text{Boc}\cdot\text{NH}\cdot\text{CHR}^1\cdot\text{CO}\cdot\text{NH}\cdot\text{CHR}^2\cdot\text{CO}\cdot\text{NH}_2} \xrightarrow{\text{i}} \underset{(67)}{\text{Boc}\cdot\text{NH}\cdot\text{CHR}^1\cdot\text{CO}\cdot\text{NH}\cdot\text{CHR}^2\cdot\overset{+}{\text{N}}\text{H}_3\ \text{CF}_3\text{CO}_2^-}$$

Reagents: i, $PhI(OCOCF_3)_2/MeCN/H_2O$

Scheme 13

An interesting insertion of an olefinic dipeptide into a renin substrate has been described.[146] These renin inhibitors contain the amino acid 5-(*S*)-amino-7-methyl-3-(*E*)-octanoic acid, which is synthesized by the route outlined in Scheme 14. This synthesis utilizes a Wittig reaction between the phosphorane derived from the phosphonium salt (68) and Boc-leucine aldehyde. The intermediate (69) is then converted to the *E*-isomer of compound (70), which is subsequently incorporated into the renin inhibitors.

Dehydropeptides[147,148] have also been of interest, and an unusual route[148] to dehydrodipeptides is shown in Scheme 15. In this reaction the benzyloxycarbonyl or trifluoroacetyl amino acid amide (71) is reacted with the pyruvic acid (72). Azeotropic distillation brings about condensation of these components, and subsequent rearrangement yields the protected dehydropeptide (73).

Replacement of the acidic function of glutamic acid with a tetrazolyl group provides isosteric replacement with a group of similar pK_a. To this end[149] glutamine-containing peptides may be dehydrated by treatment with DCCI in pyridine to give cyanoaminobutyric acid derivatives, which may then be reacted with tri-*N*-butyltin azide, giving the tetrazolyl-substituted peptide.

[144] F. Bonelli, A. Pessi, and A. S. Verdini, *Int. J. Pept. Protein Res.*, 1984, **24**, 553.
[145] M. Chorev, S. A. MacDonald, and M. Goodman, *J. Org. Chem.*, 1984, **49**, 821.
[146] R. L. Johnson, *J. Med. Chem.*, 1984, **27**, 1351.
[147] I. Ojima, N. Yoda, M. Yatabe, T. Tanaka, and T. Kogure, *Tetrahedron*, 1984, **40**, 1255.
[148] M. Makowski, B. Rzeszotarska, Z. Kubica, and P. Wieczorek, *Liebigs Ann. Chem.*, 1984, 920.
[149] J. Dubois, S. Bory, M. Gaudry, and A. Marquet, *J. Med. Chem.*, 1984, **27**, 1230.

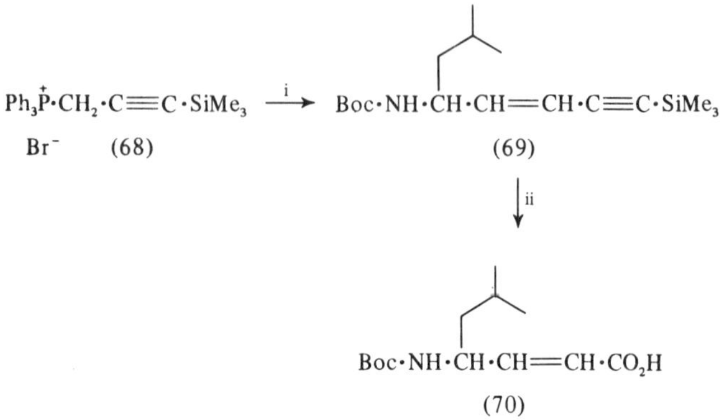

Reagents: i, (*a*) Bu^{n}Li/THF, (*b*) Boc·NH·CH·CHO; ii, (*a*) (cyclohexyl)$_2$BH, (*b*) H_2O_2/NaOH

Scheme 14

Reagents: i, Tos·OH/azeotropic distillation (R = Ph or H, X = Z or Tfa)

Scheme 15

Peptides containing thioamides have also continued to be of interest, and a number of thionating agents have been prepared.[150,151] Generally, the thionating agents have the structure (74), and compounds (74a)–(74c) will thionate most amides and lactams at room temperature, using THF as the solvent.[150] Dimethoxyethane may also be used but generally results in a longer reaction time. The scope of compound (74d) appears to be limited, as this compound is claimed to have a particularly bad odour; however, it is useful for the preparation of dithioesters. Compounds (74b) and (74c) may also be used for this, using dimethoxyethane as the solvent. Compounds (74e) and (74f) are claimed to have good solubility and to be better than other thionating reagents; they are also easier to prepare and may be used under particularly mild conditions to give clean reaction products. A simple dipeptide dissolved in THF was thionated in 90 min at room temperature, giving an 86% yield.

Dithioesters of type (75) derived from amino acids may be prepared and used for the preparation of specifically monothionated peptides.[152] Unfortunately,

[150] B. Yde, N. M. Yousif, U. Pedersen, I. Thomsen, and S.-O. Lawesson, *Tetrahedron*, 1984, **40**, 2047.

[151] M. Yokoyama, Y. Hasegawa, H. Hatanaka, Y. Kawazoe, and T. Imamoto, *Synthesis*, 1984, 827.

[152] G. Lajoie, F. Lepine, S. Lemaire, F. Jolicoeur, C. Aube, A. Turcotte, and B. Belleau, *Int. J. Pept. Protein Res.*, 1984, **24**, 316.

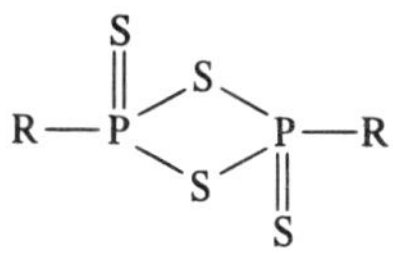

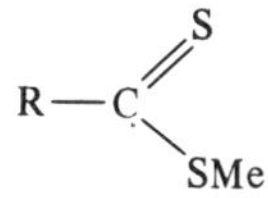

(75)

(74) a; R = 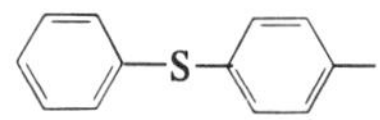

b; R = 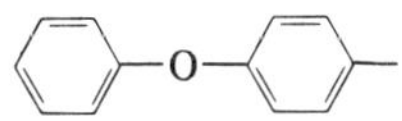

c; R = MeO— 

d; R = MeS—

e; R = 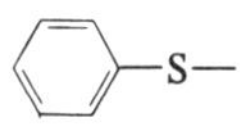

f; R = Me— 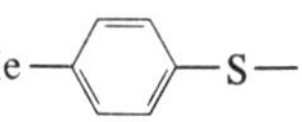

however, completely racemic products were isolated, and it is therefore doubtful whether this route will be useful for the preparation of thionated peptides.

4 Appendix I: A List of Syntheses Reported in 1984

The syntheses are listed under the name of the peptide to which they relate, as in previous years.

Peptide	*Ref.*
Natural Peptides, Proteins, and Partial Sequences	
Acetylcholine receptor	
Acetylcholine receptor (151–169)	153
Torpedo californica acetylcholine receptor α (125–147)	154
Adrenocorticotropic hormone (ACTH)	
Ostrich and turkey ACTH	155
Cyclic ACTH (5–14) analogue	156
Cyclic ACTH (5–10) analogue	157

153 M. A. Juillerat, T. Barkas, and S. J. Tzartos, *FEBS Lett.*, 1984, **168**, 143.
154 D. J. McCormick and M. Z. Atassi, *Biochem. J.*, 1984, **224**, 995.
155 D. Yamashiro, C. L. Ho, and C. H. Li, *Int. J. Pept. Protein Res.*, 1984, **23**, 42.
156 I. K. Liepkaula, A. A. Skujins, P. J. Romanovskis, E. A. Porunkevich, M. P. Ratkevich, and G. I. Chipens, *Bioorg. Khim.*, 1984, **10**, 1326.
157 I. K. Liepkaula, A. A. Skujins, P. J. Romanovskis, E. A. Porunkevich, M. P. Ratkevich, and G. I. Chipens, *Bioorg. Khim.*, 1984, **10**, 807.

Peptide	*Ref.*
ACTH (1–24)/(11–24) analogue	158
ACTH (4–10) analogue	159
ACTH analogue fragments	160
[7-Methyl-Trp[9]]β-corticotropin (1–24)	161
Adipokinetic hormone	
Tritiated locust adipokinetic hormone analogue	162
AM toxin	
AM toxin II	163
AM toxin I analogues	164
Angiotensin	
Angiotensinogen (1–14)	102
[1,8-Cysteine]angiotensin II	165
Angiotensin II analogues, substitution at residue 4	166
Apamin	
Apamin analogues	101
Aplysia peptides	
Aplysia peptides	167
Aspartame	
Analogues of α-Asp-Gly-Gly-OMe, sweet peptides	168
Aspartyl dipeptide sweeteners	169
Aspartame	170
Trifluoroacetyl-L-aspartylanilides	171
Atrial natriuretic peptide	
Atrial natriuretic peptides	172

[158] I. V. Syskov, P. J. Romanovskis, I. A. Vosekalna, A. A. Skujins, M. P. Ratkevich, B. S. Kataev, E. A. Porunkevich, and G. I. Chipens, *Bioorg. Khim.*, 1984, **10**, 618.
[159] K. I. Ratmanova, T. N. Bocharova, and L. A. Andreeva, *Bioorg. Khim.*, 1984, **10**, 564.
[160] I. V. Systov, P. J. Romanovskis, and G. I. Chipens, *Bioorg. Khim.*, 1983, **9**, 1483.
[161] M. C. Allen, P. D. Bailey, D. E. Brundish, J. H. Jones, R. Wade, and G. T. Young, *Int. J. Pept. Protein Res.*, 1984, **24**, 529.
[162] K. Muramoto, J. Ramachandran, P. Moshitzky, and S. W. Applebaum, *Int. J. Pept. Protein Res.*, 1984, **23**, 443.
[163] T. Kozono, H. Mihara, H. Aoyagi, T. Kato, and N. Izumiya, *Int. J. Pept. Protein Res.*, 1984, **24**, 402.
[164] H. Mihara, H. Aoyagi, S. Lee, M. Waki, T. Kato, and N. Izumiya, *Int. J. Pept. Protein Res.*, 1984, **23**, 447.
[165] J. M. Matsoukas, M. N. Scanlon, and G. J. Moore, *J. Med. Chem.*, 1984, **27**, 404.
[166] G. Guillemette, M. Bernier, P. Parent, R. Leduc, and E. Escher, *J. Med. Chem.*, 1984, **27**, 315.
[167] T. Kreiner, J. B. Rothbard, G. K. Schoolnik, and R. H. Scheller, *J. Neurosci.*, 1984, **4**, 2581.
[168] Y. Ariyoshi, *Bull. Chem. Soc. Jpn.*, 1984, **57**, 3197.
[169] J. W. Tsang, B. Schmied, R. Nyfeler, and M. Goodman, *J. Med. Chem.*, 1984, **27**, 1663.
[170] Y. P. Shvachkin and S. K. Girin, *J. Gen. Chem. U.S.S.R.*, 1982, **52**, 2461.
[171] M. Rodriguez and M. Goodman, *J. Med. Chem.*, 1984, **27**, 1668.
[172] M. Sugiyama, H. Fukumi, R. T. Grammer, K. S. Misono, Y. Yabe, Y. Morisawa, and T. Inagami, *Biochem. Biophys. Res. Commun.*, 1984, **123**, 338.

Peptide	*Ref.*
α-Human atrial natriuretic polypeptide	173
Biotin	
Biotin peptides	174
Bitter peptides	
Bitter peptide BPIa, dimer and Gly_2BPIa	175
An alternative bitter peptide from β-casein related to BPIa	176
Bitter peptide BPIa, a cyclic analogue	177
Bitter peptide BPIa, N-terminal analogues	178
Bitter peptide BPIc, fragments and analogues	179
Bitter peptide BPIa, retrosequence and fragments	180
Bitter peptide BPIa, hydrophobic analogues	181
Bombesin-like peptide	
Bombesin-like peptide fragment	91
Bradykinin	
Bradykinin analogues	182
Cyclic analogues of bradykinin	142
Bradykinin potentiating peptide	
Retro–inverso analogue of bradykinin potentiating peptide	144
Cadystin	
Cadystin A and B, cadmium-binding peptides	183
Cadystin A and B	184
Calcitonin	
Salmon calcitonin (1–10)	104
Calmodulin	
Calmodulin (137–143)	185
Modified calmodulin calcium-binding domain III	186

[173] K. Kangawa and H. Matsuo, *Biochem. Biophys. Res. Commun.*, 1984, **118**, 131.
[174] H. Kondo, S. Uno, Y. Komizo, and J. Sunamoto, *Int. J. Pept. Protein Res.*, 1984, **23**, 559.
[175] I. Miyake, K. Kouge, H. Kanehisa, and H. Okai, *Bull. Chem. Soc. Jpn.*, 1984, **57**, 815.
[176] H. Kanehisa, I. Miyake, H. Okai, H. Aoyagi, and N. Izumiya, *Bull. Chem. Soc. Jpn.*, 1984, **57**, 819.
[177] I. Miyake, K. Kouge, H. Kanehisa, and H. Okai, *Bull. Chem. Soc. Jpn.*, 1984, **57**, 1163.
[178] H. Kanehisa and H. Okai, *Bull. Chem. Soc. Jpn.*, 1984, **57**, 301.
[179] H. Kanehisa, *Bull. Chem. Soc. Jpn.*, 1984, **57**, 97.
[180] T. Shigenage, K. Otagiri, H. Kanehisa, and H. Okai, *Bull. Chem. Soc. Jpn.*, 1984, **57**, 103.
[181] K. Otagiri, T. Shigenage, H. Kanehisa, and H. Okai, *Bull. Chem. Soc. Jpn.*, 1984, **57**, 90.
[182] I. E. Mutule, F. K. Mutulis, O. E. Lando, A. A. Ashmanis, V. D. Grigoryeva, N. V. Myshlyakova, V. E. Klusha, and G. I. Chipens, *Bioorg. Khim.*, 1984, **10**, 891.
[183] N. Kondo, K. Imai, M. Isobe, T. Goto, A. Murasugi, C. Wada-Nakagawa, and Y. Hayashi, *Tetrahedron Lett.*, 1984, **25**, 3869.
[184] N. Kondo, M. Isobe, K. Imai, and T. Goto, *Agric. Biol. Chem.*, 1985, **49**, 71.
[185] L. J. Van Eldik, D. M. Watterson, K.-F. Fok, and B. W. Erickson, *Arch. Biochem. Biophys.*, 1983, **227**, 522.
[186] V. Pavone, A. Di Nola, S. Andini, L. Ferrara, B. Di Blasio, E. Benedetti, and P. Pucci, *Int. J. Pept. Protein Res.*, 1984, **23**, 454.

Peptide	Ref.
Chemotactic peptides	
Chemotactic peptides	187
Chlamydocin	
Dihydrochlamydocin	188
Cholecystokin (CCK)	
CCK heptapeptide	50
Analogues of CCK heptapeptide	53
Derivatives of CCK (26–32)	189
Chymostatin	
Chymostatin analogues	190
Circumsporozoite surface protein	
Fragments of circumsporozoite surface protein	191
Ciliatine	
Peptides containing (2-aminoethyl)phosphonic acid, ciliatine	192
Collagen	
Polypeptide models of collagen	193
Conotoxin	
Conotoxin GI	194
Corticotropin-releasing factor (CRF)	
Human corticotropin-releasing factor (hCRF)	56, 195
Nine fragments of human corticotropin-releasing factor	196
Cyclosporin	
Cyclosporin analogues	197
Cyclosporin A and H	198
Cytochromes	
Cytochrome C	199

[187] C. Toniolo, G. M. Bonora, H. Showell, R. J. Freer, and E. L. Becker, *Biochemistry*, 1984, **23**, 698.

[188] U. Schmidt, A. Lieberknecht, H. Griesser, and F. Bartkowiak, *Angew. Chem.*, 1984, **96**, 310.

[189] J. D. Gardner, M. Knight, V. E. Sutliff, C. A. Tamminga, and R. T. Jensen, *Am. J. Physiol., G*, 1984, **246**, 292.

[190] I. J. Galpin, A. J. Wilby, G. A. Place, and R. J. Beynon, *Int. J. Pept. Protein Res.*, 1984, **23**, 477.

[191] D. H. Schlesinger, A. H. Cochrane, R. W. Gwadz, G. N. Godson, R. Melton, R. S. Nussenzweig, and V. Nussenzweig, *Biochemistry*, 1984, **23**, 5665.

[192] K. Yamauchi, S. Ohtsuki, and M. Kinoshita, *J. Org. Chem.*, 1984, **49**, 1158.

[193] V. Guantieri and A. M. Tamburro, *Int. J. Pept. Protein Res.*, 1984, **24**, 274.

[194] A. Reyes, J. Alford, M. McIntosh, B. M. Olivera, L. J. Cruz, and J. Rivier, *Biochemistry*, 1984, **23**, 2796.

[195] H. Yajima, N. Fujii, M. Nomizu, K. Watanabe, K. Akaji, M. Shimokura, S. Katakura, F. Shono, M. Tsuda, and A. Yoshitake, *Chem. Pharm. Bull.*, 1984, **32**, 2052.

[196] N. Fujii, M. Nomizu, K. Akaji, M. Shimokura, S. Katakura, and M. Yajima, *Chem. Pharm. Bull.*, 1984, **32**, 4786.

[197] R. M. Wenger, *Chimia*, 1984, **38**, 11.

[198] R. M. Wenger, *Helv. Chim. Acta*, 1984, **67**, 502.

[199] G. I. Tesser, P. J. H. M. Adams, P. B. W. T. Kortenaar, and H. A. Boots, *Int. J. Pept. Protein Res.*, 1984, **24**, 192.

Peptide	*Ref.*
Cytochrome P450, apoprotein model peptides	200
Deoxybouvaridin	
Deoxybouvaridin	201
Dermorphin	
Dermorphin	202
Dermorphin fragments	203
Dermorphin analogues	204
[Sar⁴]dermorphin tetrapeptide analogues	205
Dermorphin tetrapeptide	206
Dihydromauritine	
Dihydromauritine A	207
Dynorphin	
Dynorphin (1–13)	100
Alanine-containing dynorphin (1–13) analogues	208
Dynorphin B29	209
Elastase-like proteinase (ELP)	
Inhibitors of human leukocyte ELP	210
Elastin	
[D-Ala³]elastin polypentapeptide	211
Eledoisin	
Eledoisin (6–11)	121
Endorphin	
Endorphin analogues	212, 213
Enkephalin	
Enkephalin alkylamides	214

200 K. Kawasaki, M. Maeda, H. Tamura, S. Ohashi, T. Yoshimura, E. Hatayama, and H. Sakurai, *Chem. Pharm. Bull.*, 1984, **32**, 1717.

201 R. B. Bates, S. L. Gin, M. A. Hassen, V. J. Hruby, K. D. Janda, G. R. Kriek, J.-P. Michaud, and D. B. Vine, *Heterocycles*, 1984, **22**, 785.

202 K. Darlak and Z. Grzonka, *Pol. J. Chem.*, 1982, **56**, 1201.

203 A. Pastore, P. A. Temussi, T. Tencredi, S. Salvadori, and R. Tomatis, *Biopolymers*, 1984, **23**, 2349.

204 K. Darlak, Z. Grzonka, P. Krzascik, P. Janicki, and S. W. Gumulka, *Peptides,* 1984, **5**, 687.

205 S. Salvadori, G. Balboni, M. Marastoni, G. Sarto, and R. Tomatis, *Hoppe-Seyler's Z. Physiol. Chem.*, 1984, **365**, 1199.

206 S. Salvadori, G. P. Sarto, and R. Tomatis, *Eur. J. Med. Chem.*, 1983, **18**, 489.

207 R. F. Nutt, K.-M. Chen, and M. M. Joullie, *J. Org. Chem.*, 1984, **49**, 1013.

208 A. Turcotte, J.-M. Lalonde, S. St-Pierre, and S. Lemaire, *Int. J. Pept. Protein Res.*, 1984, **23**, 361.

209 P. Sanchez-Blazquez, J. K. Chang, J. Garzon, and N. M. Lee, *Neuropeptides*, 1984, **4**, 369.

210 Y. Okada, Y. Tsuda, Y. Nagamatsu, and U. Okamoto, *Int. J. Pept. Protein Res.*, 1984, **24**, 347.

211 D. W. Urry, T. L. Trapane, S. A. Wood, R. D. Harris, J. T. Walker, and K. U. Prasad, *Int. J. Pept. Protein Res.*, 1984, **23**, 425.

212 D. Yamashiro, C. H. Li, and M. Westphal, *Int. J. Pept. Protein Res.*, 1984, **24**, 520.

213 D. Yamashiro, M. Westphal, and C. H. Li, *Int. J. Pept. Protein Res.*, 1984, **24**, 516.

214 B. J. Dhotre, S. Chaturvedi, and K. B. Mathur, *Indian J. Chem., Sect. B*, 1984, **23**, 828.

Peptide	Ref.
Enkephalin analogues with C-terminal methylthioamide	215
Conformationally restrained enkephalin analogue	216
Bis-[*N*,*N*-diallyl-(D-Ala^2,Leu^5)enkephalinyl] cystine	41
[D-Ala^2]enkephalin/glucose derivative	217
Ketomethylene analogues of leucine-enkephalin	218
L- and D-[Phe(2-OH)]^1-leucine-enkephalin	118
Met- and [D-Ala^2]Met-enkephalin	219
Monothionated analogues of enkephalin	152
Monothio analogues of leucine-enkephalin	220
Radioiodinated [D-Ala^2,Met^5]enkephalin	221
Retro methionine-enkephalin analogues	222
Retro–inverso enkephalins	143
Epidermal growth factor (EGF)	
EGF (14–31) and (20–31)	223
53-Residue peptide with EGF activity	224
Ferredoxin	
2Fe2S ferredoxin model	225
Fibrin	
Fibrin amino-terminal peptide	226
Fibrinogen	
Fibrinogen γ (400–411)	227
Gastrin	
Gastrin, C-terminal peptide analogues	228
Shortened little-gastrin I peptides	229

[215] K. Rolka, M. Kruszynski, and G. Kupryszewski, *Acta Pharm. Suec.*, 1984, **21**, 173.
[216] T. Deeks, P. A. Crooks, and R. D. Waigh, *J. Pharm. Sci.*, 1984, **73**, 457.
[217] B. Filippi, L. Biondi, F. Filira, and R. Rocchi, *Farmaco Ed. Sci.*, 1983, **38**, 713.
[218] R. G. Almquist, C. M. Olsen, E. T. Uyeno, and L. Toll, *J. Med. Chem.*, 1984, **27**, 115.
[219] T. Plucinski, K. Rolka, L. Baran, E. Przegalinski, and G. Kupryszewski, *Pol. J. Chem.*, 1983, **57**, 887.
[220] K. Clausen, M. Thorsen, S.-O. Lawesson, and A. F. Spatola, *J. Chem. Soc., Perkin Trans. 1*, 1984, 785.
[221] C. W. T. Yeung, *Int. J. Pept. Protein Res.*, 1984, **24**, 636.
[222] Y. P. Shvashkin, A. A. Shishkina, A. P. Smirnova, V. P. Fedotov, T. I. Ivanenko, V. I. Gudoshnikov, G. I. Bushueva, and L. A. Batrameeva, *Bioorg. Khim.*, 1983, **9**, 1492.
[223] A. Komoriya, M. Hortsch, C. Meyers, M. Smith, H. Kanety, and J. Schlessinger, *Proc. Natl. Acad. Sci. U.S.A.*, 1984, **81**, 1351.
[224] H. Yajima, K. Akaji, N. Fujii, K. Hayashi, K. Mizuta, M. Aono, and M. Moriga, *J. Chem. Soc., Chem. Commun.*, 1984, 1103.
[225] N. Ueyama, S. Ueno, M. Nakata, and A. Nakamura, *Bull. Chem. Soc. Jpn.*, 1984, **57**, 984.
[226] V. I. Unkovskii, G. A. Vasil'eva, and R. P. Evstigneeva, *J. Gen. Chem. U.S.S.R.*, 1982, **52**, 1690.
[227] E. F. Plow, A. H. Srouji, D. Meyer, G. Marguerie, and M. H. Ginsberg, *J. Biol. Chem.*, 1984, **259**, 5388.
[228] J. Martinez, R. Magous, M.-F. Lignon, J. Laur, B. Castro, and J.-P. Bali, *J. Med. Chem.*, 1984, **27**, 1597.
[229] W. Goehring, L. Moroder, G. Borin, A. Lobbia, J. P. Bali, and E. Wuensch, *Hoppe-Seyler's Z. Physiol. Chem.*, 1984, **365**, 83.

Peptide	*Ref.*
N-(2-Chloroethyl)-*N*-nitrosocarbamoyl gastrin analogues	230
Gastrointestinal polypeptide	
Gastrointestinal polypeptide	231
Glucagon	
Glucagon analogues	232
Mammalian glucagon	98
Glycophorin	
Glycophorin A analogues	233
Gramicidin	
Dimeric gramicidin A	234
Gramicidin S analogues	235
Gramacidin S macro-ring analogue	236
[D-^{2}H-Phe]gramicidin S	237
Gramicidin S2 and S3	238
Gratisin	
Gratisin	239
Growth-hormone-releasing factor (GRF)	
GRF biotinyl derivatives	240
hpGRF-40-OH	241
Human-GRF and two omission analogues	107
GRF	242
Human-GRF	243
(GRF, somatocrinin), (23—44)	244
GRF deletion analogue	245

[230] H. Sueli-Vargha and K. Medzihradszky, *Int. J. Pept. Protein Res.*, 1984, **23**, 650.
[231] S. A. St-Pierre, *Clin. Biochem.*, 1984, **17**, 68.
[232] J. Sueiras-Diaz, V. A. Lance, W. A. Murphy, and D. H. Coy, *J. Med. Chem.*, 1984, **27**, 310.
[233] B. Ferrari and A. A. Pavia, *Int. J. Pept. Protein Res.*, 1984, **24**, 488.
[234] L. A. Fonina, A. M. Demina, S. V. Sychev, V. T. Ivanov, and Y. Hlavacek, *Bioorg. Khim.*, 1984, **10**, 1073.
[235] S. Ando, T. Kato, and N. Izumiya, *Int. J. Pept. Protein Res.*, 1985, **25**, 15.
[236] S. Ando, H. Takiguchi, and N. Izumiya, *Bull. Chem. Soc. Jpn.*, 1983, **56**, 3781.
[237] K. Tanimura, T. Kato, M. Waki, S. Lee, Y. Kodera, and N. Izumiya, *Bull. Chem. Soc. Jpn.*, 1984, **57**, 2193.
[238] S. Nozaki and I. Muramatsu, *Bull. Chem. Soc. Jpn.*, 1985, **58**, 331.
[239] M. Tamaki, *Bull. Chem. Soc. Jpn.*, 1984, **57**, 3210.
[240] N. Fujii, M. Shimokura, H. Yajima, F. Shono, M. Tsuda, and A. Yoshitake, *Chem. Pharm. Bull.*, 1984, **32**, 1200.
[241] N. Fujii, M. Shimokura, H. Yajima, F. Shono, M. Tsuda, and A. Yoshitake, *Chem. Pharm. Bull.*, 1984, **32**, 1193.
[242] M. Wakimasu, C. Kitada, H. Yukimasa, T. Sugimoto, and S. Kobayashi, *Takeda Kenkyusho Ho*, 1983, **42**, 209.
[243] N. Fujii, M. Shimokura, M. Nomizu, H. Yajima, F. Shono, M. Tsuda, and A. Yoshitake, *Chem. Pharm. Bull.*, 1984, **32**, 520.
[244] N. Fujii, M. Shimokura, K. Akaji, S. Kiyama, and H. Yajima, *Chem. Pharm. Bull.*, 1984, **32**, 510.
[245] N. Ling, A. Baird, W. B. Wehrenberg, N. Ueno, T. Munegumi, and P. Brazeau, *Biochem. Biophys. Res. Commun.*, 1984, **123**, 854.

Peptide	*Ref.*
GRF analogues	246
Hybrid GRF/PHI heptacosapeptide amide	247
Haemoglobin	
Haemoglobin/myoglobin 'active-centre' model peptide	248
β-Chain human haemoglobin (67–69)	103
Heat-stable enterotoxin	
Heat-stable enterotoxin	249, 250
Heat-stable enterotoxin fragments	251–253
Hexa-acetylcelenamide	
Hexa-acetylcelenamide A	254
Histone	
Histone H1 (152–184)	255
Histone H1 (152–164)/(165–172)	256
Hylambatin	
Hylambatin (frog-skin peptide)	54
Immunoglobulin	
Ig V-H (27–68)	112
IgM hypervariable-region peptides	257
Immunostimulating peptide	
Immunostimulating peptide FK156	258
Immunostimulating peptide FK156 analogue	259
Inhibin-like peptide	
Human inhibin-like peptide (1–31)	260

[246] D. H. Coy, W. A. Murphy, J. Sueiras-Diaz, E. J. Coy, and V. A. Lance, *J. Med. Chem.*, 1985, **28**, 181.
[247] N. Fujii, W. Lee, M. Shimokura, and H. Yajima, *Chem. Pharm. Bull.*, 1984, **32**, 739.
[248] N. A. Kazakova, V. A. Radyukhin, V. M. Luzgina, E. I. Filippovich, N. V. Kamyshan, E. V. Kudryavtseva, and R. P. Evstigneeva, *J. Gen. Chem. U.S.S.R.*, 1982, **52**, 1681.
[249] R. A. Houghten, J. M. Ostresh, and F. A. Klipstein, *Eur. J. Biochem.*, 1984, **145**, 157.
[250] S. Yoshimura, M. Miki, H. Ikemura, S. Aimoto, Y. Shimonishi, T. Takeda, Y. Takeda, and T. Miwatani, *Bull. Chem. Soc. Jpn.*, 1984, **57**, 125.
[251] H. Ikemura, H. Watanabe, S. Aimoto, Y. Shimonishi, S. Hara, T. Takeda, Y. Takeda, and T. Miwatani, *Bull. Chem. Soc. Jpn.*, 1984, **57**, 2550.
[252] H. Ikemura, S. Yoshimura, S. Aimoto, Y. Shimonishi, S. Hara, T. Takeda, Y. Takeda, and T. Miwatani, *Bull. Chem. Soc. Jpn.*, 1984, **57**, 2543.
[253] S. Yoshimura, T. Takao, H. Ikemura, S. Aimoto, Y. Shimonishi, S. Hara, T. Takeda, Y. Takeda, and T. Miwatani, *Bull. Chem. Soc. Jpn.*, 1984, **57**, 1381.
[254] U. Schmidt and J. Wild, *Angew. Chem.*, 1984, **96**, 996.
[255] L. I. Maryash, V. K. Burichenko, and V. A. Shibnev, *Bioorg. Khim.*, 1984, **10**, 467.
[256] L. I. Maryash, V. K. Burichenko, and V. A. Shibnev, *Bioorg. Khim.*, 1984, **10**, 459.
[257] P. P. Chen, R. A. Houghten, S. Fong, G. H. Rhodes, T. A. Gilbertson, J. H. Vaughan, R. A. Lerner, and D. A. Carson, *Proc. Natl. Acad. Sci. U.S.A.*, 1984, **81**, 1784.
[258] H. Takeno, S. Okada, S. Yonishi, K. Hemmi, O. Nakaguchi, Y. Kitaura, and M. Hashimoto, *Chem. Pharm. Bull.*, 1984, **32**, 2932.
[259] H. Takeno, S. Okada, K. Hemmi, M. Aratani, Y. Kitaura, and M. Hashimoto, *Chem. Pharm. Bull.*, 1984, **32**, 2925.
[260] D. Yamashiro, C. H. Li, K. Ramasharma, and M. R. Sairam, *Proc. Natl. Acad. Sci. U.S.A.*, 1984, **81**, 5399.

Peptide	*Ref.*
Insulin	
[Asn^{B10}]insulin (human)	261
[Gln^{B30}]insulin	262
[Leu^{A19}]insulin	263
[D-Trp^{A1}]insulin	264
Biotinyl insulins	265
Des-[Phe^{B1}, Trp^{B2}] insulin	266
Des-(B26–B30)-insulin analogues	267
Human insulin fragments (B9–B14) and (B15–B20)	268
Bovine insulin fragment (B9–B30)/(B20–B30)	116
Insulin-like growth factor	
Insulin-like growth factor I	269
Interferon	
Interferon α1 (24–81)/(111–166)	270
C-Terminal fragment of human leucocyte interferon αF	44
Human leukocyte interferon (116–137)	99
Kinins	
Renal vasodilators, tri-/tetra-peptide analogues of kinins	271
Kyotorphin	
Kyotorphin analogue	272
lac-Repressor	
lac-Repressor (53–58)	273
Octapeptide fragments of the *lac*-repressor	274

[261] G. T. Burke, G. Schwartz, and P. G. Katsoyannis, *Int. J. Pept. Protein Res.*, 1984, **23**, 394.

[262] Y. P. Shvachkin, S. P. Krasnoshchekova, A. M. Nikitina, E. N. Voluiskaya, V. P. Fedotov, and A. T. Ivanova, *Bioorg. Khim.*, 1984, **10**, 708.

[263] K. Kitagawa, H. Ogawa, G. T. Burke, J. D. Chanley, and P. G. Katsoyannis, *Biochemistry*, 1984, **23**, 4444.

[264] P. W. Schiller and R. Geiger, *J. Protein Chem.*, 1983, **2**, 279.

[265] K. Hofmann, W. J. Zhang, H. Romovacek, F. M. Finn, A. A. Bothner-By, and P. K. Mishra, *Biochemistry*, 1984, **23**, 2547.

[266] B. Dong and K.-J. Lei, *Acta Biochim. Biophys. Sin.*, 1984, **16**, 679.

[267] S.-Q. Zhu, D.-J. Zhuang, and J.-H. Zhu, *Acta Biochim. Biophys. Sin.*, 1984, **16**, 7.

[268] L. A. Kolomeitseva, N. A. Kocharova, L. D. Ibragimova, V. F. Krivtsov, and Y. P. Shvachkin, *J. Gen. Chem. U.S.S.R.*, 1982, **52**, 372.

[269] D. Schalch, D. Reismann, C. Emler, R. Humbel, C. H. Li, M. Peters, and E. Lau, *Endocrinology*, 1984, **115**, 2490.

[270] T. Leist and R. M. Thomas, *Biochemistry*, 1984, **23**, 2541.

[271] F. R. Pfeiffer, P. A. Chambers, E. E. Hilbert, P. W. Woodward, and D. M. Ackerman, *J. Med. Chem.*, 1984, **27**, 325.

[272] K. Rolka, E. Oslislok, J. Krupa, M. Kruszynski, L. Baran, E. Przegalinski, and G. Kupryszewski, *Pol. J. Pharmacol. Pharm.*, 1983, **35**, 473.

[273] M. V. R. Rao, M. Atreyi, V. S. Chauhan, S. Kumar, and S. G. Kumar, *Indian J. Chem., Sect. B*, 1984, **6**, 353.

[274] V. A. Shibnev, A. K. Gazumyan, and M. P. Finegenova, *Bioorg. Khim.*, 1983, **9**, 1497.

Peptide	*Ref.*
Luteinizing-hormone-releasing hormone (LHRH)	
Salmon LHRH	275
[Trp^7]LHRH	276
LHRH analogues	277
LHRH antagonists	278
LHRH analogues containing 3-(2-benzimidazolyl)alanine	279
LHRH analogues containing β-(3-pyridyl)-D-alanine	280
LHRH, hydrophobic aza-Gly analogues	281
Photoreactive derivative of the gonadotropin-releasing-hormone agonist [D-Lys^6]LHRH (1–9)-NHEt	282
Lysozyme	
[Cys^3,Cys^{97}]phage T4 lysozyme	283
M13 coat protein	
Leader peptide from M13 coat protein	284
Mastopyran	
Mastopyran	59
Melanin-concentrating hormone (MCH)	
MCH	285
Salmon MCH	286
Melanotropin (MSH)	
Biotin-labelled MSH derivatives	287
Cyclic MSHs containing modified disulphide bridge	288

[275] B. Breton, A. Motin, O. Kah, F. Lemenn, S. Geoffre, G. Precigoux, and P. Chambolle, *C. R. Acad. Sci., Ser. III*, 1984, **299**, 383.

[276] K. Folkers, C. Y. Bowers, H.-M. Shieh, Y.-Z. Liu, S.-B. Xiao, P.-F. L. Tang, and J.-Y. Chu, *Biochem. Biophys. Res. Commun.*, 1984, **123**, 1221.

[277] J. J. Nestor, jun., R. Tahilramani, T. L. Ho, G. I. McRae, and B. H. Vickery, *J. Med. Chem.*, 1984, **27**, 1170.

[278] M. Kovacs, I. Mezo, B. Flerko, I. Teplan, and K. Nikolics, *Biochem. Biophys. Res. Commun.*, 1984, **118**, 351.

[279] J. J. Nestor, jun., B. L. Horner, T. L. Ho, G. H. Jones, G. I. McRae, and B. H. Vickery, *J. Med. Chem.*, 1984, **27**, 320.

[280] K. Folkers, T. Kubiak, and J. Stepinski, *Int. J. Pept. Protein Res.*, 1984, **24**, 197.

[281] T. L. Ho, J. J. Nestor, jun., G. I. McRae, and B. H. Vickery, *Int. J. Pept. Protein Res.*, 1984, **24**, 79.

[282] K. Nikolics, I. Teplan, and J. Ramachandran, *Int. J. Pept. Protein Res.*, 1984, **24**, 430.

[283] L. J. Perry and R. Wetzel, *Science*, 1984, **226**, 555.

[284] A. E. Shinnar and E. T. Kaiser, *J. Am. Chem. Soc.*, 1984, **106**, 5006.

[285] B. C. Wilkes, V. J. Hruby, W. C. Sherbrooke, A. M. De L. Castrucci, and M. E. Hadley, *Biochem. Biophys. Res. Commun.*, 1984, **122**, 613.

[286] K. Okamoto, K. Yasumura, K. Fujitani, Y. Kiso, H. Kawauchi, I. Kawazoe, and H. Yajima, *Chem. Pharm. Bull.*, 1984, **32**, 2963.

[287] D. N. Chaturvedi, J. J. Knittel, V. J. Hruby, A. M. De L. Castrucci, and M. E. Hadley, *J. Med. Chem.*, 1984, **27**, 1406.

[288] M. Lebl, W. L. Cody, B. C. Wilkes, V. J. Hruby, A. M. De L. Castrucci, and M. E. Hadley, *Collect. Czech. Chem. Commun.*, 1984, **49**, 2680.

Peptide	Ref.
Cyclic MSH analogues	105, 106, 140
[Tyr^4]-α-MSH analogues	289
Tritiated Ac-[Nle^4,D-Phe^7]-α-MSH (4–11)-NH_2	290
N-(2-Chloroethyl)-*N*-nitrosocarbamoyl-α-MSH analogues	230
Myoglobin	
Myoglobin fragment	291
Neurokinin	
Neurokinin α/β	292, 293
Neurotensin	
[Trp^{11}] neurotensin	294
Dehydroamino acid analogues of neurotensin	295
Oxytocin	
Oxytocin analogues	296
Carbaoxytocin analogues	106
Carba-des-amino-Pen^1,X^6-oxytocin analogues	297
[$Cystine^6$] oxytocin	298
[*allo*-Thr^4] oxytocin	299
^{13}C-Enriched oxytocin analogues	300
Tritiated desaminocarbaoxytocin analogues	301
P2 protein	
P2 protein (66–70), (66–78), and (70–78)	302

[289] B. C. Wilkes, T. K. Sawyer, V. J. Hruby, and M. E. Hadley, *Int. J. Pept. Protein Res.*, 1984, **23**, 621.
[290] B. C. Wilkes, V. J. Hruby, H. I. Yamamura, K. Akiyama, A. M. De L. Castrucci, M. E. Hadley, J. R. Andrews, and Y.-P. Wan, *Life Sci.*, 1984, **34**, 977.
[291] P.-T. Shi, J. P. Riehm, P. E. E. Todd, and S. J. Leach, *Mol. Immunol.*, 1984, **21**, 489.
[292] K. Folkers, Y.-A. Lu, and S. Rosell, *Biochem. Biophys. Res. Commun.*, 1984, **118**, 405.
[293] E. Munekata, M. Okada, S. Kimura, Y. Sugita, I. Kanazawa, T. Matsudo, and M. Otsuka, *Chem. Lett.*, 1984, 1013.
[294] C. Labbe-Jullie, P. Kitabgi, C. Granier, J.-P. Vincent, and J. Van Rietschoten, *Biochimie*, 1983, **65**, 534.
[295] C. Labbe-Jullie, D. Blanot, J.-L. Morgat, P. Kitabgi, F. Checler, J.-P. Vincent, C. Granier, and J. Van Rietschoten, *Biochimie*, 1983, **65**, 553.
[296] Z. Prochazka, M. Lebl, T. Barth, J. Hlavacek, A. Trka, M. Budesinsky, and K. Jost, *Collect. Czech. Chem. Commun.*, 1984, **49**, 642.
[297] M. Lebl, T. Barth, L. Servitova, J. Slaninova, and P. Hrbas, *Collect. Czech. Chem. Commun.*, 1984, **49**, 2012.
[298] J. W. Van Nispen, J. A. J. Hannink, M. S. Schoffelmeer, W. P. A. Janssen, J. P. Polderdijk, and H. M. Greven, *Recl. Trav. Chim. Pays-Bas*, 1984, **103**, 68.
[299] K. Bankowski, A. Misicka, and S. Drabarek, *Pol. J. Chem.*, 1983, **57**, 1045.
[300] M. Blumenstein, V. J. Hruby, V. Viswanatha, and D. Chaturvedi, *Biochemistry*, 1984, **23**, 2153.
[301] M. Lebl, T. Barth, D. J. Crankshaw, B. Cerny, E. E. Daniel, A. K. Grover, and K. Jost, *Collect. Czech. Chem. Commun.*, 1984, **49**, 1921.
[302] M. Suzuki, K. Kitamura, K. Uyemura, Y. Ogawa, S. Nozaki, and I. Muramatsu, *Prog. Clin. Biol. Res.*, 1984, **146**, 487.

Peptide	*Ref.*
PHI	
PHI and related analogues	303
Phosphonopeptides	
Phosphonopeptides	304, 305
Pilin	
Pilin (1–12) analogue	306
Polisteskinin	
Polisteskinin analogues	182
Preproglucagon	
Angler-fish preproglucagon I and II (AGFI and AGFII)	307
Protohaemin IX	
Derivatives of protohaemin IX	308
Quisqualic acid	
Quisqualic acid analogues	309
Ragweed allergen	
Ragweed allergen Ra^5	310
Red-pigment-concentrating hormone	
[Thr^6]red-pigment-concentrating hormone	311
Renin precursor	
Renin precursor	312
Ribonuclease (RNase)	
RNase analogue	137
RNase analogues, substitution at residue 119	136
RNase (1–118)/(111–124) complex	313
Bovine RNase (1–118)/(111–124)	135
RNase (92–94) and (113–115)	314
C-Terminal decapeptide of bovine pancreatic RNase A	315

303 K. Nokihara, C. Yanaihara, K. Iguchi, S. Fukata, M. Tanaka, T. Mochizuki, K. Tatemoto, J. M. Lundberg, V. Mutt, and N. Yanaihara, *J. Am. Chem. Soc.*, 1984, **106**, 7909.
304 P. F. Alewood, S. Palma, and R. B. Johns, *Aust. J. Chem.*, 1984, **37**, 425.
305 J. Rachon, *Synthesis*, 1984, 219.
306 E. A. Worobec, W. Paranchych, J. M. R. Parker, A. K. Taneja, and R. S. Hodges, *J. Biol. Chem.*, 1985, **260**, 938.
307 I. J. Galpin and C. A. Leach, *Tetrahedron*, 1984, **40**, 4415.
308 V. A. Radyukhin, N. A. Kazakova, E. I. Filippovich, and R. P. Evstigneeva, *J. Gen. Chem. U.S.S.R.*, 1982, **52**, 377.
309 B. W. Bycroft, S. R. Chhabra, R. J. Grout, and P. J. Crowley, *J. Chem. Soc., Chem. Commun.*, 1984, 1156.
310 A. M. Choudhury and L. Goodfriend, *J. Protein Chem.*, 1983, **2**, 321.
311 D. Yamashiro, S. W. Applebaum, and C. H. Li, *Int. J. Pept. Protein Res.*, 1984, **23**, 39.
312 G. Evin, J. Devin, J. Menard, and P. Corvol, *Proc. Natl. Acad. Sci. U.S.A.*, 1984, **81**, 48.
313 M. S. Doscher, P. D. Martin, and B. F. P. Edwards, *J. Mol. Biol.*, 1983, **166**, 685.
314 G. T. Montelione, E. Arnold, Y. C. Meinwald, E. R. Stimson, J. B. Denton, S.-G. Huang, J. Clardy, and H. A. Scheraga, *J. Am. Chem. Soc.*, 1984, **106**, 7946.
315 Y. Okada, N. Teno, J. Miyao, Y. Mori, and M. Irie, *Chem. Pharm. Bull.*, 1984, **32**, 4585.

Peptide	*Ref.*
Scorpion toxin	
Antigenic-site peptides of a scorpion toxin	316
Secretin	
Chicken secretin	55
Secretin fragments and analogues	317
Somatostatin	
[Dehydro-Phe]somatostatin analogue	318
Cyclic hexapeptide analogues of somatostatin	319
Cyclic pseudo-hexapeptide analogues of somatostatin	141
Substance K	
Substance K	320
Substance P	
Substance P	321
[Arg^{11}]substance P analogues	322
Biotinylated analogues of substance P	323
Dehydrophenylalanine substance P (6–11)	324
Radiolabelled substance P analogues	325
Substance P analogues	326
Substance P antagonists	327
Tentoxin	
Tentoxin	328
Thioredoxin	
Thioredoxin fragment	111
Thymic peptides	
Thymic peptide hormone/polyamine conjugates	329

[316] C. Granier, E. Bahraoui, J. Van Rietschoten, H. Rochat, and M. El Ayeb, *Int. J. Pept. Protein Res.*, 1984, **23**, 187.
[317] W. Koenig, M. Bickel, K. Karch, V. Tettz, and R. Uhmann, *Peptides*, 1984, **5**, 189.
[318] S. F. Brady, D. W. Cochran, R. F. Nutt, F. W. Holly, C. D. Bennett, W. J. Paleveda, P. E. Curley, B. H. Arison, R. Saperstein, and D. F. Veber, *Int. J. Pept. Protein Res.*, 1984, **23**, 212.
[319] R. M. Freidinger, D. S. Perlow, W. C. Randall, R. Saperstein, B. H. Arison, and D. F. Veber, *Int. J. Pept. Protein Res.*, 1984, **23**, 142.
[320] H. Nawa, M. Doteuchi, K. Igano, K. Inouye, and S. Nakanishi, *Life Sci.*, 1984, **34**, 1153.
[321] M. Bienert, S. Katzwinkel, and J. Schwarz, *Z. Chem.*, 1983, **23**, 375.
[322] S. Sofuku, K. Torigoe, and I. Muramatsu, *Chem. Pharm. Bull.*, 1984, **32**, 4144.
[323] S. Lavielle, G. Chassaing, J. C. Beaujouan, Y. Torrens, and A. Marquet, *Int. J. Pept. Protein Res.*, 1984, **24**, 480.
[324] T. Wasiak and W. Koziolkiewicz, *Pol. J. Chem.*, 1983, **57**, 861.
[325] M. Bienert, H. E. Schmidt, A. Ehrlich, K. Forner, E. Klauschenz, J. Furkert, R. Rathsack, and H. Niedrich, *Pharmazie*, 1984, **39**, 97.
[326] J. Hoerig and H. Schultheiss, *Eur. J. Pharmacol.*, 1984, **105**, 65.
[327] G. Koeller, M. Bienert, H. Niedrich, J. Bergmann, I. Mezoe, and P. Oehme, *Pharmazie*, 1984, **39**, 65.
[328] R. Jacquier and J. Verducci, *Tetrahedron Lett.*, 1984, **25**, 2775.
[329] H.-M. Shieh, D. Campbell, and K. Folkers, *Biochem. Biophys. Res. Commun.*, 1984, **122**, 21.

Peptide	*Ref.*
Thymosin β4, β8, and β9	330
Thymosin β9 (16–25)	331
Desacetyl-thymosin β4	332
Thyrotropin-releasing hormone (TRH)	
TRH analogues	82
Transforming growth factor I	
Rat transforming growth factor I	333
Triostin	
Triostin A	334
Nortriostin A	335
Tryptophyllins	
Tryptophyllins, tetra- and penta-peptide analogues	336
Tuftsin	
Tuftsin, elongated analogues	337
Acetylenic analogues of tuftsin	338
[3,4-Dehydroproline^3]tuftsin	339
Tuftsinyltuftsin	340
Urogastrone	
Urogastrone (h-EGF)	139
Urogastrone (32–48) and (33–48)	341
Vasointestinal polypeptide (VIP)	
VIP	120
VIP (14–28)	119
VIP (22–28)	342
Vasopressin	
Conformationally restrained vasopressin analogue	343
[Cystine^6]arginine vasopressin	298
Position-9 analogues of arginine and lysine vasopressin	344

[330] T. Abiko and H. Sekino, *Chem. Pharm. Bull.*, 1984, **32**, 228.
[331] T. Abiko and H. Sekino, *J. Appl. Biochem.*, 1983, **5**, 446.
[332] T. Abiko and H. Sekino, *Chem. Pharm. Bull.*, 1984, **32**, 4497.
[333] J. P. Tam, H. Marquardt, D. F. Rosberger, T. W. Wong, and G. J. Todaro, *Nature (London)*, 1984, **309**, 376.
[334] M. Shin, K. Inouye, and H. Ōtsuka, *Bull. Chem. Soc. Jpn.*, 1984, **57**, 2203.
[335] M. Shin, K. Inouye, N. Higuchi, and Y. Kyogoku, *Bull. Chem. Soc. Jpn.*, 1984, **57**, 2211.
[336] G. Perseo and R. De Castiglione, *Int. J. Pept. Protein Res.*, 1984, **24**, 155.
[337] D. Konopinska, B. Kazanowska, and J. Boguslawska-Jaworska, *Int. J. Pept. Protein Res.*, 1984, **24**, 267.
[338] G. Auger and D. Blanot, *Int. J. Pept. Protein Res.*, 1984, **24**, 60.
[339] A. A. Amoscato, G. F. Babcock, and K. Nishioka, *Peptides*, 1984, **5**, 489.
[340] D. Konopinska, V. A. Najjar, and M. Callery, *Pol. J. Chem.*, 1982, **56**, 1063.
[341] E. R. Mably, M. D. Hollenberg, S. St. Pierre, and G. J. Moore, *Proc. West. Pharmacol. Soc.*, 1984, **27**, 273.
[342] A. Fournier, J. K. Saunders, and S. St. Pierre, *Peptides*, 1984, **5**, 169.
[343] G. Skala, C. W. Smith, C. J. Taylor, and J. H. Ludens, *Science*, 1984, **226**, 443.
[344] A. Buku, D. Gazis, and I. L. Schwartz, *Int. J. Pept. Protein Res.*, 1984, **23**, 551.

Peptide	*Ref.*
Vasopressin analogues	345, 346
Viroidin	
Analogues of viroidin	347
Sequential Oligo- and Poly-peptides	
Z-(Aib)$_n$-OBu^t (n = 3–5)	348
Block copolymers containing γ-benzyl- or γ-methyl-D,L-glutamate and butadiene	349
Copolymer of vinyl benzoic acid and phenylalanine ethyl ester	350
Oligomers of cysteinylcysteine	351
Poly{Glu[OBzl-(4-OR)]}	352
Poly{Glu(OBzl),Gln(glucopyranosyl)}	353
Poly{D-Glu(OMe),D-Glu(Ohexyl)}	354
Poly(Gly-Pro-Ile) and poly(Gly-Ile-Pro)	355
Poly{Hyp(acyl)}	356
Poly(X-Leu) (X = Arg or His)	357
(Leu-Gly)$_n$ oligomers	358
Boc-(Leu$_3$-Pro$_2$-Gly)$_n$-OBzl (n = 4, 6, 8, 10, or 12)	359
Nps-(Leu$_2$-Ala)$_n$-OEt and Nps-(Met$_2$-Leu)$_n$-OEt (n = 1–6)	16
Oligoleucine derivatives containing proline or glycyl-*N*-(2,4-dimethoxybenzyl)-L-leucine	360
Oligoleucine peptides containing a tertiary peptide bond	361

[345] M. Bressan, F. Marchiori, and G. Valle, *Int. J. Pept. Protein Res.*, 1984, **23**, 104.
[346] L. Balaspiri, G. Toth, M. Toth, F. Sirokman, K. Kovacs, and F. A. Laszlo, *Acta Chim. (Budapest)*, 1984, **115**, 223.
[347] J. U. Kahl, G. P. Vlasov, A. M. Sellrger, and Th. Wieland, *Int. J. Pept. Protein Res.*, 1984, **23**, 543.
[348] G. M. Bonora, C. Mapelli, C. Toniolo, R. R. Wilkening, and E. S. Stevens, *Int. J. Biol. Macromol.*, 1984, **6**, 179.
[349] T. Hayashi, G.-W. Chen, and A. Nakajima, *Polym. J.*, 1984, **16**, 739.
[350] L. Andersson, B. Sellergren, and K. Mosbach, *Tetrahedron Lett.*, 1984, **25**, 5211.
[351] S. Capasso, L. Mazzarella, T. Tancredi, and A. Zagari, *Biopolymers*, 1984, **23**, 1085.
[352] K. Hanabusa, M. Sato, H. Shirai, K. Takemoto, and E. Iizuka, *J. Polym. Sci. Polym. Lett.*, 1984, **22**, 559.
[353] Y. Kuroyanagi, H. Kobayashi, M. Seno, M. Ishida, N. Tominaga, T. Akaike, M. Sakamoto, and G. Ebert, *Int. J. Biol. Macromol.*, 1984, **6**, 266.
[354] J. Watanabe, Y. Fukuda, R. Gehani, and I. Uematsu, *Macromolecules*, 1984, **17**, 1004.
[355] A. M. Tamburro, V. Guantieri, D. Carbrol, H. Broch, and D. Vasilescu, *Int. J. Pept. Protein Res.*, 1984, **24**, 627.
[356] T. Kawasaki and T. Komai, *Polym. J.*, 1983, **15**, 743.
[357] B. Barbier, A. Caille, and A. Brack, *Biopolymers*, 1984, **23**, 2299.
[358] T. Muro, Y. Tominaga, and S. Okada, *Agric. Biol. Chem.*, 1984, **48**, 1231.
[359] M. Narita, S. Nagasawa, J.-Y. Chen, H. Sato, and Y. Tanaka, *Makromol. Chem.*, 1984, **185**, 1069.
[360] M. Narita, T. Fikunaga, A. Wakabayashi, K. Ishikawa, and H. Nakano, *Int. J. Pept. Protein Res.*, 1984, **23**, 306.
[361] M. Narita, K. Ishikawa, H. Nakano, and S. Isokawa, *Int. J. Pept. Protein Res.*, 1984, **24**, 14.

Peptide	Ref.
$(Leu)_n$-Phe-polystyrene resin (n = 3, 6, 9, or 15)	362
Ac-(Lys-Leu-Glu-Ala-Leu-Glu-Gly)$_n$-Lys-OH (n = 1–5)	363
Poly{Lys(Boc)}$_n$homo-oligopeptides	364
Poly{Lys(D, L-Ala)}	365
Poly(Lys-Glu) and poly(Lys-Gln)	366
Sequential polypeptides containing methionine and alanine or leucine	367
Tfa-(Nva)$_n$-OMe (n = 1–5), Tfa-(Aib)$_n$-OBu^t (n = 2, 4, or 5)	368
Oligopeptides containing tri- or tetra-prolyl sequences	369
H-(Val-Pro-D-Ala-Val-Gly)$_n$-Val-OMe	211
Enzyme Substrates and Inhibitors	
Conformationally restrained ACE inhibitors	370
ACE inhibitors (review)	371
ACE inhibitors	372–374
ACE inhibitor (Hoe 498)	375
ACE inhibitor L681176	376, 377
ACE inhibitors, ketomethylene peptides	378, 379

362 M. Narita, Y. Tomotake, S. Isokawa, T. Matsuzawa, and T. Miyauchi, *Macromolecules*, 1984, **17**, 1903.

363 S. Y. M. Lau, A. K. Taneja, and R. S. Hodges, *J. Biol. Chem.*, 1984, **259**, 13 253.

364 C. Toniolo, G. M. Bonora, I. Lusher, and C. H. Schneider, *Int. J. Pept. Protein Res.*, 1984, **23**, 47.

365 F. Hudecz, P. Kovacs, S. Kutassi-Kovacs, and J. Kajtar, *Colloid Polym. Sci.*, 1984, **262**, 208.

366 M. V. R. Rao, M. Atreyi, V. S. Chauhan, and S. Kumar, *Int. J. Pept. Protein Res.*, 1984, **24**, 48.

367 R. Katakai and Y. Iizuka, *J. Chem. Soc., Perkin Trans 1*, 1984, 2665.

368 G. M. Bonora, C. Toniolo, B. Di Blasio, V. Pavonne, C. Pedone, E. Benedetti, I. Lingham, and P. Hardy, *J. Am. Chem. Soc.*, 1984, **106**, 8152.

369 K. Sato, M. Tanaka, and T. Hayase, *Bull. Chem. Soc. Jpn.*, 1984, **57**, 1679.

370 H. N. Weller, E. M. Gordon, M. B. Rom, and J. Pluscec, *Biochem. Biophys. Res. Commun.*, 1984, **125**, 82.

371 C. R. W. Edwards and P. L. Padfield, *Lancet*, 1985, **1**, 30.

372 J. W. Ryan and A. Chung, *Adv. Exp. Med. Biol., B*, 1983, **156**, 1133.

373 J. T. Suh, J. W. Skiles, B. E. Williams, R. D. Youssefyeh, H. Jones, B. Loev, E. S. Neiss, A. Schwab, W. S. Mann, A. Khandwala, P. S. Wolf, and I. Weinryb, *J. Med. Chem.*, 1985, **28**, 57.

374 T. Ogihara, Y. Koyama, and Y. Kumahara, *Sogo Rinsho*, 1984, **33**, 367.

375 V. Teetz, R. Geiger, R. Henning, and H. Urbach, *Arzneim.-Forsch.*, 1984, **34**, 1399.

376 O. D. Hensens and J. M. Liesch, *J. Antibiot.*, 1984, **37**, 466.

377 L. Huang, G. Rowin, J. Dunn, R. Sykes, R. Dobna, B. A. Mayles, D. M. Gross, and R. W. Burg, *J. Antibiot.*, 1984, **37**, 462.

378 S. Natarajan, E. M. Gordon, E. F. Sabo, J. K. Godfrey, H. M. Weller, J. Pluscec, M. B. Rom, and D. W. Cushman, *Biochem. Biophys. Res. Commun.*, 1984, **124**, 141.

379 E. M. Gordon, S. Natarajan, J. Pluscec, H. N. Weller, J. D. Godfrey, M. B. Rom, E. F. Sabo, J. Engebrecht, and D. W. Cushman, *Biochem. Biophys. Res. Commun.*, 1984, **124**, 148.

Peptide	*Ref.*
Peptide β-lactams, ACE inhibitor	380
ACE inhibitor (phosphinoyl-peptide derivative)	381
Thiol-containing ACE inhibitors	382
Phosphinic acid inhibitors of aspartic peptidases	383
Truncated bovine trypsin inhibitor (Kunitz)	138
Inhibitors of bacterial cell-wall biosynthesis	384, 385
Fluorogenic substrate for calpain I/II	386
Azapeptide chymotrypsin inhibitors	387
Azapeptide elastase inhibitors	388
Inhibitors of carboxypeptidase A	389
Fluorogenic substrate for chymotrypsin	390
Fluorescent substrate of cAMP-dependent protein kinase	391
Elastase inhibitors, peptidyl carbamate esters	392
Elastase inhibitors, phosphoryl dipeptides	393
Porcine pancreatic elastase inhibitors	394
Inhibitors of human spleen fibriolytic proteinase (SFP)	210
Fluorogenic enkephalinase substrates	395
Inhibitors of enkephalin-degrading enzymes	396

[380] C. J. Wharton, R. Wrigglesworth, and M. Rowe, *J. Chem. Soc., Perkin Trans. 1*, 1984, 29.

[381] T. Morikawa, K. Takada, T. Kimura, S. Sakakibara, M. Kurauchi, Y. Ozawa, C. Eguchi, S. Hashimoto, and Y. Yukari, *Biochem. Biophys. Res. Commun.*, 1984, **119**, 1205.

[382] M.-C. Fournie-Zaluski, E. Lucas, G. Waksman, and B. P. Roques, *Eur. J. Biochem.*, 1984, **139**, 267.

[383] P. A. Bartlett and W. B. Kezer, *J. Am. Chem. Soc.*, 1984, **106**, 4282.

[384] T. Loennechen, T. Bergan, L. K. Sydens, and A. J. Aasen, *Acta Chem. Scand., Ser. B*, 1984, **38**, 647.

[385] E. A. Hagen, T. Bergan, and A. J. Aasen, *Acta Chem. Scand., Ser. B*, 1984, **38**, 5.

[386] T. Sasaki, T. Kikuchi, N. Yumoto, N. Yoshimura, and T. Murachi, *J. Biol. Chem.*, 1984, **259**, 12 489.

[387] B. F. Gupton, D. L. Carroll, P. M. Tuhy, C.-M. Kam, and J. C. Powers, *J. Biol. Chem.*, 1984, **259**, 4279.

[388] J. C. Powers, R. Boone, D. L. Carroll, B. F. Gupton, C.-M. Kam, N. Nishino, M. Sakamoto, and P. M. Tuhy, *J. Biol. Chem.*, 1984, **259**, 4288.

[389] R. E. Galardy and Z. P. Kortylewicz, *Biochemistry*, 1984, **23**, 2083.

[390] E. Sato, M. Miyakawa, and Y. Kanaoka, *Chem. Pharm. Bull.*, 1984, **32**, 336.

[391] H. Kondo, K. Takaki, R. Kuroki, A. Tada, K. Fukumoto, and J. Sunamoto, *Bull. Chem. Soc. Jpn.*, 1984, **57**, 2957.

[392] K. Tsuji, B. J. Agha, M. Shinogi, and G. A. Digenis, *Biochem. Biophys. Res. Commun.*, 1984, **122**, 571.

[393] L. Poncz, T. A. Gerken, D. G. Dearborn, D. Grobelny, and R. E. Galardy, *Biochemistry*, 1984, **23**, 2766.

[394] W. C. Groutas, W. R. Abrams, M. C. Theodorakis, A. M. Kasper, S. A. Rude, R. C. Badger, T. D. Ocain, K. E. Miller, M. K. Moi, M. J. Brubaker, K. S. Davis, and M. E. Zandler, *J. Med. Chem.*, 1985, **28**, 204.

[395] R. S. Rush, M. Mitas, J. C. Powers, T. Tanaka, and L. B. Hersh, *Arch. Biochem. Biophys.*, 1984, **231**, 390.

[396] R. Bouboutou, G. Waksman, J. Devin, M.-C. Fournie-Zaluski, and B. P. Roques, *Life Sci.*, 1984, **35**, 1023.

Peptide	Ref.
Suicide inhibitor of GABA-transaminase	397
Substrate for erythrocyte protein carboxymethyltransferase	398
N-Carboxymethylpeptide inhibitors of a rat brain metalloendopeptidase	399
Lysine oxidase substrate, H-Ala_2-Lys-X-Ala_2-OH	400
Monoamine oxidase inhibitor prodrug	401
Substrate for pepsin and aspartyl proteases	402
Peptidyl dipeptidase inhibitor	403
Post proline cleaving enzyme inhibitor, Z-Val-Pro-H	404
Protein kinase C substrate	405
Proteinase inhibitors (review)	406
Renin inhibitors	407
Renin inhibitors containing 5(*S*)-amino-7-methyl-3(*E*)-octenoic acid	146
Renin inhibitors, aldehydic peptides	408
Renin inhibitors containing 2-substituted statine	409
Human renin inhibitors	410
Tripeptide thiobenzyl esters as substrates for serine proteases	411
Serine proteinase inhibitors	412–414

[397] J. P. Burkhart, G. W. Holbert, and B. W. Metcalf, *Tetrahedron Lett.*, 1984, **25**, 5267.

[398] E. D. Murray, jun. and S. Clarke, *J. Biol. Chem.*, 1984, **259**, 10 722.

[399] T. G. Chu and M. Orlowski, *Biochemistry*, 1984, **23**, 3598.

[400] H. M. Kagan, M. A. Williams, P. R. Williamson, and J. M. Anderson, *J. Biol. Chem.*, 1984, **259**, 11 203.

[401] I. A. McDonald, J. M. Lacoste, P. Bey, J. Wagner, M. Zreika, and M. G. Palfreyman, *J. Am. Chem. Soc.*, 1984, **106**, 3354.

[402] B. M. Dunn, B. Kammermann, and K. R. McCurry, *Anal. Biochem.*, 1984, **138**, 68.

[403] M. P. Filatova, N. A. Krit, G. V. Gavrilova, N. V. Beschastnaya, and S. Reissmann, *Bioorg. Khim.*, 1984, **10**, 632.

[404] H. Yokosawa, M. Nishikata, and S. Ishii, *J. Biochem.*, 1984, **95**, 1819.

[405] C. A. O'Brian, D. S. Lawrence, E. T. Kaiser, and I. B. Weinstein, *Biochem. Biophys. Res. Commun.*, 1984, **124**, 296.

[406] T. Koide, *Taisha (Rinji. Zokan)*, 1984, **21**, 995.

[407] B. J. Leckie, J. Grant, A. Hallett, M. Hughes, D. M. Jones, M. Szelke, and M. Tree, *Scott. Med. J.*, 1984, **29**, 125.

[408] J.-A. Fehrentz, A. Heitz, B. Castro, C. Cazaubon, and D. Nisato, *FEBS Lett.*, 1984, **167**, 273.

[409] D. F. Veber, M. G. Bock, S. F. Brady, E. H. Ulm, D. W. Cochran, G. M. Smith, B. I. LaMont, R. M. Dipardo, M. Poe, R. M. Freidinger, B. E. Evans, and J. Boger, *Biochem. Soc. Trans.*, 1984, **12**, 956.

[410] T. Kokubu, K. Hiwada, Y. Sato, T. Iwata, Y. Imamura, R. Matsueda, Y. Yabe, H. Kogen, M. Yamazaki, Y. Iijima, and Y. Baba, *Biochem. Biophys. Res. Commun.*, 1984, **118**, 929.

[411] J. W. Harper, R. R. Cook, C. J. Roberts, B. J. McLaughlin, and J. C. Powers, *Biochemistry*, 1984, **23**, 2995.

[412] H. Ehorn, G. Wagner, J. Stuerzebecher, P. Walsmann, and F. Markwardt, *Pharmazie*, 1983, **38**, 581.

[413] B. Voigt, G. Wagner, P. Walsmann, J. Stuerzebecher, and F. Markwardt, *Pharmazie*, 1983, **38**, 835.

[414] H. Vieweg and G. Wagner, *Pharmazie*, 1984, **39**, 82.

Peptide	*Ref.*
Serine protease inhibitors, peptide boronic acids	415
Inhibitors of trypsin-like serine protease	416
Vinylogous amino acid esters as thiol protease inhibitors	417
E64, a thiol protease inhibitor	418
Substrates for thermitase	419
Thermitase inhibitors	420
Thermitase inhibitor, Ac-$(Ala)_n$-X-CH_2Cl	421
Thrombin inhibitors	422–425
Inhibitors of trypsin, thrombin, and plasmin	426
Kunitz soybean trypsin inhibitor, photoreactive derivative	427
Lima bean trypsin inhibitor, trinitrophenyl derivative	428
Penicillopepsin substrate	429
Enzyme-activated inhibitors of pyridoxyl phosphate-dependent enzymes	430
Inhibitor of vitamin K-dependent carboxylase	431
Vitamin K-dependent carboxylase, diastereoisomeric γ-substituted glutamic acid-containing substrates	432
Vitamin K-dependent carboxylase substrate	149
Glycopeptides	
Glycopeptide synthesis (review)	433
Galactosylated tripeptides	434

[415] C. A. Kettner and A. B. Shenvi, *J. Biol. Chem.*, 1984, **259**, 15 106.
[416] H. R. Lijnen, M. Uytterhoeven, and D. Collen, *Thromb. Res.*, 1984, **34**, 431.
[417] R. P. Hanzlik and S. A. Thompson, *J. Med. Chem.*, 1984, **27**, 711.
[418] T. Amamoto, T. Okazaki, T. Komurasaki, K. Hanada, and S. Omura, *Biochem. Biophys. Res. Commun.*, 1984, **118**, 117.
[419] G. Jahreis and S. Fittkau, *J. Prakt. Chem.*, 1984, **326**, 35.
[420] S. Fittkau and G. Jahreis, *J. Prakt. Chem.*, 1984, **326**, 48.
[421] G. Jahreis, K. Smalla, and S. Fittkau, *J. Prakt. Chem.*, 1984, **326**, 41.
[422] R. Kikumoto, Y. Tamao, T. Tezuka, S. Tonomura, H. Hara, K. Ninomiya, A. Hijikata, and S. Okamoto, *Biochemistry*, 1984, **23**, 85.
[423] L. Svendsen, M. Brogli, G. Lindeberg, and K. Stocker, *Thromb. Res.*, 1984, **34**, 457.
[424] G. Wagner, B. Voigt, and H. Vieweg, *Pharmazie*, 1984, **39**, 226.
[425] G. Wagner, B. Voigt, and C. Pfeiffer, *Pharmazie*, 1984, **39**, 315.
[426] H. Horn and G. Wagner, *Pharmazie*, 1983, **38**, 889.
[427] D. Gazis, I. L. Schwartz, B. Lammek, and Z. Grzonka, *Int. J. Pept. Protein Res.*, 1984, **23**, 78.
[428] S. Takahashi, T. Yamamura, M. Kamo, and K. Satake, *Chem. Lett.*, 1984, **127**.
[429] T. Hofmann, R. S. Hodges, and M. N. G. James, *Biochemistry*, 1984, **23**, 635.
[430] P. Casara, K. Jund, and P. Bey, *Tetrahedron Lett.*, 1984, **25**, 1891.
[431] E. Guibe, P. Decottignies-Le Marechal, P. Le Marechal, and R. Azerad, *FEBS Lett.*, 1984, **177**, 265.
[432] S. Bory, J. Dubois, M. Gaudry, and A. Marquet, *Int. J. Pept. Protein Res.*, 1984, **24**, 505.
[433] H. Kunz, *Nachr. Chem. Tech. Lab.*, 1984, **32**, 11.
[434] K. Dill, M. E. Daman, R. L. Batstone-Cunningham, M. Denarie, and A. A. Pavia, *Carbohydr. Res.*, 1983, **124**, 11.

Peptide	Ref.
Glucopyranosylactoyl-Ala-D-Glu-NH_2	435
Cluster ligand for galactose-/NAG-specific lectin	436
N-Tripeptidyl-D-glucosamine	437
N-(Aspartyl)-β-D-glucopyranosylamine, α- and β-amides	438
Glycosylthiocarbamoyl peptides	439
Glycosylated asparagine peptides	440
O-Glycopeptides	32
Immunostimulating dipeptide/saccharide	441
Lipophilic thiomuramyl dipeptides	442
Muramyl-Ala-D-Glu-NH_2 carbohydrate analogue	443
Penicillin-binding protein, photoreactive peptidoglycan analogue	444
Quinonyl muramyl dipeptides	445
Thiomuramyl dipeptides	446
Thiomuramyl dipeptides	447
Miscellaneous Peptides	
Aib peptides	448
α-Alkyl-leucine methyl esters	449
δ-(L-α-Aminoadipoyl)-cysteinyl-*N*-hydroxy-D-valine	450
o- and *p*-aminobenzoic acid peptides	451
m-Aminomethylbenzoic acid peptides	452

435 A. Hasegawa, M. Kiso, and I. Azuma, *Carbohydr. Res.*, 1983, **122**, 99.
436 R. T. Lee, P. Lin, and Y. C. Lee, *Biochemistry*, 1984, **23**, 4255.
437 M. Oya and T. Negishi, *Bull. Chem. Soc. Jpn.*, 1984, **57**, 439.
438 M. Tamura, H. Nishizaki, C. Miyazaki, and H. Okai, *Bull. Chem. Soc. Jpn.*, 1984, **57**, 3167.
439 Y. P. Abashev, T. M. Andronova, S. E. Zurabyan, V. P. Malkova, I. B. Sorokina, and A. Y. Khorlin, *Bioorg. Khim.*, 1984, **10**, 18.
440 M. Tamura and H. Okai, *Carbohydr. Res.*, 1984, **133**, 207.
441 M. Guinand, A. Francon, M.-J. Vacheron, and G. Michel, *Eur. J. Biochem.*, 1984, **143**, 359.
442 A. Hasegawa, E. Seki, Y. Hioki, M. Kiso, and I. Azuma, *Carbohydr. Res.*, 1984, **131**, 61.
443 H. Okumura, Y. Tokushima, I. Saiki, I. Azuma, M. Kiso, and A. Hasegawa, *Carbohydr. Res.*, 1983, **122**, 87.
444 V. Aran, A. Rodriguez-Tebar, and D. Vazquez, *Eur. J. Biochem.*, 1984, **144**, 613.
445 S. Kobayashi, T. Fukuda, H. Yukimasa, M. Fujino, I. Azuma, and Y. Yamamura, *Bull. Chem. Soc. Jpn.*, 1984, **57**, 3182.
446 A. Hasegawa, Y. Hioki, M. Kiso, H. Okumura, and I. Azuma, *Carbohydr. Res.*, 1983, **123**, 183.
447 A. Hasegawa, E. Seki, Y. Hioki, M. Kiso, and I. Azuma, *Carbohydr. Res.*, 1984, **129**, 271.
448 G. De Pieri, A. Signor, G. M. Bonora, and C. Toniolo, *Int. J. Biol. Macromol.*, 1984, **6**, 35.
449 U. Schoellkopf, U. Busse, R. Kilger, and P. Lehr, *Synthesis*, 1984, 271.
450 R. L. Baxter, G. A. Thompson, and A. I. Scott, *J. Chem. Soc., Chem. Commun.*, 1984, 32.
451 F. H. C. Stewart, *Aust. J. Chem.*, 1983, **36**, 1629.
452 F. H. C. Stewart, *Aust. J. Chem.*, 1983, **26**, 2511.

Peptide	*Ref.*
Oligomers of *N*-(2-aminoethyl)glycine	453
Ala_3 isonicotinyl hydrazide, antimicrobial activity	454
Analgesic peptides containing tyrosine	455
Antitumour agents, nitrosourea, and nitrogen mustard amino acid derivatives	456
Model arginine/glycine peptides	457
A monocylic β-lactam tripeptide	458
Benzyl-penicilloylated *eicosa*-L-lysine conjugate with cholestane	459
4-Bromo-1,8-naphthaloyl dipeptides	460
Diastereoisomeric cyclo(-L- or D-Ala-Gly-Pro-L- or D-Phe-$)_2$	461
Antitumour cyclic hexapeptides RA	462
Cyclo(-Ala-Gly-D-Phe-Pro-$)_3$	463
Cyclic biscystine peptides	464
Cyclic cystine tetrapeptides containing proline	465
Cyclo{-Lys(acryloyl)-Sar-}	466
Cyclo(-Phe-Pro-D-Ala-$)_2$	467
Cyclo(-Val-Lac-$)_3$	468
Ten-membered cyclotripeptides	469
Bridged cyclodipeptides	470

[453] E. P. Heimer, H. E. Gallo-Torres, A. M. Felix, M. Ahmad, T. J. Lambros, F. Scheidl, and J. Meienhofer, *Int. J. Pept. Protein Res.*, 1984, **23**, 203.
[454] H. Brueckner, G. Jung, R. G. Werner, and K.-R. Appel, *Arzneim.-Forsch.*, 1983, **33**, 1630.
[455] M. Maeda, S. Okusada, and K. Kawasaki, *Chem. Pharm. Bull.*, 1984, **32**, 4157.
[456] M. Rodriguez, J.-L. Imbach, and J. Martinez, *J. Med. Chem.*, 1984, **27**, 1222.
[457] N. V. Avdyukova, A. G. Shirokova, and L. B. Radina, *J. Gen. Chem. U.S.S.R.*, 1982, **52**, 156.
[458] J. E. Baldwin, E. P. Abraham, R. M. Adlington, M. J. Crimmin, L. D. Field, G. S. Jayatilake, R. L. White, and J. J. Usher, *Tetrahedron*, 1984, **40**, 1907.
[459] I. F. Luescher, *J. Med. Chem.*, 1984, **27**, 1502.
[460] A. M. El-Naggar, M. R. Zaher, B. M. Haroun, and A. A. Salem, *Indian J. Chem., Sect. B*, 1982, **21**, 1106.
[461] K. D. Kopple, K. N. Parameswaran, and J. P. Yonan, *J. Am. Chem. Soc.*, 1984, **106**, 7212.
[462] H. Itokawa, K. Takeya, N. Mori, T. Sonobe, N. Serisawa, T. Hamanaka, and S. Mihashi, *Chem. Pharm. Bull.*, 1984, **32**, 3216.
[463] J. P. Degelaen, P. Pham, and E. R. Blout, *J. Am. Chem. Soc.*, 1984, **106**, 4882.
[464] R. Kishore and P. Balaram, *J. Chem. Soc., Chem. Commun.*, 1984, 778.
[465] A. Ravi and P. Balaram, *Tetrahedron*, 1984, **13**, 2583.
[466] A. Mori and Y. Imanishi, *Polymer*, 1984, **25**, 1837.
[467] G. Kartha, K. K. Bhandary, K. D. Kopple, A. Go, and P.-P. Zhu, *J. Am. Chem. Soc.*, 1984, **106**, 3844.
[468] H. Kunz and H.-G. Lerchen, *Angew. Chem.*, 1984, **96**, 798.
[469] F. Pinnen, G. Zanotti, and G. Lucente, *Tetrahedron Lett.*, 1984, **25**, 5201.
[470] Y. Sato, C. Shin, S. Sumiya, Y. Nakajima, and J. Yoshimura, *Bull. Chem. Soc. Jpn.*, 1984, **57**, 1265.
[471] C.-G. Shin, Y. Yonezawa, and T. Yamada, *Chem. Pharm. Bull.*, 1984, **32**, 3934.

Peptide	*Ref.*
Dehydropeptides	147, 148, 471–473
Electron-transfer peptide complexes	474, 475
Peptidyl derivatives of 5-fluorouracil as prodrugs	476
(^{3}H)-D,L-γ-Carboxyglutamic acid	477
Glp-His-Gly-OH	478
Derivatives of γ-L-glutamyl taurine	479
S-Substituted glutathione derivatives	480
Histidine and arginine tetrapeptides	481
3(*E*)-Hydroxyiminomethyl derivatives of tyrosine	482
δ-*N*-Hydroxyornithine derivatives	483, 484
Ketomethylene analogues of peptides	485
Peptide lipid with NADH activity	486
(^{2}H)-Methionine derivatives	487
Methotrexate monoamides	488
Methotrexate analogues containing L-homocysteic and L-cysteic acids in place of glutamate	489
Methotrexate monoesters	490
2-Methoxy-6,9-dichloroacridinyl fluorescent-label peptides	491

[472] C.-G. Shin, *Kagaku Kogyo Bessatsu*, 1984, **28**, 189.
[473] C.-G. Shin, Y. Yonezawa, and T. Yamada, *Chem. Pharm. Bull.*, 1984, **32**, 2825.
[474] S. S. Isied and A. Vassilian, *J. Am. Chem. Soc.*, 1984, **106**, 1726.
[475] S. S. Isied and A. Vassilian, *J. Am. Chem. Soc.*, 1984, **106**, 1732.
[476] W. D. Kingsbury, J. C. Boehm, R. J. Mehta, S. F. Grappel, and C. Gilvarg, *J. Med. Chem.*, 1984, **27**, 1447.
[477] Y. Ishikawa and R. E. Wuthier, *J. Labelled Comp. Radiopharm.*, 1984, **21**, 507.
[478] Y. P. Shvachkin, A. P. Smirnova, V. P. Fedotov, and G. I. Bushueva, *J. Gen. Chem. U.S.S.R.*, 1982, **52**, 1054.
[479] J. Gulyas, A. Horvath, A. Furka, F. Sevestyen, and L. Feuer, *Int. J. Pept. Protein Res.*, 1984, **24**, 367.
[480] S. J. Addison, B. D. M. Cunningham, E. N. Gate, P. Z. Shah, and M. D. Threadgill, *J. Chem. Soc., Perkin Trans. 1*, 1985, 75.
[481] A. M. El-Naggar, M. R. Zaher, and S. A. El-Ghaffar, *Indian J. Chem., Sect. B*, 1983, **22**, 392.
[482] Z. Arnold, *Pol. J. Chem.*, 1982, **56**, 1021.
[483] B. H. Lee, G. J. Gerfen, and M. J. Miller, *J. Org. Chem.*, 1984, **49**, 2418.
[484] R. K. Olsen, K. Ramasamy, and T. Emery, *J. Org. Chem.*, 1984, **49**, 3527.
[485] R. L. Johnson and R. B. Miller, *Int. J. Pept. Protein Res.*, 1984, **23**, 581.
[486] Y. Murakami, J. Kikuchi, K. Nishida, and A. Nakano, *Bull. Chem. Soc. Jpn.*, 1984, **57**, 1371.
[487] K. Ramalingam and R. W. Woodward, *J. Labelled Comp. Radiopharm.*, 1984, **21**, 563.
[488] D. J. Antonjuk, D. K. Boadle, H. T. A. Cheung, and T. Q. Tran, *J. Chem. Soc., Perkin Trans. 1*, 1984, 1989.
[489] A. Rosowsky, R. A. Forsch, J. H. Freisheim, R. G. Moran, and M. Wick, *J. Med. Chem.*, 1984, **27**, 600.
[490] A. Rosowsky, R. A. Forsch, C.-S. Yu, H. Lazarus, and G. P. Beardsley, *J. Med. Chem.*, 1984, **27**, 605.
[491] V. A. Shibnev, M. P. F. Nogenova, A. K. Gazumyan, A. I. Poletaev, and L. I. Marjash, *Bioorg. Khim.*, 1984, **10**, 610.

Peptide	Ref.
Mosquito-attractant/-repellent peptide	492
Polylysine conjugates of methotrexate	493
7-Hydroxymethotrexate conjugates	494
Di- and tri-lysine methotrexate derivatives	495
Opiate-receptor mimetic peptide	46
Phage-inactivating lysine derivatives	496
Derivatives of the phosphonic acid analogue of serine	497
Alkylaminomethane phosphonic acid derivatives	498
Proline-containing tripeptides	499
Ser(phosphono) derivatives	500
Polypeptide liquid crystals	501
Polypeptide membranes	502
N-Nitrosodipeptides	503
Restricted GABA analogues	504
Nucleopeptides	505
Nucleopeptide fragments	506
$[^{17}O]$Carbonyl-enriched peptides	507
Thiacyclols	508
threo-γ-Methyl- and *threo*-γ-fluoro-glutamate dipeptides	509

[492] Y. Okada, S. Iguchi, T. Hirai, Y. Goto, M. Yagyu, and H. Yajima, *Chem. Pharm. Bull.*, 1984, **32**, 4608.
[493] A. Rosowsky, R. A. Forsch, J. H. Freisheim, J. Galivan, and M. Wick, *J. Med. Chem.*, 1984, **27**, 888.
[494] J. J. McGuire, P. Hsieh, and J. R. Bertino, *Biochem. Pharmacol.*, 1984, **33**, 1355.
[495] R. A. Forsch and A. Rosowsky, *J. Org. Chem.*, 1984, **49**, 1305.
[496] M. Kondo, K. Miyazaki, Y. Yada, H. Horimoto, M. Samoto, and A. Murata, *Agric. Biol. Chem.*, 1984, **48**, 1263.
[497] B. Lejczak, P. Kafarski, M. Soroka, and P. Mastalerz, *Synthesis*, 1984, 577.
[498] R. Tyka, *Synthesis*, 1984, 218.
[499] A. M. Tamburro, V. Guantieri, and A. Scatturin, *Int. J. Biol. Macromol.*, 1984, **6**, 241.
[500] P. F. Alewood, J. W. Perich, and R. B. Johns, *Aust. J. Chem.*, 1984, **37**, 429.
[501] I. Uematsu, and Y. Uematsu, *Kagaku Kogyo Bessatsu*, 1984, **28**, 214.
[502] A. Takizawa, *Kagaku Kogyo Bessatsu*, 1984, **28**, 222.
[503] B. C. Challis, J. R. Milligan, and R. C. Mitchell, *J. Chem. Soc., Chem. Commun.*, 1984, 1050.
[504] R. D. Allan and H. W. Tran, *Aust. J. Chem.*, 1984, **37**, 1123.
[505] O. N. Ryabtseva, G. A. Korshunova, Y. A. Semiletov, and Y. P. Shvachkin, *J. Gen. Chem. U.S.S.R.*, 1983, **53**, 2176.
[506] C. Schattenkerk, C. T. J. Wreesmann, M. J. De Graff, G. A. Van Der Marel, and J. H. Van Boom, *Tetrahedron Lett.*, 1984, **25**, 5197.
[507] A. Steinschneider and D. Fiat, *Int. J. Pept. Protein Res.*, 1984, **23**, 591.
[508] G. Zanotti, F. Pinnen, G. Lucente, S. Cerrini, W. Fedeli, and F. Mazza, *J. Chem. Soc., Perkin Trans. 1*, 1984, 1153.
[509] S. Bory, J. Dubois, M. Gaudry, A. Marquet, L. Lacombe, and S. Weinstein, *J. Chem. Soc., Perkin Trans. 1*, 1984, 475.

5 Appendix II: Amino Acid Derivatives Useful in Synthesis

The list of derivatives is divided into two groups, the coded and the uncoded amino acids. The following unusual abbreviations are employed: Dppe, diphenylphosphinoethyl; Hcy, homocysteine; Mpt, dimethylphosphinothioyl; OBzl·SO_3H, 4-sulphobenzyl ester; OMTM, methylthiomethyl ester; O(2-Pet), 2-(2-pyridyl)-ethyl ester; O(4-Pet), 2-(4-pyridyl)ethyl ester; OTAT, thiazoline-2-thione ester; Pbm, bis-[2-methylsulphonyl)ethyl] phosphate; Pym, 2-oxo-1-pyrrolidinyl; Pyoc, 2-pyridylethoxycarbonyl.

Compound	*M.p./°C*	$[\alpha]_D^{*1}$	*Conc./g 100 cm^{-3}*	*Solvent*	*Ref.*
Coded Amino Acids					
Alanine					
Boc-Ala-ODppe	Oil	−13	2	CH_2Cl_2	33
$Cl^-H_2^+$-Pyoc-Ala-OH	151–152	−5.4	1	DMF	18
Dnpy-Ala-OH	195	−26.4	1	DMF	22
Dpp-Ala-OH	152–154	−21.4	1	MeOH	23
Dpp-Ala-OMe	114–115	−31.6	1	MeOH	23
H_2^+-Ala-OBzl·SO_3^-	234–236	−8.3	1	DMSO	34
Nps-Ala-ONSu	149–150	−59.3	1	DMF	16
Trt-Ala-OBt	145–147	+52.7	2	$CHCl_3$	74
Z-Ala-ODppe	Oil	−7.1	2	CH_2Cl_2	33
Z-Ala-OMTM	Oil	−60.6	1	95% EtOH	29
Z-Ala-O(2 Pet)	Oil	−3.5	0.8	$CHCl_3$	31
Z-Ala-OTAT	163–165	−120.0	2	$CHCl_3$	77
Arginine					
Boc-Arg(H_2^+)-OBzl·SO_3^-	142–149	−34.9	1	30% HOAc	34
Dnpy-Arg-OH	176	−3.5	1	DMF	22
Z-Arg(Mbs)-OBu^t	59	−6.6	1	DMF	298
Asparagine					
Dnpy-Asn-OH	130–132	+30.5	1	DMF	22
H_2^+-Asn-OBzl·SO_3^-	240–245	+9.1	1	30% HOAc	34
Nps-Asn-OH	168–169	−114.4	4	DMF	16
Aspartic acid					
Dnpy-Asp-OH	112	−69.5	1	DMF	22
Nps-Asp(OBzl)-ONSu	89–91	−44.7	1	DMF	16
Z-Asp(OMTM)-OMTM	Oil	−24.5	1	95% EtOH	29
Cysteine					
Boc-Cys(Bzl)-NHMe	123	–	–	–	465
Boc-Cys(4-Me-Bzl)-OH	76–77	−41.7	1	HOAc	335
Boc-Cys(Pym)-OH	113–114	−28.3	0.5	H_2O	40
Boc-Cys(Pym)-ONp	100–101	−40.1	0.1	MeOH	40
$Cl^-H_2^+$-Cys(Pym)-OH	174–175	+4.7	1	H_2O	40
$Cl^-H_2^+$-Cys(Pym)-OMe	70–72	−5.2	1	MeOH	40
Fmoc-Cys(Bzl)-OTcp	152	−52	1	DMF	83

Compound	*M.p./°C*	$[\alpha]_D^{*1}$	*Conc./g 100 cm^{-3}*	*Solvent*	*Ref.*
H_2^+-Cys(Trt)-OBzl·SO_3^-	186–194	+20.9	1	HMPA	34
Mpt-Cys(Mpt)-OMob	Oil	−5.12	1	MeOH	41
Glutamine					
Dnpy-Gln-OH	225–226	−83.5	1	DMF	22
H_2^+-Gln-OBzl·SO_3^-	175–185	+4.3	1	30% HOAc	34
Nps-Gln-ONSu	147–148	−56.0	1	DMF	16
Glutamic acid					
Dnpy-Glu-OH	218–226	−71.5	1	DMF	22
Fmoc-Glu-OTcp	172–174	−26	1	DMF	83
H_2^+-Glu(OBu^t)-OBzl·SO_3^-	209–215	+17.1	1	DMSO	34
Nps-Glu(OBzl)-ONSu	85–87	−43.4	1	DMF	16
Z-Glu(OMe)-OPfp	59–60	−17.9	1	EtOAc	82
Z-Glu(OMTM)-OMTM	Oil	−41.7	1	95% EtOH	29
Glycine					
Boc-Gly-ODppe	Oil	–	–	–	33
Dnpy-Gly-OH	220	–	–	–	22
Dpp-Gly-OBzl	117–118	–	–	–	23
Dpp-Gly-OH	133–134	–	–	–	23
Dpp-Gly-OMe	110–112	–	–	–	23
Fmoc-Gly-OTcp	146	–	–	–	83
H_2^+-Gly-OBzl·SO_3^-	265–270	–	–	–	34
Nps-Gly-OMTM	Oil	–	–	–	29
Trt-Gly-OBt	137–140	–	–	–	74
Z-Gly-ODppe	Oil	–	–	–	33
Z-Gly-O(2 Pet)	Oil	–	–	–	31
Histidine					
Dnpy-His-OH	255–257	–	–	–	22
Fmoc-His(π-Bum)-OH	175–176	−7.5	1	AcOH	47
Z-His(π-Bum)-OH	186–187	−10.7	1	AcOH	47
Isoleucine					
Dnpy-Ile-OH	240–245	−114.0	1	DMF	22
Dpp-Ile-OBzl	106–108	−30.7	1	MeOH	23
Dpp-Ile-OH	92	−7.2	1	MeOH	23
Dpp-Ile-OH-DCHA	129–131	+5.0	1	MeOH	23
Fmoc-Ile-OTcp	155	−37	1	DMF	83
H_2^+-Ile-OBzl·SO_3^-	248–268	–	–	–	34
Trt-Ile-OBt	87–90	+3.4	2	$CHCl_3$	74
Z-Ile-O(2 Pet)	Oil	+7.5	2.1	$CHCl_3$	31
Leucine					
Boc-Leu-ODppe	Oil	+43.5	2	CH_2Cl_2	33
Boc-Leu-O(2 Pet)	Oil	−7.9	2.0	$CHCl_3$	31
$Cl^-H_2^+$-Pyoc-Leu-OH	152	−9.5	1	DMF	18
Dnpy-Leu-OH	210–212	−53.5	1	DMF	22
Dpp-Leu-OBzl	100–102	−27.8	1	MeOH	23

Compound	*M.p./°C*	$[\alpha]_D^{*1}$	*Conc./g 100 cm^{-3}*	*Solvent*	*Ref.*
Dpp-Leu-OH	135–136	−18.6	1	MeOH	23
Dpp-Leu-OMe	102	−31.5	1	MeOH	23
Fmoc-Leu-OTcp	136	−43	1	DMF	83
H_2^+-Leu-OBzl·SO_3^-	243–244	−6.5	1	30% HOAc	34
Nps-Leu-OH	108–109	−101.8	2	DMF	16
Nps-Leu-ONSu	Oil	−61.6	1	DMF	16
Trt-Leu-OBt	135–136	−35.0	2	$CHCl_3$	74
Z-Leu-O(2 Pet)	Oil	−6.2	2.2	$CHCl_3$	31
Z-Leu-OTAT	78–79	−98.4	2	$CHCl_3$	77
Lysine					
Dnpy-Lys(Dnpy)-OH	191–192	−52.3	1	DMF	22
Dpp-Lys(Dpp)-OMe	152–153	−1.3	1	MeOH	23
Dpp-Lys(Z)-OH-DCHA	152–154	+13.3	1	MeOH	23
H_2^+-Lys(Z)-OBzl·SO_3^-	229–237	−6.9	1	DMSO	34
Nps-Lys(Z)-OH-DCHA	188–190	−29.9	0.7	DMF	16
Z-Lys(Boc)-OTAT	102–103	−66.3	1	$CHCl_3$	77
Methionine					
Dnpy-Met-OH	149–151	−61.5	1	DMF	22
Dpp-Met-OH	142–144	−13.9	1	MeOH	23
Dpp-Met-OMe	93–94	−35.8	1	MeOH	23
H_2^+-Met-OBzl·SO_3^-	227–229	+10.1	1	HMPA	34
Nps-Met-OH-DCHA	201–202	−34.2	0.7	MeOH	16
Nps-Met-OMTM	Oil	−81.8	1	95% EtOH	29
Trt-Met-OBt	142–143	+52.0	2	$CHCl_3$	74
Z-Met-OMTM	Oil	−67.8	1	95% EtOH	29
Z-Met-OTAT	99–101	−97.2	2	$CHCl_3$	77
Phenylalanine					
Boc-Phe-OMTM	Oil	−26.1	1	95% EtOH	29
Boc-Phe-OTAT	168.5–170.5	−29.8	2	$CHCl_3$	77
$Cl^-H_2^+$-Pyoc-Phe-OH	210–212	−29.6	1	DMF	18
Dnpy-Phe-OH	105–107	−150.0	1	DMF	22
Dpp-Phe-OBzl	158	−46.1	1	MeOH	23
Dpp-Phe-OH	131–133	−40.1	1	MeOH	23
Dpp-Phe-OMe	156–158	−49.7	1	MeOH	23
Fmoc-Phe-OTcp	187	−55	1	DMF	83
For-Phe-OMTM	Oil	−12.4	1	95% EtOH	29
H_2^+-Phe-OBzl·SO_3^-	241–246	+19.1	1	DMSO	34
Nps-Phe-OMTM	Oil	−21.0	1	95% EtOH	29
Pht-Phe-OMTM	52	−169.4	1	95% EtOH	29
Tfa-Phe-OMTM	63	−39.0	1	95% EtOH	29
Trt-Phe-OBt	Foam	−26.7	2	$CHCl_3$	74
Trt-Phe-OMTM	Oil	+11.0	1	95% EtOH	29
Z-Phe-ODppe	Oil	−14.5	2	CH_2Cl_2	33
Z-Phe-OMTM	Oil	−28.8	1	95% EtOH	29

Compound	*M.p./°C*	$[\alpha]_D^{*1}$	*Conc./g* 100 cm^{-3}	*Solvent*	*Ref.*
Z-Phe-O(2 Pet)	Oil	+23.5	1.9	$CHCl_3$	31
Z-Phe-O(4 Pet)	Oil	+23.5	2.1	$CHCl_3$	31
Proline					
Dpp-Pro-OBzl	80–82	−43.3	1	MeOH	23
Dpp-Pro-OH	170–173	−17.6	1	MeOH	23
H_2^+-Pro-OBzl·SO_3^-	230–237	−38.9	1	30% HOAc	34
Nps-Pro-OH-DCHA	171–172	−55.3	0.7	DMF	16
Nps-Pro-OMTM	Oil	−58.3	1	95% EtOH	29
Nps-Pro-ONSu	178–180	−50.3	1	DMF	16
Trt-Pro-OBt	Foam	−109.7	2	$CHCl_3$	74
Serine					
$Cl^-H_2^+$-Pyoc-Ser-OH	145	−3.7	1	DMF	18
Dnpy-Ser-OH	185–186	−0.87	1	DMF	22
H_2^+-Ser(Bu^t)-OBzl·SO_3^-	218 (dec.)	–	–	–	34
Z-Ser(Bu^t)-OPfp	Oil	−4.3	1	EtOAc	82
Z-Ser-OMTM	64	−20.4	1	95% EtOH	29
Z-Ser-O(2 Pet)	Oil	−43.4	1.5	$CHCl_3$	31
Threonine					
Boc-Thr(Bzl)-OPfp	85–87	−24.1	1	EtOAc	82
$Cl^-H_2^+$-Pyoc-Thr-OH	Amorph	−4.4	1	DMF	18
H_2^+-Thr(Bu^t)-OBzl·SO_3^-	167–180	−27.2	1	30% HOAc	34
Dnpy-Thr-OH	203	+25.0	1	DMF	22
H_2^+-Thr-OBzl·SO_3^-	155–165	−18.5	1	30% HOAc	34
Tryptophan					
Boc-Trp(Mts)-OH-DCHA	154–155	+19.9	1.0	MeOH	50
Dnpy-Trp-OH	191–193	−190.8	1	DMF	22
Dpp-Trp-OH	159–162	−60.5	1	MeOH	23
Dpp-Trp-OMe	153–155	−48.9	1	MeOH	23
H-Trp(Mts)-OH	211–213	−30.0	0.6	DMF	50
Nps-Trp-OMTM	Oil	−44.7	1	95% EtOH	29
Z(OMe)-Trp(Mts)-OBzl	77–79	+10.1	1.8	DMF	50
Z(OMe)-Trp(Mts)-$NHNH_2$	121–123	−41.3	0.8	DMF	50
Z(OMe)-Trp-OBzl	94–95	−16.1	0.6	DMF	50
Z-Trp-OMTM	126–127	−23.0	1	95% EtOH	29
Tyrosine					
Boc-Tyr(Boc)-OH	96–98	+10.5	2	MeOH	152
Boc-Tyr-OMTM	Oil	−15.6	1	95% EtOH	29
Boc-Tyr[$PO(OMe)_2$]-OH	Oil	−0.10	1	MeOH	510
Boc-Tyr[$PO(OMe)_2$]-ONb	Oil	−10.0	1	MeOH	510
$Cl^-H_2^+$-Pyoc-Tyr(Bzl)-OH	212	−21.4	1	DMF	18
Dnpy-Tyr-OH	234–235	−208.0	1	DMF	22
Dpp-Tyr(Dpp)-OMe	189–190	−25.2	1	MeOH	23

[510] R. M. Valerio, P. F. Alewood, R. B. Johns, and B. E. Kemp, *Tetrahedron Lett.*, 1984, **25**, 2609.

Compound	*M.p./°C*	$[\alpha]_D^{*1}$	*Conc./g 100 cm⁻³*	*Solvent*	*Ref.*
Dpp-Tyr-OH-DCHA	218–220	+9.9	1	MeOH	23
Fmoc-Tyr-OTcp	186–188	−44	1	DMF	83
H_2^+-Tyr-OBzl·SO_3^-	275–285	–	–	–	34
Trt-Tyr-OBt	Foam	−10.8	2	$CHCl_3$	74
Z-Tyr(Pbm)-OBzl	94–96	+3.1	0.8	$CHCl_3$	51
Z-Tyr(Z)-OMTM	95–97	−27.8	1	95% EtOH	29
Valine					
Boc-Val-ODppe	Oil	+91	2	CH_2Cl_2	33
Boc-Val-O(2 Pet)	Oil	+2.8	1.8	$CHCl_3$	31
$Cl^-H_2^+$-Pyoc-Val-OH	157–158	+5.4	1	DMF	18
Dnpy-Val-OH	238–239	−127.5	1	DMF	22
Dpp-Val-OH	103–104	−15.2	1	MeOH	23
Dpp-Val-OMe	119–124	−32.7	1	MeOH	23
H_2^+-Val-OBzl·SO_3^-	248–251	+4.3	1	DMSO	34
Nps-Val-ONSu	136–137	−90.7	1	DMF	16
Trt-Val-OBt	Foam	+9.3	2	$CHCl_3$	74
Z-Val-O(2 Pet)	Oil	+2.8	1.9	$CHCl_3$	31
Z-Val-O(4 Pet)	Oil	+2.5	1.9	$CHCl_3$	31

Uncoded Amino Acids

(All compounds are of the L-configuration unless specified otherwise.)

Compound	M.p./°C	$[\alpha]_D$	Conc.	Solvent	Ref.
Aminobutyric acid (Abu)					
Boc-Abu-OPfp	83–84	−32.8	1	EtOAc	82
2-Amino-4-cyanobutyric acid [Abu(CN)]					
Boc-Abu(CN)-OH	155–157	+4	0.5	$CHCl_3$	149
Boc-Abu(CN)-ONSu	113–115	–	–	–	149
Aminoethylglycine (Aeg)					
Ac-Aeg(HCl)-OH	196–198	–	–	–	453
$Br^-H_2^+$-Aeg(HBr)-OH	172–175	–	–	–	453
$Cl^-H_2^+$-Aeg(HCl)-NH_2	182	–	–	–	453
$Cl^-H_2^+$-Aeg(HCl)-OMe	190–191	–	–	–	453
$Cl^-H_2^+$-Aeg(Z)-OH	173–174	–	–	–	453
Z-Aeg(Boc)-OH	90–95	–	–	–	453
Z-Aeg(HCl)-OH	176–177	–	–	–	453
Z-Aeg(Z)-OH	80–83	–	–	–	453
Z-Aeg(Z)-NH_2	117–118	–	–	–	453
Alanine					
Fmoc-D-Ala-OTcp	164–165	+39	1	DMF	83
Aminosuccinic acid (Asu)					
Fmoc-Asu-OTcp	190–192	−37	1	DMF	83
Cyclohexylalanine (Cha)					
Boc-Cha-OPfp	75–77	−23.9	1	EtOAc	82
γ-Aminobutyric acid (Gaba)					
Boc-Gaba-OH-DCHA	132–134	–	–	–	186

Compound	*M.p./°C*	$[\alpha]_D^{*1}$	*Conc./g 100 cm^{-3}*	*Solvent*	*Ref.*
Homocysteine (Hcy)					
Boc-Hcy(CH_2CH_2-CO_2Bu^t)-OH	129–130	+15.6	1.8	EtOH	288
Boc-Hcy(Dmb)-OH-DCHA	105–107	+20.1	1.1	EtOH	105
H-Hcy($CH_2CH_2CO_2$$Bu^t$)-OH	–	−0.85	2	HOAc	288
H-Hcy(Dmb)-OH	235–237	+27.4	0.25	6M HCl	105
Leucine					
Boc-D-Leu-OPfp	53–55	+31.7	1	EtOAc	82
Methylhistidine					
Boc-His(π-Me)-OH	–	+13.8	1	MeOH	136
Boc-His(τ-Me)-OH	–	+13.3	1	MeOH	136
Phenylalanine					
Boc-D-Phe(2-OH)-OH-DCHA	205	−25.4	1.85	DCHA	118
Boc-Phe(2-OH)-OH-DCHA	203	+28.8	1.78	MeOH	118
Tos-$O^-H_2^+$-Phe-H semicarbazone	200–202	+42.8	2	MeOH	190
Z-Phe-H semicarbazone	142–143	−4.1	1	MeOH	190
Phenylglycine (Phg)					
Fmoc-Phg(4-OH)-OH	155–157	−44	1	DMF	83
Pyroglutamic acid (Glp)					
Z-D-Glp-OPfp	81–82	+40.1	1	EtOAc	82
Z-Glp-OPfp	80–82	+40.4	1	EtOAc	82
Valine					
Fmoc-D-Val-OTcp	150–151	+41	1	DMF	83
Vinylglycine					
Z-Vinylglycine-OMe	Oil	−12.4	0.5	MeOH	511

6 Appendix III: Purification Methods

Methods for the purification of protected peptides and proteins are given; the list also includes references to the purification of free peptides and separation of diastereoisomers.

Technique	*Ref.*
High-performance Liquid Chromatography	
H.p.l.c. analysis of Dns-amino acids	512

[511] S. Hanessian and S. P. Sahoo, *Tetrahedron Lett.*, 1984, **25**, 1425.
[512] S. Weinstein and S. Weiner, *J. Chromatogr.*, 1984, **303**, 244.

Technique	*Ref.*
H.p.l.c. separation of amino acid enantiomers using precolumn derivatization	513
H.p.l.c. estimation of amino acids using automated precolumn derivatization	514
H.p.l.c. separation on a chiral stationary phase	515
H.p.l.c. separation of enantiomers (review)	516
H.p.l.c. separation of enantiomers	517–519
H.p.l.c. of LHRH on a radially compressed column	520
H.p.l.c. of cyclic α-MSH analogues	521
Resolution of human mercapt- and non-mercapt-albumin by h.p.l.c.	522
H.p.l.c. separation of peptides on a polystyrene-resin column	523
H.p.l.c. analysis of phenylthiohydantoins	524
H.p.l.c. of synthetic peptides	525
H.p.l.c. of low-molecular-weight proteins	526
Reversed-phase h.p.l.c. separation of four monoiodoinsulins	527
Purification of radioiodinated peptides using Sep-Pak cartridges for h.p.l.c.	528
Gas–Liquid Chromatography	
G.c. retention of enantiomers on two stationary phases	529
Separation of enantiomeric *N*-methylamino acids	530
G.c. resolution of β-hydroxyamino acids	531
G.c. estimation of lysinoalanine	532

[513] R. H. Buck and K. Krummen, *J. Chromatogr.*, 1984, **315**, 279.
[514] R. H. Buck and K. Krummen, *J. Chromatogr.*, 1984, **303**, 238.
[515] N. Oi and H. Kitahara, *Bunseki Kagaku*, 1984, **33**, 386.
[516] D. W. Armstrong, *J. Liq. Chromatogr.*, 1984, **7** (Suppl. 2), 353.
[517] N. Nimura, A. Toyama, and T. Kinoshita, *J. Chromatogr.*, 1984, **316**, 547.
[518] M. H. Engel and S. A. Macko, *Anal. Chem.*, 1984, **56**, 2598.
[519] N. Oi, M. Nagase, and Y. Sawada, *J. Chromatogr.*, 1984, **292**, 427.
[520] J. Rivier, R. McClintock, R. Galyean, and H. Anderson, *J. Chromatogr.*, 1984, **288**, 303.
[521] M. Lebl, W. L. Cody, and V. J. Hruby, *J. Liq. Chromatogr.*, 1984, **7**, 1195.
[522] M. Sogami, S. Nagoka, J. Eva, M. Honda, and K. Noguchi, *Int. J. Pept. Protein Res.*, 1984, **24**, 96.
[523] R. Sasagawa, L. H. Ericsson, D. C. Teller, K. Titani, and K. A. Walsh, *J. Chromatogr.*, 1984, **307**, 29.
[524] D. H. Schlesinger, *Methods Enzymol.*, 1983, **91**, 494.
[525] G. J. Putterman, M. B. Spear, and F. Perini, *J. Liq. Chromatogr.*, 1984, **7**, 341.
[526] T. Isobe, Y. Kurosu, Y.-I. Fang, N. Ishioka, H. Kawasaki, N. Takai, and T. Okuyama, *J. Liq. Chromatogr.*, 1984, **7**, 1101.
[527] B. S. Welinder, S. Linde, B. Hansen, and O. Sonne, *J. Chromatogr.*, 1984, **298**, 41.
[528] J. J. Miller, G. S. Schultz, and R. S. Levy, *Int. J. Pept. Protein Res.*, 1984, **24**, 112.
[529] R. Charles and K. Watabe, *J. Chromatogr.*, 1984, **298**, 253.
[530] W. A. Koenig, I. Benecke, N. Lucht, E. Schmidt, J. Schulze, and S. Sievers, *J. Chromatogr.*, 1983, **279**, 555.
[531] H. Su and K. Bernauer, *J. Chromatogr.*, 1984, **301**, 461.
[532] W. Bueser and H. F. Erbersdobler, *J. Chromatogr.*, 1984, **303**, 234.

Technique	*Ref.*
Other Chromatographic Methods	
Biospecific sorbent for aminopeptidases	533
Countercurrent distribution of synthetic peptides	525
Multi-layer-coil countercurrent chromatography	534
Rotating-coil countercurrent distribution	535
Purification of CCKPZ using a coil-planet centrifuge	536
CCKPZ purification using a coil-planet centrifuge (CCC)	537
Liquid-chromatographic separation of diastereoisomers	538
Separation of Dns-amino acid by reversed-phase t.l.c.	539
Peptide isolation using fast protein liquid chromatography (f.p.l.c.)	540

[533] L. A. Lyublinskaya, M. P. Yusupova, T. I. Vaganova, N. M. Ivanova, and V. M. Stepanov, *Bioorg. Khim.*, 1984, **10**, 1490.
[534] J. L. Sandlin and Y. Ito, *J. Liq. Chromatogr.*, 1984, **7**, 323.
[535] Y. Ito and R. Bhatnagar, *J. Liq. Chromatogr.*, 1984, **7**, 257.
[536] M. Knight, C. A. Tamminga, Y. Ito, J. D. Gardner, and T. N. Chase, *J. Chromatogr.*, 1984, **301**, 277.
[537] M. Knight, A. M. Kask, and C. A. Tamminga, *J. Liq. Chromatogr.*, 1984, **7**, 351.
[538] Z. Iskandarani, R. L. Smith, and D. J. Pietrzyk, *J. Liq. Chromatogr.*, 1984, **7**, 111.
[539] N. Grinberg and S. Weinstein, *J. Chromatogr.*, 1984, **303**, 251.
[540] C. Vita, D. Dalzoppo, S. Patti, and A. Fontana, *Int. J. Pept. Protein Res.*, 1984, **24**, 104.

3

Analogue and Conformational Studies on Peptide Hormones and Other Biologically Active Peptides

BY J. S. DAVIES

1 Introduction

Structure–activity studies on biologically active peptides have been an area of immense activity in academia and industry with the ultimate goal of understanding the regulatory mechanisms, the substrate–receptor interactions, and enabling the development of super-active analogues or mimics as potential chemotherapeutic agents. This vast effort has contributed greatly to our understanding of mechanisms, but in only a few cases has a peptide or analogue achieved commercial success.[1] This is most probably due to the inherent structure of peptides and the ability of the host system to complicate the delivery of a peptide to its 'active site' of biological activity.

Precise mimicking of the active peptide form is already a prominent feature of research activity. Success depends on obtaining meaningful information about the natural conformation of peptides especially the receptor-bound conformation, so precise physico-chemical and analogue studies are essential. The introduction of non-peptide modifications introduces new challenges in synthesis, since preservation of molecular shape and topochemistry will be paramount.

The emphasis in this chapter, therefore, will be on papers published during 1984 that highlight the developments described above. The effects of introducing backbone modifications, restricted conformational forms, and side-chain modifications are reviewed, and an attempt is made to survey the increasing use being made of computer modelling. This latest development should greatly assist in targeting the research effort into the most fruitful areas of study.

2 Peptide-backbone Modifications

The excellent coverage of the literature up to 1982 by A. F. Spatola[2] is a very helpful entry into this field as the review subdivides the literature coverage into peptide-bond replacement categories. To bring order into a vast field of

[1] M. Bodanszky in his foreword to 'Principles of Peptide Synthesis', Springer Verlag, Berlin, Heidelberg, 1984.

[2] A. F. Spatola in 'Chemistry and Biochemistry of Amino Acids, Peptides and Proteins', ed. B. Weinstein, 1983, Vol. 7, p. 267.

endeavour, Spatola suggested that peptide-bond surrogate groups should be identified by the nomenclature ψ [ ] with the group within the brackets defining the nature of the peptide-bond replacement. The Table summarizes the peptide-bond replacements known in 1982.

Table

ψ [CH_2S]	ψ [CONR]	ψ [CH_2(*R*)SO] or ψ [CH_2(*S*)SO]
ψ [NHCO]	ψ [COO]	ψ [CH_2SO_2]
ψ [NHCONH]	ψ [CSNH]	ψ [(*R*)-CH(Me)S] or ψ [(*S*)-CH(Me)S]
ψ [$COCH_2$]	ψ [CH_2NH]	ψ [(*Z*)-CH=CH] or ψ [(*E*)-CH=CH]
ψ [COS]	ψ [CH_2CH_2]	ψ [CONHO]
		ψ [C(=CH_2)CH_2]

Of equal interest to this field are modifications such as: —NHN(R)CO— (α-aza), —NHB(R)CO— (α-bora), —NHC(R^1R^2)CO— (α-substituted), —NHC(=CHR)CO— (dehydro), —NHCHCON— (lactam), —HNC_6H_4CO— (phenyl), —$NHCH_2$CH(R)-
└X┘
CO (β-amino acids), —NHCHR(CH_2)CO— (β-homo-amino acids), —D-NHCH(R)-CO— (replacement of L-residue by D-form).

Having reviewed the field, Spatola concludes, 'Generalizations regarding the efficacy of particular modifications are not yet justified or justifiable. While the degree of isoterism of modifications certainly varies, the constantly changing importance of steric, electronic, geometric, and lipophilic parameters makes it virtually impossible to cite one modification as inherently superior to another, and it is likely that all of the various types described are often combinations of these modifications and will prove efficacious, depending on the particular parent peptide chosen.'

It is against this background, and the coverage in other reviews[3] on analogue design, that the present chapter will annually assess the published literature.

ψ [CSNH] Analogues. – The development of the Lawesson reagent to introduce the thioamide unit has provided further examples for assessing the resulting activity. Lawesson and co-workers[4] have synthesized the thionated analogues H-Pro-Leu-Glyψ[CSNH]-NH_2, H-Pro-Leuψ[CSNH]Gly-NH_2, and H-Proψ[CSNH]-Leu-Gly-NH_2 of melanostatin, and have also replaced the amide bond in the sweetener aspartame. The aspartame analogue[5] H-Aspψ[CSNH] Phe-OMe, however, showed a reduced sweetening power at the receptor.

[3] G. R. Marshall in 'The Chemical Regulation of Biological Mechanisms', ed. A. M. Creighton and S. Turner, Special Publication No. 42, The Royal Society of Chemistry, London, 1982, p. 279.

[4] M. Thorsen, B. Yde, U. Pedersen, K. Clausen, and S.-O. Lawesson, *Tetrahedron*, 1983, **39**, 3429.

[5] B. Yde, I. Thomsen, M. Thorsen, K. Clausen, and S.-O. Lawesson, *Tetrahedron*, 1983, **39**, 4121.

Structure–activity studies on dermorphin endothiotetrapeptides[6] were also made possible by the use of the Lawesson reagent. Taking H-Tyr-D-Ala-Phe-Gly-NH_2 as standard, thioamide substitution at Gly^4 or Phe^3 meets bond requirements for μ-receptors and shows significant antinociceptive effects after i.c.v. administration in mice. The *in vitro* opioid activities of a series of Leu-enkephalin thioamide analogues[7] have been determined in μ- and δ-receptor selective bioassays and binding assays. Potency enhancement was only achieved when the thioamide occurred at positions 2–3 and 4–5, with the former showing a 3–5 times higher preference for δ- over μ-receptors than natural Leu-enkephalin. In a separate study[8] on Leu-enkephalins, it is also confirmed that activities depend on the backbone position of the thioamide function. In this study, position 2–3 was again the only position for insertion of the thioamide to bring enhancement of potency, which could well reflect a stronger resistance to proteolytic cleavage in this position.

ψ[NHCO] Retro–Inverso Analogues. – Interest continues in preserving the stereochemistry of the overall peptide but inverting the direction of the amide bonds. Analogues of somatostatin with extended retro–inverso modifications[9] at the six central residues exhibit low potency in growth-hormone inhibition tests *in vitro*. retro-Leu^5-enkephalin has been synthesized[10] *via* stepwise coupling in solution and has been shown to stimulate prolactin secretion in the rat.

Enhanced resistance to hydrolysis by the angiotensin-converting enzyme (ACE) is obtained[11] by the retro-inverso analogue (1) of the bradykinin-potentiating peptide BPP_{9a}. Synthesis was achieved by solid-phase techniques on Sheppard's polyamide resin combined with the use of [bis(trifluoroacetoxy)iodo] benzene to give good conversion of C-terminal $CONH_2$ to an amine on the resin.

Glp-Trp-Pro-Arg-Pro-NHCHNHCOCH(CH_2Ph)CO-Ala-Pro-OH
(*RS*)
(CH_2)$_4$$NH_2$ (on the NHCH carbon)

(1)

ψ[$COCH_2$]-Ketomethylene Analogues. – Following up on their studies on ketomethylene analogues as ACE inhibitors, Almquist *et al.* have incorporated[12] ketomethylene groups between both Tyr-Gly and Gly-Gly in Leu-enkephalin. The extra synthetic demand required in the incorporation of non-amide units is well exemplified in this work. On testing, H-Tyrψ[$COCH_2$]Glyψ[$COCH_2$]Gly-Phe-Leu

6 S. Salvadori, M. Marastoni, G. Balboni, R. Tomatis, and G. Sarto, *Farm. Ed. Sci.*, 1984, **39**, 316.

7 K. Clausen, A. F. Spatola, C. Lemieux, P. W. Schiller, and S.-O. Lawesson, *Biochem. Biophys. Res. Commun.*, 1984, **120**, 305.

8 G. Lajoie, F. Lépine, S. Lemaire, F. Jolicoeur, C. Aubé, A. Turcotte, and B. Belleau, *Int. J. Pept. Protein Res.*, 1984, **24**, 316.

9 P. Pallai, S. Struthers, M. Goodman, J. Rivier, and W. Vale, *Biopolymers*, 1983, **22**, 2523.

10 Y. P. Shvachkin, A. A. Shishkina, A. P. Smirnova, V. P. Fedotov, T. I. Ivanenko, G. I. Bushueva, and L. A. Batrameeva, *Zh. Obshch. Khim.*, 1984, **54**, 1907.

11 F. Bonelli, A. Pessi, and A. S. Verdini, *Int. J. Pept. Protein Res.*, 1984, **24**, 553.

12 R. G. Almquist, C. M. Olsen, E. T. Uyeno, and L. Toll, *J. Med. Chem.*, 1984, **27**, 115.

had less than 0.0001 times the opiate receptor-binding ability but had 10 times the analgesic activity of Leu-enkephalin. It is concluded that the somewhat surprising *in vivo* activity may be due to the ketomethylene analogue being 10 000 times more stable to peptidase degradation.

ψ[CH_2S]-Thiomethylene Analogues. – A conformational comparison[13] using n.m.r. techniques has been made between cyclo(-Gly-Pro-Gly-D-Phe-Pro-) and cyclo(-Gly-Proψ[CH_2S]Gly-D-Phe-Pro-). ^{1}H and ^{13}C n.m.r. evidence confirms the retention of β- and γ-turns, but more flexibility is seen around the proline bond. The effect of combining both the insertion of a CH_2S group and the cyclic conformational restriction has been surveyed *via* the synthesis[14] of the somatostatin analogue cyclo(-Proψ[CH_2S]Phe-D-Trp-Lys-Thr-). Solid-phase techniques with diphenylphosphoryl azide for ring closure achieved the synthesis, but the analogue only achieved 6% of the activity of the all-amide cyclic hexapeptide.

ψ[CONR]-*N*-Alkylated Analogues. – Analogues of vasopressin where proline^7 has been replaced by *N*-methylalalanine (sarcosine) and where position 1 contains either deaminopenicillamine or β-mercapto-β,β-cyclopentylmethylenepropionic acid have been investigated[15] for their inhibitory effects. In general they did not show a consistent pattern, but milk ejection and antidiuretic activities were depressed whereas pressor antagonism together with antagonism in the uterus *in vitro* were maintained. Using ^{1}H n.m.r. and *X*-ray data, conformational changes on *N*-methylation and its effect on chemotactic response have been rationalized.[16] Selective *N*-methylation of gramicidin S derivatives has yielded[17] [Me-Orn$^{22'}$,D-Me-Phe$^{44'}$]gramicidin S, which has the same antimicrobial activity as gramicidin S itself.

Alkylation of a terminal-amide bond in Met-enkephalins increases analgesic potency. A terminal *N*-dimethylamide gives an analogue[18] 250 times more potent than Met-enkephalin; [D-Ala^2, Met^5]enkephalin with a C-terminal *N*-methylamide was about 70 times more potent.[19]

Azapeptides. – Although the aza analogue (2) of the ACE inhibitor enalaprilat (3) differs[20] markedly in conformation (*X*-ray data), they are both potent inhibitors.

[13] A. F. Spatola, L. M. Gierasch, and A. L. Rockwell, *Biopolymers*, 1983, **22**, 147.
[14] T. W. Gero, A. F. Spatola, I. T. Aleman, and A. V. Schally, *Biochem. Biophys. Res. Commun.*, 1984, **120**, 840.
[15] D. Gazis, I. L. Schwartz. B. Lammek, and Z. Grzonka, *Int. J. Pept. Protein Res.*, 1984, **23**, 78.
[16] P. W. Jeffs, S. L. Heald, D. F. Chodosh, and D. S. Eggleston, *Int. J. Pept. Protein Res.*, 1984, **24**, 442.
[17] M. Kawai, M. Ohya, Y. Butsugan, K. Saito, and T. Higashijima, *Chem. Lett.*, 1984, 1835.
[18] B. J. Dhotre, S. Chaturvedi, and K. B. Mathur, *Indian J. Chem., Sect. B*, 1984, **23**, 828; B. J. Dhotre and K. B. Mathur, *ibid.*, p. 1231.
[19] T. Plucinski, K. Rolka, L. Baran, E. Przegalinski, and G. Kupryszewski, *Pol. J. Chem.*, 1983, **57**, 887.
[20] W. J. Greenlee, E. D. Thorsett, J. P. Springer, A. A. Patchett, E. H. Ulm, and T. C. Vassil, *Biochem. Biophys. Res. Commun.*, 1984, **122**, 791.

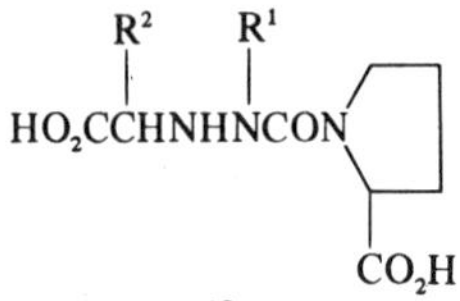

(2)

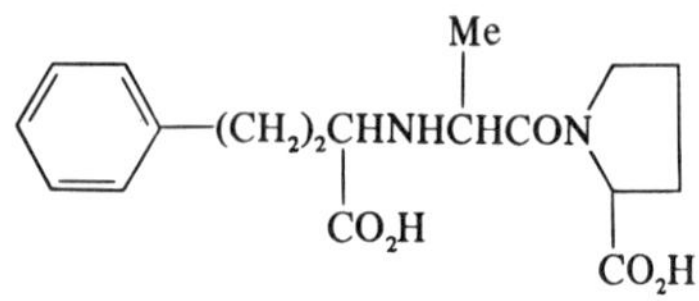

(3)

Lack of a requirement for a basic amine group and distortion of the semicarbazide system due to the presence of a proline ring are put forward as explanations. Further improvement in potency of LHRH analogues has been achieved[21] by a combination of azaglycine insertions at position 10 and hydrophobic D-residues[22] at position 6. [6-(3-Benzimidaz-2-yl)-D-Ala, 10-azaGly] LHRH, [6-(3 -5, 6-dimethylbenzimidazol-2-yl)-D-Ala,10-azaGly] LHRH, and [6-{3-(2-naphthyl)-D-Ala},10-azaGly] LHRH increased the potency by 40, 190, and 230 times that of LHRH, respectively.

α,α-Dialkylated Glycine Analogues. – The effect of introducing conformational restraint into the peptide backbone through incorporation of α,α-dialkylated glycine residues has been analysed.[23] Incorporation[24] of the α-aminoisobutyric acid (Aib) group at positions 2 and 3 in Leu-enkephalinamide also shows intramolecular H-bonding patterns that agree with a β-turn structure centred at positions 2,3 and 3,4, respectively. Analogues incorporating the cyclic unit Acc, 1-aminocyclopentane-1-carboxylic acid, at positions 2 and 3 show similar n.m.r. patterns to the acyclic acid. Incorporation of two residues of Aib or of Acc in the same molecule favours consecutive β-turn conformation.

Phosphinic Acid Analogues. – Current interest in the use of phosphorus analogues as antimicrobial agents has initiated[25] a synthesis of tripeptidic phosphinic acids such as (4).

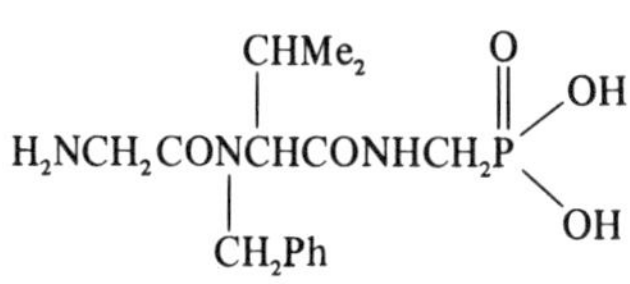

(4)

[21] T. L. Ho, J. J. Nestor, jun., G. I. McRae, and B. H. Vickery, *Int. J. Pept. Protein Res.*, 1984, **24**, 79.

[22] J. J. Nestor, jun., B. L. Horner, T. L. Ho, G. H. Jones, G. I. McRae, and B. H. Vickery, *J. Med. Chem.*, 1984, **27**, 320.

[23] E. Benedetti, C. Toniolo, P. Hardy, V. Barone, A. Bavaso, B. DiBlasio, P. Grimaldi, F. Lelj, V. Pavone, C. Pedone, G. M. Bonara, and I. Lingham, *J. Am. Chem. Soc.*, 1984, **106**, 8146; G. M. Bonara, C. Toniolo, B. DiBlasio, V. Pavone, C. Pedone, E. Benedetti, I. Lingham, and P. Hardy, *ibid.*, p. 8152.

[24] R. Kishore and P. Balaram, *Indian J. Chem., Sect. B*, 1984, **23**, 1137.

[25] J. Rachon, *Synthesis*, 1984, 219.

3 Conformationally Restricted Bridged Analogues

Somatostatin. – The success of the Merck group, using all the modern tools available to drug designers, in producing a highly active cyclic hexapeptide analogue[26] of somatostatin has resulted in an explosion of interest in this approach to conformationally restricting bioactive peptides to overcome rapid metabolism *in vivo*. The same group of workers have now modified[27] the original highly active analogue cyclo(-Pro-Phe-D-Trp-Lys-Thr-Phe-) (5) to include a dehydro-Phe residue at the position equivalent to Phe^7, the new analogue being produced by a combination of solid-phase and solution methodology. Although n.m.r. studies show comparable backbone conformations for (5) and its dehydro analogue, the latter only has 10% of the potency of the saturated form. It is implied that flexibility in the conformation of Phe^7 is needed to allow it to adopt a different conformation at the receptor. Replacing the Pro in (5) with open-chain *N*-methyl analogues retains[28] high potency, but constraining the backbone of the cyclic hexapeptide with lactams as bridged analogues drastically lowers the potency. These bridged analogues enforce a *trans* peptide bond at the 6-position in contrast to the *cis* form presumed to be present in the *N*-Me- and Pro-containing compounds. Increasing the flexibility[14] by incorporating a $-CH_2S-$ moiety lowers potency, which could also be due to loss of amide H-bonding.

Restricting conformational mobility by bridging *via* a disulphide bond between Cys^6 and $penicillamine^{11}$ gives the cyclic somatostatin analogue[29] D-Phe-Cys-Try-D-Trp-Lys-Thr-Pen-Thr-NH_2 (6). This, the most potent of a series of analogues synthesized, exhibited 7800 times the potency of somatostatin in its affinity for μ-opiate receptors. Various other cyclic analogues have been shown[30] to give somatostatin activity.

Enkephalins. – Conformationally restricting the flexibility of enkephalins by linking the 2- and 5-positions through disulphide links has produced[31] useful molecules for n.m.r. comparisons and activity correlations. H-Tyr-D-Cys-Gly-Phe-D- or L-Cys-NH_2, which exhibits moderate selectivities for μ-opioid receptors, and the penicillamine analogue H-Tyr-D-Pen-Gly-Phe-D- or L-Cys-NH_2, which exhibits δ-opioid receptor selectivity, do seem to have very similar overall conformations, but close study shows that the penicillamine analogue does give additional rigidity, which might explain the receptor selectivity.

26 D. F. Veber in 'Peptides: Synthesis, Structure, Function', Proc. 7th Am. Pept. Symp., ed. D. H. Rich and E. Gross, Pierce Chemical Co., 1981, p. 685.

27 S. F. Brady, D. W. Cochran, R. F. Nutt, F. W. Holly, C. D. Bennett, W. J. Palevada, P. E. Curley, B. H. Arison, R. Saperstein, and D. F. Veber, *Int. J. Pept. Protein Res.*, 1984, **23**, 212.

28 R. M. Friedinger, D. S. Perlow, W. C. Randall, R. Saperstein, B. H. Arison, and D. F. Veber, *Int. J. Pept. Protein Res.*, 1984, **23**, 142.

29 J. T. Pelton, K. Gulya, V. J. Hruby, S. P. Duckles, and H. I. Yamamura, *Proc. Natl. Acad. Sci. U.S.A.*, 1985, **82**, 236.

30 A. Friedrich, W. Koenig, V. Teetz, R. Geiger, and J. Sandow, Ger. Offen. DE 3,303,348 (*Chem. Abstr.*, 1985, **102**, 7095).

31 H. I. Mosberg and P. W. Schiller, *Int. J. Pept. Protein Res.*, 1984, **23**, 462.

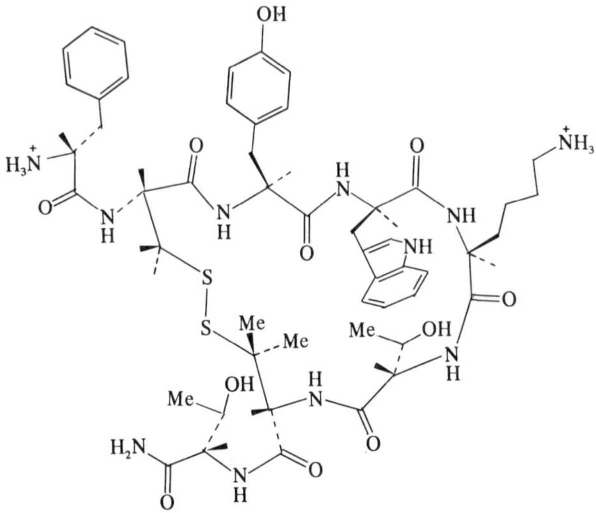

(6)

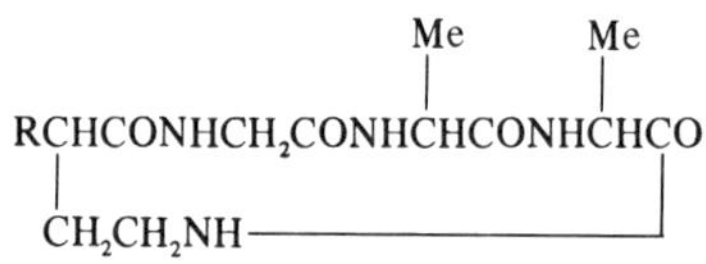

(7) R = H-Tyr-NH
(8) R = Me

The conformational restricting offered by bridging *via* an α,γ-diaminobutyric acid residue in the 2-position, which has already been shown to give high activity in the molecule (7), has now been subject[32] to computional molecular mechanics using the cyclic analogue (8) as a model. One extensive low-energy conformational region has been identified, indicating a possible significance for a Gly^3-Phe^4 type II′ bend in the enkephalins as per earlier models.[33]

ACTH Analogues. – Cyclic compounds[34] formed by bridging between C-terminal glycine and the ϵ-amino group of Lys^5 to give [Lys^5,cyclo(-Gly^{10}→Lys^5-)]ACTH-(5–10) and [Lys^5,(Gly^{11}→Lys^5)]ACTH(5–11) are 2–3 order of magnitude more active than their linear counterparts. Cyclization was effected by diphenylphosphoryl azide. Synthesis of a cyclic decapeptide ACTH(5–14) analogue has also been reported.[35]

[32] D. Hall and N. Pavitt, *Biopolymers*, 1984, **23**, 1441.
[33] F. H. Clarke, H. Jaggi, and R. A. Lovell, *J. Med. Chem.*, 1978, **21**, 600.
[34] I. K. Liepkaula, A. A. Skujins, P. J. Romanovskis, E. A. Porunkevich, M. P. Ratkevich, and G. I. Chipens, *Bioorg. Khim.*, 1984, **10**, 807.
[35] I. K. Liepkaula, A. A. Skujins, P. Y. Romanovskis, E. A. Porunkevich, M. P. Ratkevich, and G. I. Chipens, *Bioorg. Khim.*, 1984, **10**, 1326.

Melanotropins. – In an authoritative review[36] on recent developments on superpotent and prolonged analogues of melanotropins, the bridged cyclic disulphide structure (9), [Cys^4,Cys^{10}]α-MSH, and analogues are highlighted as potent agonists in the frog skin assay system.

However, a cyclic bridge is not the only criterion to consider for potency. The amino acid residues Lys^{11} and Pro^{12} and to a lesser extent Val^{13} of the C-terminal tripeptide sequence also contribute to potency.[37]

Bridging the backbone *via* carba analogues gives rise to a highly active analogue (10) of MSH, which was prepared[38] by solid-phase methodology that included

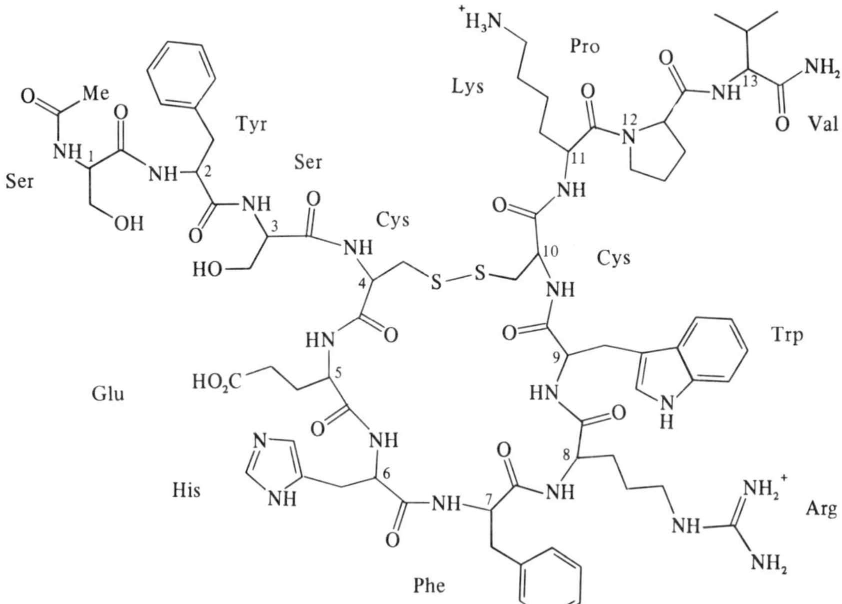

(9)

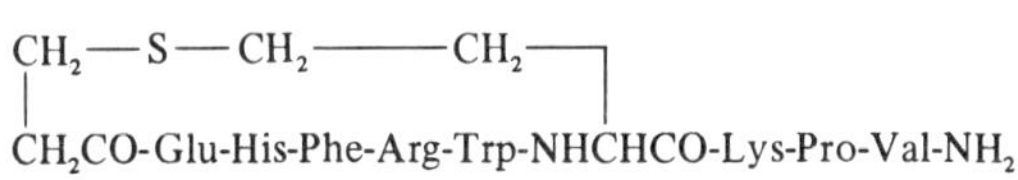

(10)

[36] V. J. Hruby, B. C. Wilkes, W. L. Cody, T. K. Sawyer, and M. E. Hadley, *Pept. Protein Rev.*, 1984, **3**, 1.

[37] W. L. Cody, B. C. Wilkes, B. J. Muska, V. J. Hruby, A. M. de L. Castrucci, and M. E. Hadley, *J. Med. Chem.*, 1984, **27**, 1186.

[38] M. Lebl and V. Hruby, *Tetrahedron Lett.*, 1984, **25**, 2067; M. Lebl, W. L. Cody, B. C. Wilkes, V. J. Hruby, A. M. de L. Castrucci, and M. E. Hadley, *Collect. Czech. Chem. Commun.*, 1984, **48**, 2680.

cyclization with DCCI/HOBt while the peptide was still on the resin. Similarly, four carba analogues of oxytocin could be prepared.

Nature also produces[39] a cyclic bisulphide antagonist hormone, melanin-concentrating hormone (MCH), as found in teleost fishes. The bisulphide link in the peptide H-Asp-Thr-Met-Arg-Cys-Met-Val-Gly-Arg-Val-Tyr-Arg-Pro-Cys-Trp-Glu-Val-OH is at a different position to α-MSH, and when synthesized[39] using solid-phase techniques the product gave full agonist activity in melanosome dispersion but had only 1/600th of the potency of α-MSH.

Angiotensin. – A cyclic analogue of angiotensin,[40] containing arginine in the cyclic structure Lys-Arg-Val-Tyr-Ile-His-Pro-Phe, has been synthesized and shown to have the depressor effect characteristic of bradykinin, but it inhibits the effects of angiotensin.

Bradykinin. – Cyclization using DCCI/HONSu has given[41] a cyclic analogue (11) of bradykinin, in which the bridge occurs between C-terminal arginine and the side chain of Lys[2], while the same cyclization conditions gave (12), which is a cyclic analogue of kallidin (lysylbradykinin). Both analogues exhibit prolonged hypotensive activity.

Carba Analogues of Oxytocin. – Deaminopenicillamine provides[42] a suitable bridge link for restricting the conformation, to give analogues (13) and (14), the former having a strong agonist effect in the *in vivo* uterotonic assay and being a weak antagonist in pressor assay.

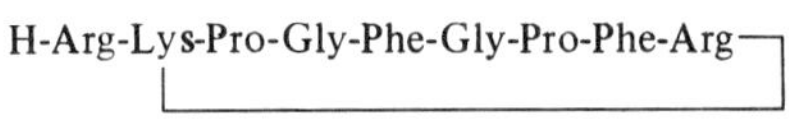

(11)

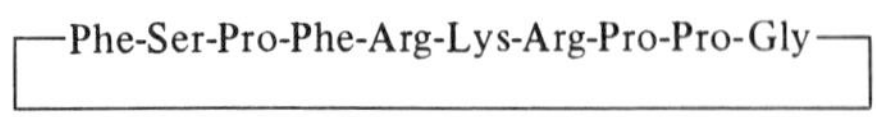

(12)

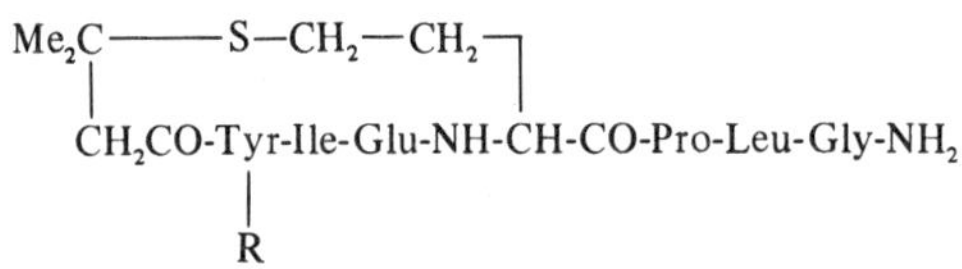

(13) R = H
(14) R = Me

[39] B. C. Wilkes, V. J. Hruby, A. M. de L. Castrucci, W. C. Sherbrooke, and M. E. Hadley, *Science*, 1984, **224**, 1111.

[40] G. Chipens, J. Ancans, D. Berga, I. Vosekalna, N. Mishlyakova, and A. Krikis in 'Chemistry of Peptides and Proteins', Vol. 2, ed. W. Voelter, E. Bayer, Yu. A. Ovchinnikov, and E. Wunsch, W. de Gruyter and Co., Berlin, 1984, p. 261.

[41] F. K. Mutulis, G. I. Chipens, O. E. Lando, and I. E. Mutule, *Int. J. Pept. Protein Res.*, 1984, **23**, 235.

[42] M. Lebl, T. Barth, L. Servitova, J. Slaninova, and P. Hrbas, *Collect. Czech. Chem. Commun.*, 1984, **49**, 2012.

Analogue (14) inhibited *in vitro* and *in vivo* uterotonic activity of oxytocin as well as the action of Lys-vasopressin. The synthesis and properties of [2-(3,5-$^3H_2$)-tyrosine,4-Glu] deamino-1-carbaoxytocin have also been reported.[43]

4 Dehydroamino Acid Analogues

In recent years conformational restriction through the insertion of α,β-didehydroamino acid residues of precise geometry has given many potent bioactive compounds. More work has been reported[44,45] on incorporating such residues into peptides and on determining their configuration. Similarly, studies have been reported[46] on gramicidin S analogues containing α,β-didehydroalanine, while both analogues containing dehydrophenylalanine at position 2 or 3 of the 6–11 fragment of substance P gave lowering of blood pressure in rats.[47]

5 Other Insertions

The Gly-Gly moiety in Leu-enkephalin has been replaced by a γ-aminobutyric acid residue to give an analogue with hypotensive activity.[48]

6 Mimics of β-Turns of Bioactive Peptides

C.d. data confirm[49] that the model dipeptide sequence (15) should be a reasonably restricted model for β-turns. Since it is reasonably accessible synthetically,

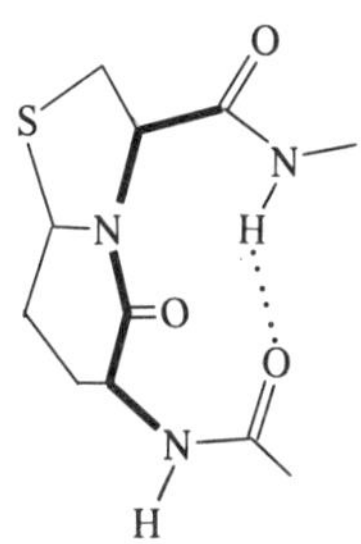

(15)

[43] M. Lebl. T. Barth, D. J. Crankshaw, B. Cerny, E. E. Daniel. A. K. Grover, and K. Jost, *Collect. Czech. Chem. Commun.*, 1984, **49**, 1921.
[44] C. Shin, Y. Yonezawa, and T. Yamada, *Chem. Pharm. Bull. (Jpn.)*, 1984, **32**, 2825.
[45] T. Kozono, H. Mihara, H. Aoyagi, T. Kato, and N. Izumyia, *Int. J. Pept. Protein Res.*, 1984, **24**, 402.
[46] S. Ando, T. Kato, and N. Izumyia, *Int. J. Pept. Protein Res.*, 1985, **25**, 15.
[47] T. Wasiak and W. Koziolkiewicz, *Pol. J. Chem.*, 1983, **57**, 861.
[48] A. V. Il'ina, Yu. A. Davidovich, and S. V. Rogozhin, *Zh. Obsch. Khim.*, 1984, **54**, 2385.
[49] U. Nagai and K. Sato, *Tetrahedron Lett.*, 1985, **26**, 647.

it will be interesting to see the result of its incorporation into a longer peptide. Detailed conformational work[50] using i.r. and 270 MHz n.m.r. spectroscopy of a bridged bisulphide bond as in the models Boc-Cys-Pro-X-Cys-NHMe, where X = Gly, L-Ala, D-Ala, Aib, or L-Leu, indicates a Pro-X β-turn stabilized by a transannular 4 → 1 H-bond involving the Cys(4)NH in all the analogues.

A conformational-energy study[51] to investigate β-turn preferences in tetrapeptides with different amino acid configurations revealed that the L-Ala[1,4] favours a β-turn in Ac-L-Ala-D-Ala-L-Pro-L-Ala-NHMe whereas D-Ala residues in the same positions do not. When L-Pro-L-Ala was substituted in the 2,3-positions, the effect was shown to be just the opposite of the previous examples when either L- or D-Ala was considered for the 1,4-positions.

7 Enzyme Inhibitors

The design of efficient inhibitors of enzymes involved in key control processes in the biological system is now clearly regarded as a prime source of chemotherapeutic agents, with compounds such as SQ 20881 (a nonapeptide), captopril, MK 421 (enalapril), and MK 422 (enalaprilat) already leading the way. This is an area where sensitive configurational and conformation design work is clearly benefiting from modern computer aids and physical methods, to achieve the best fit on the enzyme site.

Angiotensin-converting-enzyme (ACE) Inhibitors. – In an excellent example[52] of the use of molecular graphics for design work, active bicyclic mimetrics of ACE inhibitor captopril have been designed by comparing the arrays of the amide, carboxy, and methane thiol groups in the three bicyclic systems (16), (17), and (18) with the same functions in favoured conformations of captopril. Enlarging the size of these bicyclic systems to the 7/6 series to relieve internal strain has provided[53] another series of very active ACE inhibitors *in vivo* and *in vitro*. Extending the *N*-carboxymethyl dipeptide design feature in enalapril

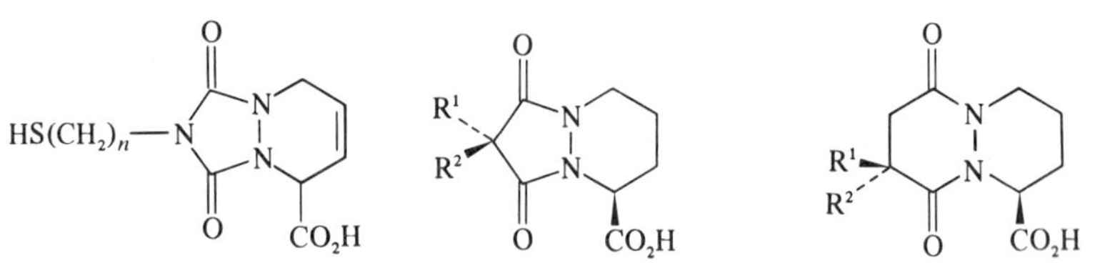

(16) n = 1, 2, or 3 (17) R^1, R^2 = Me or $HSCH_2$ (18) R^1, R^2 = H or $HSCH_2$

[50] A. Ravi and P. Balaram. *Tetrahedron*, 1984, **40**, 2577.
[51] M. Kawai, K. Sato, and U. Nagai, *Int. J. Pept. Protein Res.*, 1984, **24**, 607.
[52] C. H. Hassall, A. Krohn, C. J. Moody, and W. A. Thomas, *J. Chem. Soc., Perkin Trans. 1*, 1984, 155.
[53] M. R. Attwood, R. J. Francis, C. H. Hassall, A. Krohn, G. Lawton, I. L. Natoff, J. S. Nixon, S. Redshaw, and W. A. Thomas, *FEBS Lett.*, 1984, **165**, 201.

MK 421 into acyltripeptides and larger peptides,[54] *e.g.* BzNHCH(CH_2Ph)$(CH_2)_n$-CH(CO_2H)-Ala-Pro-OH ($n = 0$ or 1), gives no increase in potency, so the effect of having the possibility of additional binding interactions did not bear fruit. A series of β-lactam analogues of the tripeptide Phe-Ala-Pro, a substrate/inhibitor of ACE, showed[55] no inhibition of the enzyme.

Renin Inhibitors. – A potent inhibitor (19) of renin has been produced[56] by incorporating the statine residue (3*S*, 4*S*)-4-amino-3-hydroxy-6-methylheptanoic acid and a carefully designed mimic of Leu-Phe-NH_2 to give (19). A K_1 value of 3×10^{-8}M against human kidney renin shows it to be the most potent renin inhibitor reported, having fewer than five amino acid residues.

Linear and cyclic peptides containing the His^6-Pro^7-Phe^8-His^9 sequence of renin's substrate (angiotensinogen) have been shown[57] to be effective inhibitors of the enzyme. A β-turn-like structure involving this tetrapeptide region is regarded as a possible inhibition conformation.

Other Proteinase Inhibitors. – The natural chymotrypsin inhibitor chymostatin can be mimicked[58] by a series of tripeptide aldehydes, Z-Arg-X-Phe-H, where X = Leu, Ile, or Val.

Potent slow-binding inhibitors of aspartic peptidases are obtained[59] if tetrahedral intermediate mimics of the key step in peptidic hydrolysis are incorporated

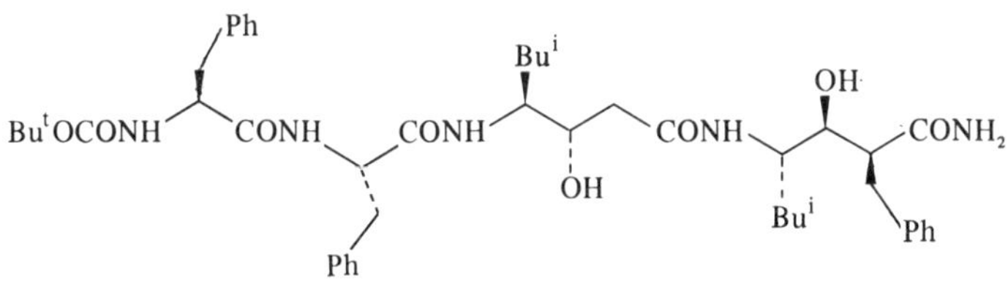

(19)

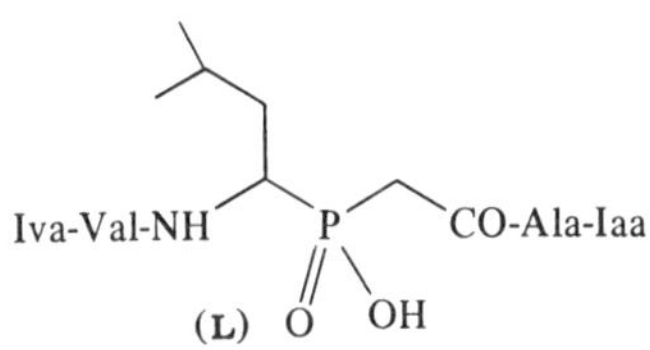

(20)

[54] W. J. Greenlee, P. L. Allibone, D. S. Perlow, A. A. Patchett, E. H. Ulm, and T. C. Vassil, *J. Med. Chem.*, 1985, **28**, 434.

[55] C. J. Wharton, R. Wrigglesworth, and M. Rowe, *J. Chem. Soc., Perkin Trans. 1*, 1984, 29.

[56] M. G. Bock, R. M. DiPardo, B. E. Evans, K. E. Rittle, J. S. Boger, R. M. Friedinger, and D. F. Veber, *J. Chem. Soc., Chem. Commun.*, 1985, 109.

[57] I. Liepina, G. V. Nikiforovich, and A. C. M. Paiva, *Biochem. Biophys. Res. Commun.*, 1984, **122**, 700.

[58] I. J. Galpin, A. H. Wilby, G. A. Place, and R. J. Beynon, *Int. J. Pept. Protein Res.*, 1984, **23**, 477.

[59] P. A. Bartlett and W. B. Kezer, *J. Am. Chem. Soc.*, 1984, **106**, 4282.

into model substrates. Phosphorous-containing statine analogues such as (20) give rise to one of the most potent inhibitors of pepsin known.

8 Side-chain Interactions Studied by Residue Substitution or Deletion

Analogue design by replacement or deletion of residues has been one of the traditional methods of investigating structure–activity relationships. It remains an active field of endeavour.

The biological activity *in vitro* of human-growth-hormone-releasing factor hGRF(1–44)NH_2 with the structure H-Tyr-Ala-Asp-Ala-Ile-Phe-Thr-Asn-Ser-Tyr-Arg-Lys-Val-Leu-Gly-Gln-Leu-Gly-Gln-Leu-Ser-Ala-Arg-Lys-Leu-Leu-Gln-Asp-Ile-Met-Ser-Arg-Gln-Gln-Gly-Glu-Ser-Asn-Gln-Glu-Arg-Gly-Ala-Arg-Ala-Arg-Leu-NH_2 can still be maintained[60] by deletion of the C-terminal residues down to hGRF-(1–21)NH_2, and the minimal core to maintain full intrinsic activity is hGRF-(3–21)NH_2.

Having established that the N-terminal tetrapeptide sequence of dermorphin is the minimum sequence for opioid activity, pharmacological properties of varying sequences within this tetrapeptide have been investigated.[61] Amongst the arginine analogues tested, H-Tyr-D-Arg-Phe-Sar-OH showed the highest activities in long-lasting analgesia, being up to 50 times as active as morphine. Other modifications to the tetrapeptide sequence, which include C- and N-terminal modifications, give rise[62] to $H_2NC(=NH)$-Tyr-D-Ala-Phe-Sar-NHCHMePh-D, which is 650 and 950 times as active as morphine in two *in vitro* tests and is believed to be a μ-type agonist.

Replacement of residues in the natural 19–24 sequence of ACTH to give analogues[63] containing hexaglycyl, hexaphenylalanyl, hexaglutamic acid, or hexalysyl residues gives c.d. curves characteristic of random-coil conformations. ACTH(11–24)tetradecapeptide containing hexalysine in the 19–24 region manifested higher steroidogenic activity than the natural sequence, while other analogues had less activity, suggesting a functional requirement for a correct 19–24 sequence for activity. It has also been shown[64] that ACTH(17–24) interacts specifically with angiotensin II receptors, which is also proved from pressor and steroidogenic activity.

60 N. Ling, A. Baird, W. B. Wehrenberg, N. Ueno, T. Munegumi, and P. Brezeau, *Biochem. Biophys. Res. Commun.*, 1984, **123**, 854.

61 Y. Sasaki, M. Matsui, M. Taguchi, K. Suzuki, S. Sakurada, T. Sato, T. Sakurada, and K. Kisara, *Biochem. Biophys. Res. Commun.*, 1984, **120**, 214.

62 S. Salvadori, G. Balboni, M. Marastoni, G. Sarto, and R. Tomatis, *Hoppe-Seyler's Z. Physiol. Chem.*, 1984, **365**, 1199.

63 I. V. Syskov, P. J. Romanovskis, I. A. Vosekalna, A. A. Skujins, M. P. Ratkevich, B. S. Kataev, E. A. Porunkevich, and G. I. Chipens, *Bioorg. Khim.*, 1984, **10**, 618.

64 I. V. Syskov, G. G. Kublis, A. A. Skuin'sh, B. S. Kataev, E. A. Porunkevich, G. A. Afanas'eva, Z. P. Pudane, P. Ya. Romanovskii, and G. I. Chipens, *Biochem. Acad. Sci. U.S.S.R.*, 1983, **48**, 436.

Further modifications to the clinically useful synthetic corticotropin synacthen to include a 7-methyl-Trp^9 residue give rise to a hyperactive analogue.[65]

In order to examine the question of why the C-terminal Trp-Met-Asp-Phe-NH_2 tetrapeptide of the gastrins contains all the physiological properties of the natural hormone but only low potency, N-terminal extension of the tetrapeptide has been studied[66] to find a possible site for additional functional information. In a series of gastrin analogues Pyr-$(Glu)_x$-Ala-Tyr-Gly-Trp-Nle-Asp-Phe-NH_2 (x = 0, 1, 2, 3, or 4) the potency of the C-terminal peptides reaches a level similar to natural minigastrin and little gastrin when the third glutamic acid residue is included. So this 'acidic' region of gastrin could well carry the information for additional recognition or for an improved transport mechanism. It has also been shown[67] that deletion of the C-terminal phenylalanine residue from the natural sequence gives peptide analogues that competitively inhibit the binding of labelled human gastrin to its receptors.

Deletion of the C-terminal phenylalanine residue in cholecystokinin CCK(27–33) to give CCK(27–32)NH_2 gives rise to the most potent peptidic antagonist[68] of the cholecystokinin receptor, and this peptide also inhibits gastrin-induced acid stimulation in the rat. Residues within Ac-CCK-7 (the shortest sequence giving biological activity) have been replaced[69] to achieve better structure–activity understanding. In the sequence Ac-Tyr$(SO_3H)^2$-Met^3-Gly^4-Trp^5-Met^6-Asp^7-Phe^8-NH_2 all analogues with *O*-sulphate esters of Ser, Thr, or Hyp in position 7 proved to be up to three times more potent than CCK-8 *in vitro*. A D-Ala^4 replacement gave rise to the first example of prolonged activity *in vivo*.

A comprehensive replacement programme[70] at position 4 of the potent antidiuretic antagonist [1-(β-mercapto-β,β-pentamethylene propionic acid), 2-D-Ile, 4-Val] Arg-vasopressin (21) shows that all analogues involving any one of

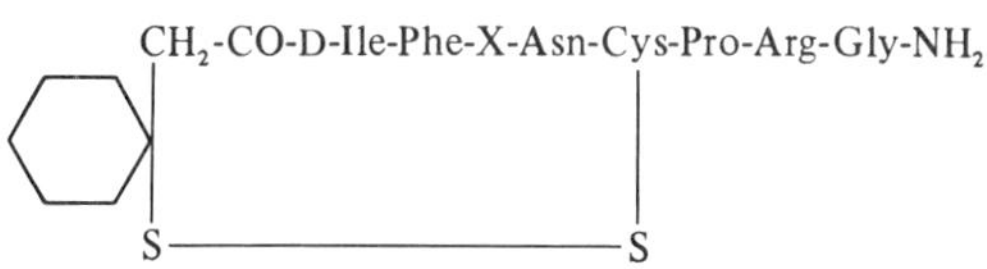

(21)

[65] M. C. Allen, P. D. Bailey, D. E. Brundish, J. H. Jones, R. Wade, and G. T. Young, *Int. J. Pept. Protein Res.*, 1984, **24**, 529.

[66] L. Moroder, G. Borin, A. Lobbia, J.-P. Bali, and E. Wunsch in 'Chemistry of Peptides and Proteins', Vol. 2, ed. W. Voelter, E. Bayer, Yu. A. Ovchinnikov, and E. Wunsch, W. de Gruyter and Co., Berlin, 1984, p. 255; W. Gohring, L. Moroder, G. Borin, A. Lobbia, J.-P. Bali, and E. Wunsch, *Z. Physiol. Chem.*, 1984, **365**, 83.

[67] J. Martinez, R. Magous, M.-F. Lignon, J. Laur, B. Castro, and J.-P. Bali, *J. Med. Chem.*, 1984, **27**, 1597.

[68] J. Martinez, C. Briet, F. Winternitz, B. Castro, V. Mutt, and J. D. Gardner in 'Peptides', Proc. 8th Am. Pept. Symp., ed. V. Hruby and D. H. Rich, Pierce Chemical Co., 1984, p. 673.

[69] B. Penke, F. Hajnal, J. Lonovics, G. Holzinger, T. Kadar, G. Telegdy, and J. Rivier, *J. Med. Chem.*, 1984, **27**, 845.

[70] M. Manning, E. Nawrocka, A. Misicka, A. Olma, W. A. Klis, J. Seto, and W. H. Sawyer, *J. Med. Chem.*, 1984, **27**, 423.

Abu, Ile, Thr, Ala, Ser, Nva, Glu, Leu, Cha, Lys, Asn, and Orn, but not Phe at position 4, are antidiuretic and vasopressor antagonists.

Synthesis and biological testing[71] of stereoisomeric analogues of FK-156 (22), an immunostimulating microbial metabolite, revealed that a D-Glu configuration and an L-configuration at the diaminopimelic acid residue are essential for activity whereas the configurations of Lac and Ala are of little importance. The minimum sequence needed for activity is deduced[72] to be N^2-(γ-D-glutamyl)-2(L),2′(D)-diaminopimelic acid.

The efficiency of carrying out multi-synthesis using solid-phase techniques[73] has provided the opportunity of introducing Ala to each of the 1–11 positions in dynorphin(1–13). Only when an Ala residue was substituted in positions 1 and 4 did the activity decrease significantly in guinea-pig ileum and mouse vas-deferens studies. When the nitrogen mustard melphalan (Mel) was incorporated[74] into the peptide chain of D-Ala^2,Leu^5-enkephalin methyl esters, replacement of Tyr^1 with Mel^1 did not result in loss of binding affinity, which suggests a need to reconsider the role of Tyr^1 in opioid activity. The *p*-chloroethyl amino group in melphalan gives rise to irreversible binding to the receptor because of the alkylating properties of this unit.

Two analogues of red-pigment-concentrating hormone (RPCH), namely [Thr^6]RPCH and [Tyr^4,Thr^6]RPCH, have been synthesized.[75] The biological response of the former analogue lends support to the conclusion that its sequence is that of adipokinetic hormone II and that replacement of Phe^4 by Tyr^4 does not reduce its adipokinetic response. The most subtle of changes in the side chain of [Thr^4]oxytocin in the form of a replacement of Thr by its *allo* form to give [*allo*-Thr^4]oxytocin gives a pronounced decrease[76] in oxytocic activity.

9 Conformational Information Derived from Physical Methods

In the period under review the advantages of using two-dimensional n.m.r. techniques have become obvious, and the broadening of the scope of *X*-ray

```
                           2
D-Lac-L-Ala-D-Glu-NHCHCO-Gly-OH
                     |
                    (CH2)3CH(NH2)CO2H
                                2'
```

(22)

[71] H. Takeno, S. Okada, K. Hemmi, M. Aratani, Y. Kitaura, and M. Hashimoto, *Chem. Pharm. Bull (Jpn.)*, 1984, **32**, 2925.

[72] H. Takeno, S. Okada, S. Yonishi, K. Hemmi, O. Nakaguchi, Y. Kitaura, and M. Hashimoto, *Chem. Pharm. Bull (Jpn.)*, 1984, **32**, 2932.

[73] A. Turcotte, J.-M. Lalonde, S. St.-Pierre, and S. Lemaire, *Int. J. Pept. Protein Res.*, 1984, **23**, 361.

[74] M. Sziics, K. DiGleria, and K. Medzihradszky, *FEBS Lett.*, 1985, **179**, 87.

[75] D. Yamashiro, S. W. Applebaum, and C. H. Li, *Int. J. Pept. Protein Res.*, 1984, **23**, 39.

[76] K. Bankowski, A. Misicka, and S. Drabarek, *Pol. J. Chem.*, 1983, **57**, 1045.

data *via* interactive computer-graphic techniques is highlighted. Solid-phase n.m.r. is also being applied to many examples, but still the greatest void is our lack of understanding of the bioactive conformation of the peptides at the receptor site. Therefore, papers that report attempts at a better understanding of the interactions between peptides and their biological environment have been brought together to form the first subparagraph in this section. Many of the papers refer to the use of more than one technique. No longer are structure–function relationships fully characterized by any one technique, but by a combination of methods. The achievements are now quite numerous, as typified by Ovchinnikov[77] and Bystrov[78] in their reviews.

External Influences on Conformation. – Owing to the difficulty of determining the conformation at the receptor using physical methods, Snell[79] has undertaken a detailed examination of 41 peptide ligands using the empirical methods of Chou and Fasman. Only four distinct conformational groupings need to be postulated to cover all the examples: β-bend, β-structure, and α-helical conformations are predicted for various groups of linear peptides, whereas disulphide-bridge-containing peptides are predicted to show common β-bend β-structure at the receptor sites. Interaction between two ACTH fragments and dioleylphosphatidylcholine membranes has been studied[80] by i.r.-attenuated total-reflection spectroscopy. ACTH(1–10) exists as a rigid anti-parallel pleated sheet in dry membranes, but in an aqueous environment it escapes from the lipid. On the other hand, ACTH(1–24) was firmly incorporated in the membrane in both dry and wet conditions, the latter conditions actually promoting peptide–lipid interaction. The helical part entering the bilayer was identified as the hydrophobic N-terminal decapeptide (message segment) with the C-terminal tetradecapeptide (address segment) remaining on the membrane surface. Peptide interactions[81] with an aqueous interfacial region within reversed micelles produced by solubilizing cyclo(-Gly-Pro-Gly-D-Ala-Pro-) in Aerosol OT [bis-(2-ethylhexyl)sodium sulphosuccinate] in heptane or octane have been studied using n.m.r., c.d., and i.r. The conformation of the peptide was perturbed predominantly by the high effective cation concentration in the aqueous core of the micelle. Complexes of Met- and Leu-enkephalin amides with 18-crown-6-ether have been studied[82] by 500 MHz n.m.r. spectroscopy in $CDCl_3$ solution to simulate two of the features of the opioid receptor: the apolar environment and the binding of the charged N atom. The NH resonances suggest a C_{10} β-turn in which the Phe^4-NH is linked to the Tyr^1-CO group. The conformations of enzyme-bound Arg-Arg-Ala-Ser-Leu and Leu-Arg-Arg-Ala-Ser-Leu-Gly, both substrates of protein kinase, have been studied[83] by two-dimensional 250 MHz n.m.r. as quaternary complexes

[77] Yu. A. Ovchinnikov, *Pure Appl. Chem.*, 1984, **56**, 1049.
[78] V. F. Bystrov, *Bioorg. Khim.*, 1984, **10**, 997.
[79] C. R. Snell, *Biochim. Biophys. Acta*, 1984, **787**, 53.
[80] H.-U. Gremlich, U.-P. Fringeli, and R. Schwyzer, *Biochemistry*, 1983, **22**, 4257.
[81] K. F. Thompson and L. M. Gierasch, *J. Am. Chem. Soc.*, 1984, **106**, 3648.
[82] C. A. Beretta, M. Parrilli, A. Pastore, T. Tancredi, and P. A. Temussi, *Biochem. Biophys. Res. Commun.*, 1984, **121**, 456.
[83] P. R. Rosevear, D. C. Fry, A. S. Mildvan, M. Doughty, C. O'Brian, and E. T. Kaiser, *Biochemistry*, 1984, **23**, 3161.

peptide enzyme–AMPPCP–$Co(NH_3)_4$

$\searrow$ $\swarrow$

Mn^{2+}

(23) AMPPCP = adenosine-β,γ-methylenetriphosphate P^{β},P^{γ}

(23). The measurements fit only extended-coil conformations for the bound peptide substrates with a minor difference in θ torsion angle at Arg_3C and ψ at Arg_2C between the penta- and hepta-peptides. An n.m.r. study[84] on N-terminal-residue ^{13}C-enriched tripeptides CF_3CO-Ala_3 and CF_3CO-Lys-Ala_2, bound to elastase, showed that in the two enzyme-inhibitor complexes obtained the $C_{\beta}H_3$ of ^{13}C-Ala is still freely rotating in the complex whereas the side chain of ^{13}C-Lys is highly immobilized. On investigating[85] the interactions between the bacterial cell-wall percursors Ac-Ala-Ala and Ac-Gly-Ala and the glycopeptide antibiotics vancomycin and ristocetin, it is revealed that replacing the Gly with Ala not only allows hydrophobic bonding but also strengthens existing hydrogen bonds.

N.M.R. Studies. – The development of two-dimensional n.m.r. at high field has simplified the task of signal correlation and hence improved the conformational information. The technique corroborates[86] the double-helix $\downarrow\uparrow\pi\pi L^5D^6$ structure of gramicidin A in dioxan. Spectral assignments have been completed[87] using both homo- and hetero-nuclear two-dimensional n.m.r. for cyclolinopeptide A, cyclo[-Pro-Pro-Phe-Phe-Leu-Ile-Ile-Leu-Val-], cyclo[-Pro-Phe-Gly-Phe-Gly-], and cyclo[-Pro-*o*-nitroBz-Gly_2-]. The same techniques[88] have proved the existence in DMSO of a $\beta II'$ turn of Phe^7-CO$\leftarrow$NHThr and a βVI turn for ThrCO$\leftarrow$HNPhe^7 in the cyclic hexapeptides cyclo[-Phe-D-Trp-Lys(Z)-Thr-X-Pro-] (X = Gly or Phe). When X = Phe the Phe-Pro bond is *cis*, but when X = Gly the Gly-Pro bond becomes *trans*. Combination of two-dimensional n.m.r. techniques and interactive computer graphics provides[89] a three-dimensional structural model for the C-terminal half of calmodulin. HC-toxin, a phytotoxin affecting certain varieties of corn, has been subjected[90] to one- and two-dimensional n.m.r. analysis at 200 and 600 MHz to reveal the conformation (24) for the cyclic tetrapeptide. High-field n.m.r. and c.d. studies reveal[91] an order conformational form as

[84] J. L. Dimicoli, H. L. Tanh, F. Toma, and S. Fermandjian, *Biochemistry*, 1984, **23**, 3173.

[85] M. P. Williamson and D. H. Williams, *Eur. J. Biochem.*, 1984, **138**, 345; D. H. Williams, *Acc. Chem. Res.*, 1984, **17**, 364.

[86] A. S. Arseniev, V. F. Bystrov, V. T. Ivanov, and Yu. A. Ovchinnikov, *FEBS Lett.*, 1984, **165**, 51.

[87] H. Kessler and R. Schuck in 'Chemistry of Peptides and Proteins', ed. W. Voelter, E. Bayer, Yu. A. Ovchinnikov, and E. Wunsch, W. de Gruyter and Co., Berlin, 1984, p. 233.

[88] H. Kessler, M. Bernd, H. Kogler, J. Zarbock, O. W. Sorenson, and G. Bodenhauser, *J. Am. Chem. Soc.*, 1983, **105**, 6944.

[89] A. Aulabough, W. P. Niemczura, T. L. Blundell, and W. A. Gibbons, *Eur. J. Biochem.*, 1984, **143**, 409.

[90] M. Pope, P. Mascagni, W. A. Gibbons, L. M. Giuffetti, and H. W. Knoche, *J. Am. Chem. Soc.*, 1984, **106**, 3863.

[91] T. Higashijima, Y. Masui, N. Chino, S. Sakakibara, H. Kita, and T. Miyazawa, *Eur. J. Biochem.*, 1984, **140**, 163.

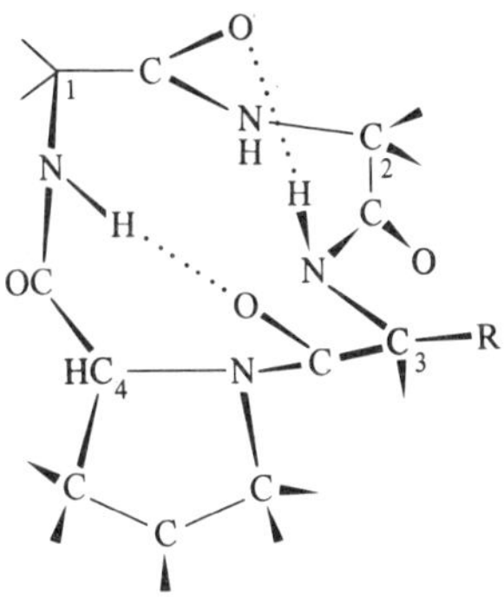

(24) R = $(CH_2)_5COCHCH_2$ (with O bridging the CH and CH_2, as drawn)

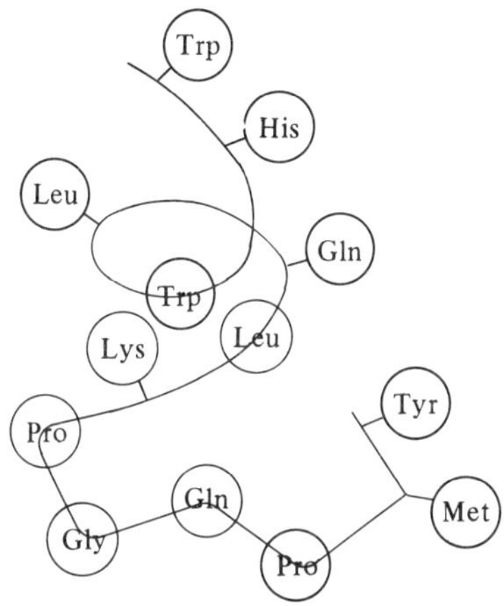

(25)

depicted in (25) for the tridecapeptide α-mating factor from yeast, *Saccharomyces cerevisiae.*

Solid-state n.m.r. spectra of polycrystalline cyclopentapeptides have been compared[92] with the solution-phase data. Cyclo[-D-Phe-Pro-Gly-D-Ala-Pro-] exists in the same single conformation in both phases, but cyclo[-D-Phe-Gly-Ala-Gly-Pro-] appears to crystallize in two different forms, which may be in dynamic equilibrium in solution. ^{13}C chemical shifts in the solid and solution phases have also been ascertained[93] for conformationally rigid polypeptides. In general[94] magic-angle-spin n.m.r. shows down-field shifts in side-chain methyl groups, which are believed to be due to van der Waals interactions. Spin–lattice relaxation rates[95] of deuteriomethyl groups in crystalline L-[β-$^2H_3$]Ala, DL-[γ-$^2H_6$]Val, and DL-[$\alpha\beta,\gamma\gamma'\delta$-$^2H_{10}$]Ile show similar dynamics as in membrane proteins. ^{13}C rapid-spin exchange in the presence of magic-angle sample spinning shows promise[96] as a means of correlating chemical shifts.

With the increased sensitivity of the modern spectrometers, ^{15}N n.m.r. offers an improved method for monitoring specific interactions with solvent *etc*. Long-range perturbations were detected[97] in the ^{15}N spectra of gramicidin S in aqueous solutions when compared with organic solvents. One- and two-dimensional natural-abundance ^{15}N n.m.r. shows[98] that gramicidin A has a greater mobility of its backbone towards the formyl end of the chain but does not

[92] M. H. Frey, S. J. Opella, A. L. Rockwell, and L. M. Gierasch, *J. Am. Chem. Soc.*, 1985, **107**, 1946.
[93] A. E. Tonelli, *Biopolymers*, 1984, **23**, 819.
[94] C. F. Brewer, *Eur. J. Biochem.*, 1984, **143**, 363.
[95] M. A. Keniry, A. Kintanar, R. L. Smith, H. S. Gutowsky, and E. Oldfield, *Biochemistry*, 1984, **23**, 288.
[96] M. H. Frey and S. J. Opella, *J. Am. Chem. Soc.*, 1984, **106**, 4942.
[97] D. H. Live, D. G. Davies, W. C. Agosta, and D. Cowburn, *J. Am. Chem. Soc.*, 1984, **106**, 1939.
[98] G. E. Hawkes, L. Y. Lian, and E. W. Randall, *J. Magn. Reson.*, 1984, **56**, 539.

necessarily indicate a random-coil conformation, since the β^6-helix includes a degree of flexibility at each peptide bond.

Cyclic depsipeptide ionophores and their analogues have always featured as pioneer molecules for n.m.r. applications. Together with *X*-ray data, solid- and solution-phase n.m.r. reveals[99] that the barium perchlorate complex of valinomycin is a flat open structure with two barium atoms per molecule, with infinite layers interconnected by perchlorate groups. An ion-binding cyclic analogue of valinomycin,[100] cyclo(-L-Ala-Gly-D-Phe-L-Pro-)$_3$, exists in a C_3 symmetric propeller structure, with its K^+ complex having a bracelet-type conformation. Replacement[101] of all D-Hyiv by D-Pro residues in hexadecavalinomycin leads to a preference for *cis* conformation of the tertiary-peptide bond in polar solvents; replacement of L-Lac by L-Pro residues, however, gives *cis* bonds in any polar medium.

Saturated-transfer n.m.r. experiments have been used[102] to study *cis–trans* proline-bond isomerism in cyclo[-Abu^i-L-Phe-D-Pro-L-Ada-] (Ada = ethylene ketal of 2-amino-10-ethoxy-8-oxodecanoic acid). Gramicidin S has served[103] as a model for relaxation analysis of heteronuclear selective nuclear Overhauser effects (NOEs), and conformation dependencies of the local NOEs have been used[104] to identify β-bends in peptides. No intramolecular hydrogen bonds or β-turns have been identified[105] from the n.m.r. spectra, including NOE studies on cyclo(-L-Phe-L-Pro-D-Ala-)$_2$. The two-*cis* backbone is the major form in solution, while in another series of diastereoisometric cyclo-octapeptides,[106] cyclo[-L- or D-Ala-Gly-L-Pro-L or D-Phe-]$_2$, the most stable conformations have *trans*-Gly-Pro bonds and C_2 symmetry in the n.m.r. average. An interesting comparison[107] of the 400 MHz spectra of [Ile^5]angiotensin II with an antagonist [Sar^1,Ile^{5,8}]-angiotensin II shows that in the former the chemical shifts for Phe aromatic protons and C_2–C_4 of His indicate shielding that is absent in the antagonist. The stacking of His/Phe side chains could, therefore, be important for activity at the receptors.

400 MHz spectra measured in [$^2H_6$]DMSO have been reported[108] for the enkephalin analogues Tyr-D-Nle-Gly-Phe-D(and L)-NleS, the C-terminal residue representing norleucine sulphonic acid. In biological tests the D^2L^5 analogue is known to be 500 times more effective and displays a preference for δ-sites. N.m.r. evidence suggests a significant population of conformer (26), which is less feasible in the D^2D^5 analogue, and is supported by lack of evidence for

99 S. Deverajan, M. Vijayan, and E. R. K. Easwaran, *Int. J. Pept. Protein Res.*, 1984, **23**, 324.
100 J. P. Degelaen, P. Pham, and E. R. Blout, *J. Am. Chem. Soc.*, 1984, **106**, 4882.
101 T. A. Balashova, L. A. Fonina, L. B. Senyania, N. V. Starovoitava, G. Ya. Avolina, V. T. Ivanov, and Yu. A. Ovchinnikov, *Bioorg. Khim.*, 1984, **10**, 437.
102 E. Haslinger, H. Kalchhauser, and P. Wolschann, *Monatshefte*, 1984, **115**, 779.
103 N. Niccolai, C. Rossi, P. Mascagni, P. Neri, and W. A. Gibbons, *Biochem. Biophys. Res. Commun.*, 1984, **124**, 739.
104 M. D. Shenderovich, G. V. Nikiforovich, and G. I. Chipens, *J. Magn. Reson.*, 1984, **59**, 1.
105 G. Kartha, K. K. Bhandary, K. D. Kopple, A. Go, and P.-P. Zhu, *J. Am. Chem. Soc.*, 1984, **106**, 3844.
106 K. D. Kopple, K. N. Parameswaran, and J. P. Yonan, *J. Am. Chem. Soc.*, 1984, **106**, 7212.
107 J. M. Matsoukas and G. J. Moore, *Biochem. Biophys. Res. Commun.*, 1984, **122**, 434.
108 S. Bajusz and A. F. Casy, *Org. Magn. Reson.*, 1984, **22**, 395.

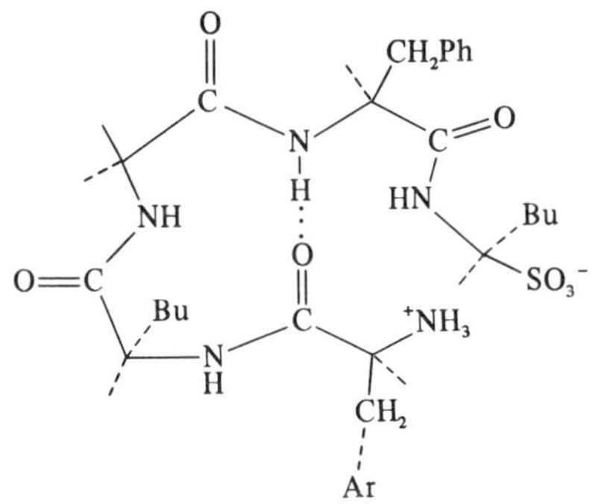

(26)

hydrogen bonding in the latter. ^{13}C spin-relaxation times have been collected[109] for the ring carbons of 26 pyrrolidine ring systems.

***X*-Ray and Related Techniques.** – An excellent example[110] of the use of *X*-ray techniques coupled with the use of interactive computer graphics has been reported for a family of 36-residue pancreatic polypeptides with some hormonal properties. Several TRH analogues[111] have been compared *via* their *X*-ray structures, and all show *trans* amide bonds. Computer graphics and a least-squares superposition fit show a surprising degree of similarity between the TRH conformations and those of the last three residues of the enkephalins. An *X*-ray determination[112] of t-Boc-Tyr-Gly-4BrPhe-Met-OH, a monoanionic Met-enkephalin, reveals an extended conformation, with the formation of dimers due to four intermolecular hydrogen bonds. The C-terminal $\alpha/3_{10}$ helical nonapeptide of alamethicin has been shown,[113] using *X*-ray data, to agree with α-helical conformation for alamethicin and is in conflict with the n.m.r. picture where a β-pleated sheet model has been proposed.[114] Crystal structures for pseudovalinomycin,[115] cyclo(-D-Hyi-Ala-Hyi-D-Val-)$_3$·$2H_2O$, hexa-*N*-methylvalinomycin,[116] cyclo(-D-MeVal-Lac-MeVal-D-Hyi-)$_3$, and model dipeptides[117] with a Pro-Ser sequence have been reported.

109 S. C. Shekar, M. B. Sankaram, and K. R. K. Easwaran, *Int. J. Pept. Protein Res.*, 1984, **23**, 166.
110 I. D. Glover, D. J. Barlow, J. E. Pitts, S. P. Wood, I. J. Tickle, T. L. Blundell, K. Tatemoto, J. R. Kimnel, A. Wollmer, W. Strassburger, and Y.-S. Zhang, *Eur. J. Biochem.*, 1984, **142**, 379.
111 E. Eckle and J. J. Stezowski, *J. Med. Chem.*, 1985, **28**, 125.
112 M. Doi, T. Ishida, T. Fujiwara, K. Tomita, T. Kimura, and S. Sakakibara. *FEBS Lett.*, 1984, **170**, 229.
113 R. Bosch, G. Jung, H. Schmitt, G. M. Sheldrick, and W. Winter, *Angew. Chem., Int. Ed.*, 1984, **23**, 450.
114 J. E. Hall, I. Vodyanoy, T. M. Balasubramanian, and G. R. Marshall, *Biophys. J.*, 1984, **45**, 233.
115 V. A. Popovich and O. I. Zaitsev, *Bioorg. Khim.*, 1984, **10**, 595.
116 V. A. Popovich and O. I. Zaitsev, *Bioorg. Khim.*, 1984, **10**, 581.
117 A. Aubry, N. Ghermani, and M. Marraud, *Int. J. Pept. Protein Res.*, 1984, **23**, 113.

I.R. and C.D. Studies. – I.r. and c.d. studies[118] on gramicidin A dimers reveal the formation of antiparallel double-helical aggregates and, at high dilution, single-stranded helices, which agrees with the model of a sliding or zipper mechanism for helix formation. Leu- and Met-enkephalins together with 17 analogues have been subjected[119] to c.d. and fluorescence spectroscopic analysis in dioxan and aqueous solutions. β-Turns are indicated in dioxan and to a lesser extent in aqueous media. C.d. spectra of four β-turn model peptides have been examined[120] in a range of solvent conditions, while sensitive changes in the i.r. frequency of terminal N—H absorptions provide[121] a means of assessing the percentage β-turn content in dipeptides. α-Aminoisobutyric acid- (Aib) containing tetrapeptides with Aib-L-Ala and L-Ala-Aib central sequences have been studied[122] to assess the amount of β-bend conformers produced. Solvent polarity effects on the secondary structure of a number of pituitary gland hormones have been studied[123] using c.d. and i.r. techniques. The same techniques reveal[124] that protection of peptide bonds and insertion of Pro residues had the same effect on conformation and solubilizing factors of oligo-L-leucine peptides.

Computational Methods. – Energy calculations[125] on the semirigid analogues Ac-[$\overline{Cys^4, Cys^{10}}$]α-$MSH_{4-10}$-$NH_2$, Ac-[$\overline{Cys^4, Cys^{10}}$]α-$MSH_{4-13}$-$NH_2$, and [$\overline{Cys^4, Cys^{10}}$]α-MSH agree fairly well with the existence of a chain-reversal structure in α-MSH_{6-9}. An equilibrium distribution of conformations for [D-Ala^2, Ser^5]enkephalin has been calculated[126] by the Monte Carlo method with Metropolis sampling; a preference for a 1—4-folded conformation is shown. Energy calculations[127] as well as n.m.r. and c.d. measurements on a conformationally rigid active cyclo analogue, Thr-Lys-Pro-Arg, have been used to describe a biologically active conformation for tuftsin; a *trans* conformation for the Pro bond is evident from the studies. A number of conformations in a limited energy range have been predicted[128] for the C-terminal heptapeptide, Pro-Arg-Arg-Pro-Tyr-Ile-Leu-

[118] S. V. Sychev, L. A. Fonina, and V. T. Ivanov, *Bioorg. Khim.*, 1984, **10**, 1080; see also S. V. Sychev and V. T. Ivanov in 'Chemistry of Peptides and Proteins', Vol. 2, ed. W. Voelter, E. Bayer, Yu. A. Ovchinnikov, and E. Wunsch, W. de Gruyter and Co., Berlin, 1984, p. 291.

[119] Z. A. Strel'tsova, *Bioorg. Khim.*, 1984, **10**, 817.

[120] M. Crisma, G. D. Fasman, H. Balaram, and P. Balaram, *Int. J. Pept. Protein Res.*, 1984, **23**, 411.

[121] G. Boussard and M. Marraud, *J. Am. Chem. Soc.*, 1985, **107**, 1825.

[122] G. De Pieri, A. Signor, G. M. Bonora, and C. Toniolo, *Int. J. Biol. Macromol.*, 1984, **6**, 35; G. M. Bonora, C. Mapelli, and C. Toniolo, *ibid.*, p. 179.

[123] A. A. Makarov, N. G. Esipova, V. M. Lobachov, and B. A. Grishkovsky, *Biopolymers*, 1984, **23**, 5; N. G. Esipova, V. M. Lobachov, V. N. Rogulenkova, A. A. Makarov, V. A. Shibnev, and M. P. Finogenova, *Mol. Biol.*, 1984, **18**, 725.

[124] M. Narita, K. Ishikawa, H. Nakano, and S. Isokawa, *Int. J. Pept. Protein Res.*, 1984, **24**, 14.

[125] S. A. Rozenblit, M. D. Shenderovich, and G. I. Chipens, *FEBS Lett.*, 1984, **170**, 315.

[126] S. Y. Yu and Y. R. Huai, *Int. J. Biol. Macromol.*, 1984, **6**, 228.

[127] G. V. Nikiforovich, I. T. Liepina, I. P. Sekacis, E. E. Liepins, B. S. Katayev, N. I. Veretennikova, and G. I. Chipens, *Int. J. Pept. Protein Res.*, 1984, **23**, 271.

[128] M. Cotrait, *Int. J. Pept. Protein Res.*, 1984, **23**, 355.

OH, of neurotensin, which is contrary to the deductions for the active C-terminal pentapeptide. Conformational-energy computations to examine the ornithine side chain[129] in gramicidin S and the ionophore[130] conformation of cyclo[-D-Ile-Lac-Ile-D-Hyi-]$_4$ agree fully with deductions from other physical methods.

[129] G. Nemethy and H. A. Scheraga, *Biochem. Biophys. Res. Commun.*, 1984, **118**, 643.
[130] V. A. Popovich, O. I. Zaitsev, and V. Z. Pletnev, *Bioorg. Khim.*, 1984, **10**, 1089.

4

Cyclic, Modified, and Conjugate Peptides

BY P. M. HARDY

1 Introduction

Although the title of this chapter has been changed, the coverage remains essentially as in earlier volumes of this series. It is hoped that the somewhat extended number of subheadings will assist the reader in locating compounds of interest in this rather heterogeneous compilation. Classification of compounds falling under more than one heading is arbitrary, but generally they are placed in the first appropriate category discussed. The section on enzyme inhibitors differs from the rest in being based on activity rather than structure, but it seemed desirable to unify this material to facilitate comparison. Protecting groups commonly used in peptide synthesis are not regarded here as conjugates, and cyclic peptides containing cystine bridges in the ring system are covered in Chapter 2. In most cases, papers that are solely biological or biochemical in content and have no organic structural or synthetic interest have not been covered.

Once again, where the papers cited refer adequately to the earlier literature on the topic, no additional references are given.

2 Cyclic Peptides

2,5-Dioxopiperazines (Cyclic Dipeptides). – Two new naturally occurring 2,5-dioxopiperazines have been reported. *Gliocladium virens* has yielded a compound (1) identified as the *N*-methyl derivative of gliovirin, which was isolated in 1983 from the same source. The latter is selectively active against members of the Oomycetes.[1] Two groups have characterized (3*Z*, 6*E*)-1-*N*-methylalbonursin (2) as a novel metabolite of *Streptomyces* species. The 6*E*-alkylidene configuration, supported by *X*-ray analysis, is unique amongst natural products of this type.[2]

The total synthesis of bicyclomycin has been reviewed,[3] and two other approaches to formation of the main skeleton of this antibiotic have been

[1] H. Yokose, N. Nakayama, C. Miyamoto, T. Furumai, H. B. Maruyama, R. D. Stipanovic, and C. R. Howell, *J. Antibiot.*, 1984, **37**, 667.

[2] A. A. Freer, D. Gardner, and J. P. Poyser, *J. Chem. Res.*, 1984, 283; D. J. Robins and M. A. Sefton, *Phytochemistry*, 1984, **23**, 200.

[3] S.-I. Nakatsuka and T. Goto, *Heterocycles*, 1984, **21**, 61.

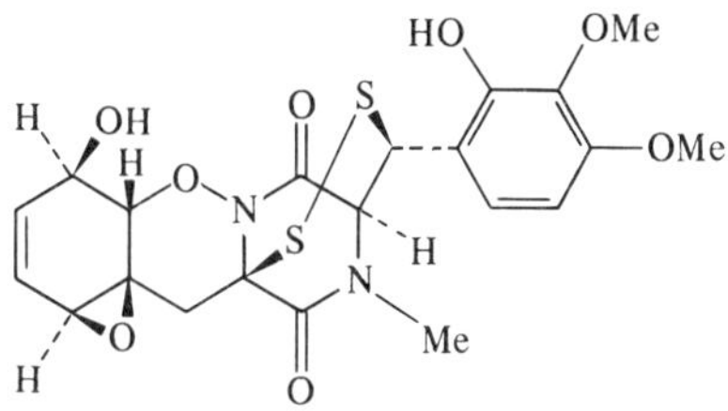

(1)

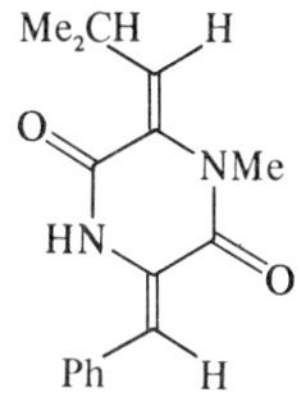

(2)

described. In one method 3,6-dimethoxypiperazinedione (3) was converted in six steps to (6) in an overall yield of 32% (Scheme 1), the key step (4) to (5) being a one-pot alkylation–cyclization.[4] In the other method cyclization followed *C*-hydroxylation of the dioxopiperazine ring using *N*-bromosuccinimide in the presence of water (Scheme 2).[5] A new versatile synthesis of δ-*N*-hydroxyornithine derivatives from glutamic acid has been developed and used in two syntheses of rhodotorulic acid (7).[6] An attempted conversion of poly(*N*-benzyl-

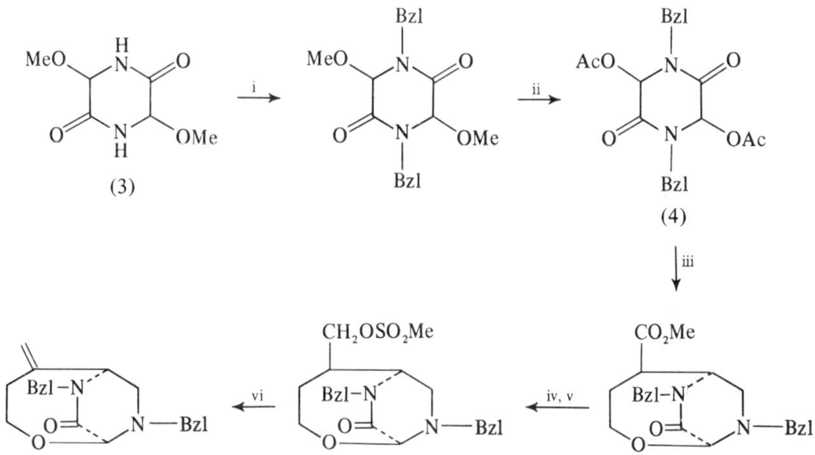

Reagents: i, NaH/DMF, $PhCH_2Cl$; ii, *p*-TsOH, Ac_2O; iii, $ZnCl_2/CH_2Cl_2$, $Me_3SiOCH_2CH_2CH{=}C(OMe)OSiMe_3$; iv, $LiAlH_4$/THF; v, $MeSO_2Cl$/pyridine; vi, Bu^tOK/DMSO

Scheme 1

[4] A. Sera, K. Itoh, H. Yamada, and R. Aoki, *Heterocycles*, 1984, **22**, 713.
[5] Y. Sato, C. Shin, S. Sumiya, Y. Yakajima, and Y. Yoshimure, *Bull. Chem. Soc. Jpn.*, 1984, **57**, 1265.
[6] B. H. Lee, G. J. Gerfen, and M. J. Miller, *J. Org. Chem.*, 1984, **49**, 2418.

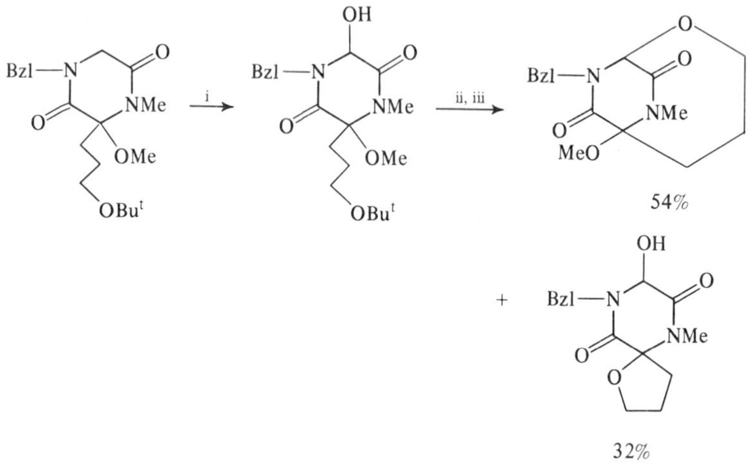

Reagents: i, NBS, $CHCl_3$–H_2O; ii, Ac_2O/NaAc; iii, CF_3CO_2H

Scheme 2

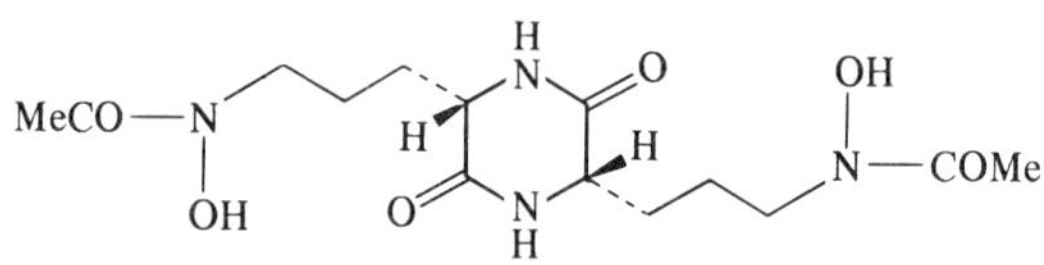

(7)

oxyalanine) into poly(*N*-hydroxyalanine) using boron trifluoroacetate in trifluoroacetic acid gave instead 1,4-dihydroxy-3,6-dimethyl-2,5-piperazinedione. This apparently occurs by fragmentation from the N-terminus of the *N*-hydroxy polymer, since no depolymerization was observed from poly(*N*-benzyloxyalanine) in trifluoroacetic acid alone.[7]

Two unsubstituted cyclic dipeptide crystal structures have been determined this year. The geometry of cyclo(-Phe-Pro-) proved to be not significantly different from that determined earlier for cyclo(-D-Phe-Pro-);[8] cyclo(-His-Met-) adopts a buckled backbone ring with the side chains folded, the imidazole ring facing the methionine side chain. Spectroscopic studies of the latter indicate that Cu^{II} adducts (but not Ni or Zn) involve additional co-ordination through the sulphur atom.[9] The acylated dioxopiperazine *N*-(phenylacetyl-Ala)-cyclo(-Phe-D-Pro-) is shown by *X*-ray to adopt a boat conformation that is appreciably more puckered than in the parent compound, with the phenylalanine side

[7] K. Shimizu, M. Hasegawa, and M. Akiyama, *Bull. Chem. Soc. Jpn.*, 1984, **57**, 495.

[8] F. Mazza, G. Lucente, F. Pinnen, and G. Zanotti, *Acta Crystallogr., Sect. C*, 1984, **40**, 1974.

[9] M. Bressan, F. Marchiori, and G. Valle, *Int. J. Pept. Protein Res.*, 1984, **23**, 104.

chain in a quasi-axial flagpole orientation.[10] The crystal structure of the lithiated bis-lactim ether of racemic cyclo(-Ala-Ala-) shows a dimeric aggregate of homochiral moieties in which two lithium atoms and two negatively charged ring nitrogens form a four-membered ring. This appears to be the first example of an organolithium dimer containing lithium atoms in chemically different surroundings; one lithium is co-ordinated by a THF molecule and an ethoxy group, the other by two THF molecules.[11]

Conformational-energy calculations on eight conformers typical of dioxopiperazine ring folding have been carried out. The results confirm considerable flexibility of the skeleton, and, as the degree of folding increases, twisted boat conformations with non-planar peptide bonds tend to be more stable.[12] The formation of the metal complex bis[cyclo(-His-His-)]-Cu^{II} has been determined potentiometrically. This complex catalyses the dismutation of superoxide anion, mimicking the active centre of Cu/Zn superoxide dismutase. On a weight basis it is three times as active as the enzyme, but a tenth as active on a molar basis.[13] A reinvestigation of the reduction of 3,6-dibenzylidene-2,5-piperazinedione with zinc and acetic acid has established the product to be (*Z*)-6-benzyl-3-benzylidene-2,5-piperazinedione. If HCl is present also, a mixture of *cis*- and *trans*-3,6-dibenzyl-2,5-piperazinedione is obtained instead.[14]

Cyclic Tetra-, Penta-, and Hexa-peptides. – The structure of Cyl-1, a novel cyclotetrapeptide from *Cylindrocladium scoparium*, has been shown by a linked-scan m.s. technique to be cyclo(-D-*O*-MeTyr-Ile-Pro-Aoe-) (Aoe = 2-amino-9,10-epoxy-8-oxodecanoic acid). Proline replaces the pipecolic acid residue seen in Cyl-2.[15] A stereoselective total synthesis of chlamydocin, cyclo(-D-Pro-Aoe-Aib-Phe-), has been reported from the aldehyde (8), which is readily available from tartaric acid. After building up to (9) (*E*/*Z* mixture), condensation with *N*-benzyloxycarbonyl-2-(dimethoxyphosphinyl)glycinate followed by asymmetric homogeneous hydrogenation catalysed by $[Rh(cod)(dipamp)]^+BF_4^-$ gave the (*S*)-amino acid derivative (10a). After conversion into (10b), this compound was incorporated into the cyclic tetrapeptide ring as earlier described. Hydrolysis

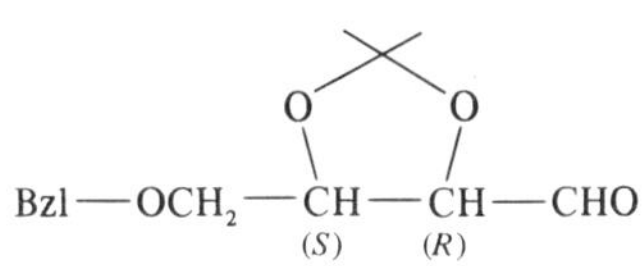

(8)

[10] C. Cerrini, W. Fedeli, G. Lucente, F. Mazza, F. Pinnen, and G. Zanotti, *Int. J. Pept. Protein Res.*, 1984, **23**, 223.
[11] D. Seeback, W. Bauer, J. Hansen, T. Laube, W. B. Schweizer, and J. D. Dunitz, *J. Chem. Soc., Chem. Commun.*, 1984, 853.
[12] J. Ciarkowski, *Biopolymers*, 1984, **23**, 397.
[13] S. Kubota and J. T. Yang, *Proc. Natl. Acad. Sci. U.S.A.*, 1984, **81**, 3283.
[14] S. M. Marcuccio and J. A. Elix, *Aust. J. Chem.*, 1984, **37**, 1791.
[15] S. Takayama, A. Isogai, M. Nakata, H. Suzuki, and A. Suzuki, *Agric. Biol. Chem.*, 1984, **48**, 839.

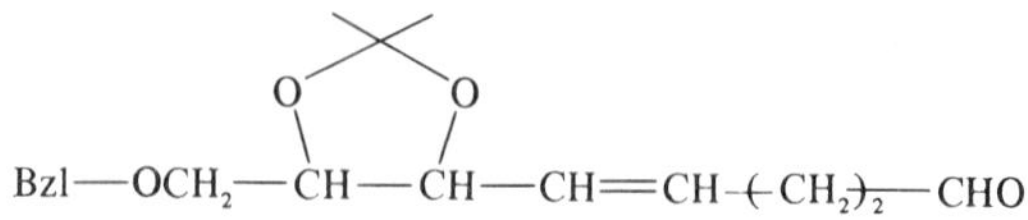

(9)

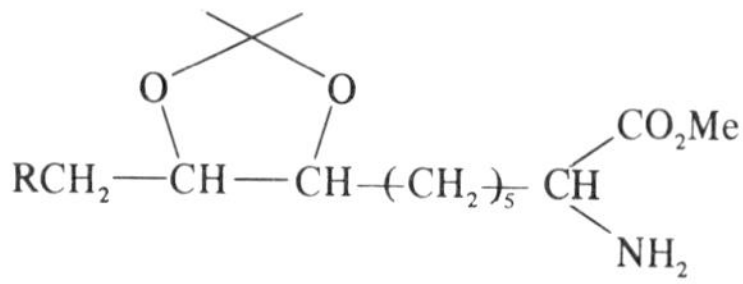

(10) a; R = OH
b; R = Cl

of the ketal, epoxidation, and oxidation of the resultant dihydrochlamydocin with DCC/DMSO/$ClCH_2CO_2H$ gave material identical with natural chlamydocin.[16]

A new synthesis of tentoxin, cyclo(-*N*-MeAla-Leu-*N*-MeΔPhe-Gly-), has been developed in which the dehydrophenylalanine residue is incorporated as such through its *N*-carboxyanhydride, and not as a precursor. Overall the yield is better than one earlier described.[17] Full details of the synthesis of three cyclotetrapeptide partial retro-inverso modified enkephalins (Volume 16 of this title, p. 352) have appeared.[18]

Upon long standing and slow evaporation, cyclo(-D-Phe-Pro-Gly-D-Ala-Pro-) in $CDCl_3$/CDCN containing $Mg(SCN)_2$ gives crystals of a 1 : 1 Mg^{II} complex, the first observed metal complex of a cyclic pentapeptide. *X*-Ray analysis indicates an infinite stack of alternating peptide and Mg^{II} moieties. The octahedrally coordinated metal ion has two peptide carbonyls as ligands, three water molecules, and the N-terminus of a thiocyanate ion. The HSO_4^- counter-ion observed is thought to have arisen by slow oxidation of NCS^-.[19] Rapid ^{13}C spin exchange of the parent cyclopentapeptide occurs in natural-abundance samples in the presence of magic-angle sample spinning. This not only allows resonances from carbons bonded to each other to be identified, but also carbon atoms that are near to each other in space even when they are in separate residues.[20]

The water-soluble cyclo(-Gly-Pro-Gly-D-Ala-Pro-) in bis-(2-ethylhexyl)sodium sulphosuccinate reversed micelles in heptane or octane adopts a conformation dominated by the counter-ion of the surfactant; the sodium ions of the surfactant effectively increase the cation concentration in the water pool so that the peptide undergoes a conformational transition to an ion-binding conformation.[21] Using energy-minimization and harmonic-analysis techniques, a number of theoretically stable vacuum configurations of cyclo(-Gly-Pro-Gly-D-Ala-Pro-)

[16] U. Schmidt, A. Lieberknecht, H. Griesser, and F. Bartkowiak, *Angew. Chem., Int. Ed. Engl.*, 1984, **23**, 318.
[17] R. Jacquier and J. Verducci, *Tetrahedron Lett.*, 1984, **25**, 2775.
[18] J. M. Berman and M. Goodman, *Int. J. Pept. Protein Res.*, 1984, **23**, 610.
[19] I. L. Karle, *Int. J. Pept. Protein Res.*, 1984, **23**, 32.
[20] M. H. Frey and S. J. Opella, *J. Am. Chem. Soc.*, 1984, **106**, 4942.
[21] K. F. Thompson and L. M. Gierasch, *J. Am. Chem. Soc.*, 1984, **106**, 3648.

have been found, one of which is similar to the observed crystal conformation. Stable complexes with lithium occur when the ion binds to the carbonyl oxygen atoms of three residues.[22] Application of the linked-atom least-squares technique to malformin A, cyclo(-D-Cys-D-Cys-Val-D-Leu-Ile-), suggests that a variety of conformational states are accessible, including one in which the peptide bond across the intramolecular disulphide bridge has a *cis* configuration. On the basis of hydrogen-bonding interactions, a model has been proposed to explain how the peptide is inactivated specifically by hydroxyproline.[23]

The structures of four cyclohexapeptides isolated from *Rubia cordifolia* have been announced (11). They have antineoplastic activity against a number of tumours.[24] Cyclo(-*N*-MeTyr-*N*-MeTyr-D-Ala-Ala-*O*,*N*-diMeTyr-Ala-), a compound related to the antitumour agent deoxybouvardin, has been synthesized. Its lack of biological activity shows that the ether link between the aromatic rings of the adjacent tyrosine residues is essential. Attempts to couple oxidatively these rings in the synthetic analogue were unsuccessful.[25] Substitution

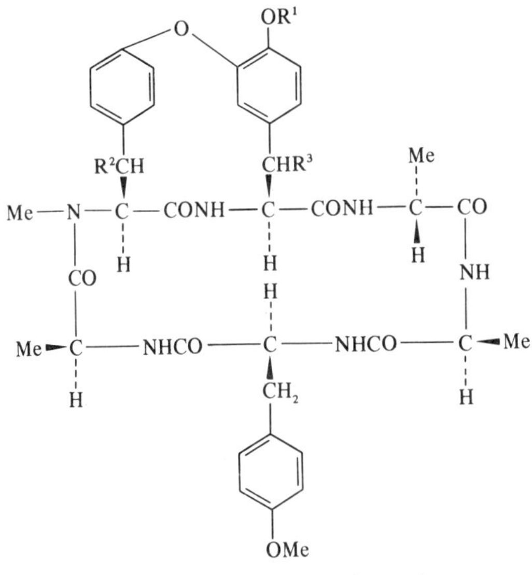

	R^1	R^2	R^3
RA 7	Me	H	H
RA 5	H	H	H
RA 4	Me	H	OH
RA 3	Me	OH	H

(11)

[22] T. E. Lynn and J. N. Kushick, *Int. J. Pept. Protein Res.*, 1984, **23**, 601.
[23] A. K. Mitra and R. Chandrasekaran, *Biopolymers*, 1984, **23**, 2513.
[24] H. Itokawa, K. Takeya, N. Mori, T. Hamanaka, T. Sonabe, and K. Mihara, *Chem. Pharm. Bull.*, 1984, **32**, 284.
[25] R. B. Bates, S. L. Gin, M. A. Hassen, V. J. Hruby, K. D. Janda, G. R. Kriek, J.-P. Michaud, and D. B. Vine, *Heterocycles*, 1984, **22**, 785.

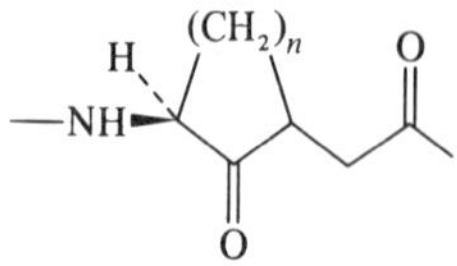

(12) $n = 2$ or 3

of proline in the somatostatin analogue cyclo(-Pro-Phe-D-Trp-Lys-Thr-Phe-) by *N*-methylalanine or sarcosine gives highly active compounds, but introduction of the lactam dipeptide unit (12) in place of Phe-Pro decreased potency.[26] Replacement of phenylalanine by *Z*-dehydrophenylalanine also decreases the potency for inhibition of growth-hormone release *in vitro* to one tenth, although n.m.r. indicates comparable backbone conformations for the two molecules.[27]

The crystal structure of cyclo(-Pro_2-Gly-Pro-Leu-Gly-)·MeOH·H_2O shows a β-turn in one half of the molecule and two *cis* peptide bonds in the other half. The paucity of hydrophilic atoms pointing towards the centre of the molecule suggests that it would lack the ability to act as an ionophore if no severe deformation is possible.[28] A second crystal form of cyclo(-Pro-Val-Phe_2-Ala-Gly-)·H_2O·3EtOH has been characterized in which the peptide-ring conformation is almost identical to the $4H_2O$ form, retaining the type II′ β-bend, an unusual feature for an L-L sequence.[29] Two molecules of cyclo(-Phe-Pro-D-Ala-)$_2$ of nearly identical conformation are found in the asymmetric unit of a monoclinic cell, each having two *cis* Phe-Pro bonds. There are no intramolecular hydrogen bonds or β-turns, and n.m.r. evidence supports a similar conformation in solution.[30] This cyclopeptide, together with cyclo(-Gly-Pro-D-Ala-)$_2$ and cyclo(-Val-Pro-Gly-)$_2$, has been studied in the crystalline powder form by ^{13}C m.a.s. n.m.r. A comparison of chemical-shift differences between the β- and γ-carbons of the proline ring suggests *cis*-Val-Pro and -Phe-Pro peptide bonds but a *trans*-Gly-Pro bond.[31]

The catalysis of hydrolysis of *p*-nitrophenyl acetates and hexanoates by cyclo(-Xxx-D-Leu-His-)$_2$ [Xxx = Cys(*S*-Acm), Ser(OBzl), Ser, or Ala] showed activities lower than that found for imidazole.[32]

Larger Cyclic Peptides. – Cyanoginosin-LA, previously called toxin BE-4 and the component of *Microcystis aeruginosa* responsible for the algal poisoning of livestock in South Africa and Australia, has been finally identified as the

[26] R. M. Friedinger, D. S. Perlow, W. C. Randall, R. Saperstein, B. H. Arison, and D. F. Veber, *Int. J. Pept. Protein Res.*, 1984, **23**, 142.
[27] S. F. Brady, D. W. Cochran, R. F. Nutt, F. W. Holly, C. D. Bennett, W. J. Paleveda, P. E. Curley, B. H. Arison, R. Saperstein, and D. F. Veber, *Int. J. Pept. Protein Res.*, 1984, **23**, 212.
[28] T. Nakashima, T. Yanane, I. Tanaka, and T. Ashida, *Acta Crystallogr., Sect. C*, 1984, **40**, 171.
[29] I. L. Karle and C. C. Chiang, *Acta Crystallogr., Sect. C*, 1984, **40**, 1381.
[30] G. Kartha, K. L. Bhandary, K. D. Kopple, A. Go, and P. P. Zhu, *J. Am. Chem. Soc.*, 1984, **106**, 3844.
[31] S. Sarkar, D. A. Torchia, K. D. Kopple, and D. L. Vanderhart, *J. Am. Chem. Soc.*, 1984, **106**, 3328.
[32] M. Kodaka, *Bull. Chem. Soc. Jpn.*, 1983, **56**, 3857.

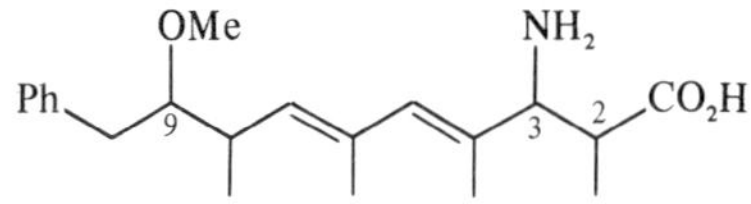

(13)

a; cyclo (-Ala-D-Thr-D-Ser-4-*cis*-Hyp-Ala-2-Methio-Trp-Leu-)
b; ,, (,, ,, ,, ,, ,, ,, -2-MeSO$_2$-Trp- ,,)
c; ,, (-Val- ,, ,, ,, ,, ,, -2-Methio-Trp- ,,)
d; ,, (-Val- ,, ,, ,, ,, ,, -2-MeSO$_2$-Trp- ,,)

(14)

cyclic hexapeptide cyclo(-D-Ala-Leu-*erythro*-β-Me-D-isoAsp-Ala-Adda-D-Glu-*N*-Me-ΔAla-). Adda is the novel β-amino acid (13); the stereochemistry at positions 2, 3, and 9 has not been established.[33]

Four analogues (14) of the virotoxins, cyclic heptapeptides from the fungus *Amanita virosa*, have been synthesized. Only poor yields of (14c) and (14d) were obtained using mixed-anhydride cyclizations with the γ-hydroxyleucine residue at the C-terminus. Two of the compounds, (14b) and (14c), bind to rabbit muscle F-actin, although less strongly than do the natural virotoxins; (14a) and (14d) were not tested.[34]

A cyclic analogue, cyclo(-Pro$_2$-Phe-Ile-Val-Arg-Gly-), of the bitter peptide BPIa isolated from a hydrolysate of casein has been synthesized. A l-succinimidyl ester cyclization gave a 10% yield of the protected macrocycle, and after deprotection the product had an extremely bitter taste. The similarity of its c.d. spectrum to that of BPIa is thought to indicate a similar molecular shape.[35] The constituents of the antifungal antibiotic iturin A_L from *Bacillus subtilis* and their relative proportions have been determined (15). Iturin A_L differs from other iturins by its high content of C_{16} acids.[36]

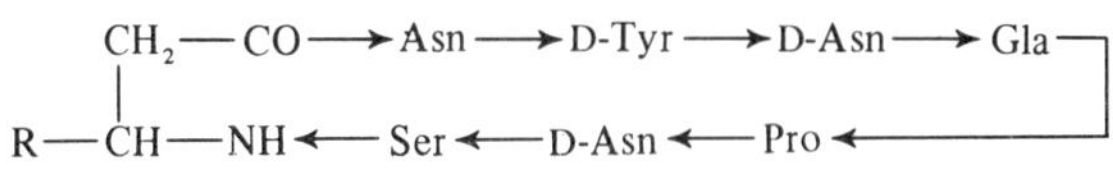

R is the side chain of:	%
3-amino-12-methyltridecanoic acid	3
3-aminotetradecanoic acid	31
3-amino-13-methyltetradecanoic acid	15
3-amino-12-methyltetradecanoic acid	9
3-amino-14-methylpentadecanoic acid	35
3-aminohexadecanoic acid	5

(15)

[33] D. P. Botes, A. A. Tuinman, P. L. Wessels, C. C. Viljoen, H. Kruger, D. H. Williams, S. Santikarn, R. J. Smith, and S. J. Hammond, *J. Chem. Soc., Perkin Trans 1*, 1984, 2311.
[34] J. U. Kahl, G. P. Vlasov, A. Seeliger, and Th. Wieland, *Int. J. Pept. Protein Res.*, 1984, **23**, 543.
[35] I. Miyake, K. Konge, H. Kanehisa, and H. Okai, *Bull. Chem. Soc. Jpn.*, 1984, **57**, 1163.
[36] H. Allgaier, G. Winkelmann, and G. Jung, *Liebigs Ann. Chem.*, 1984, 854.

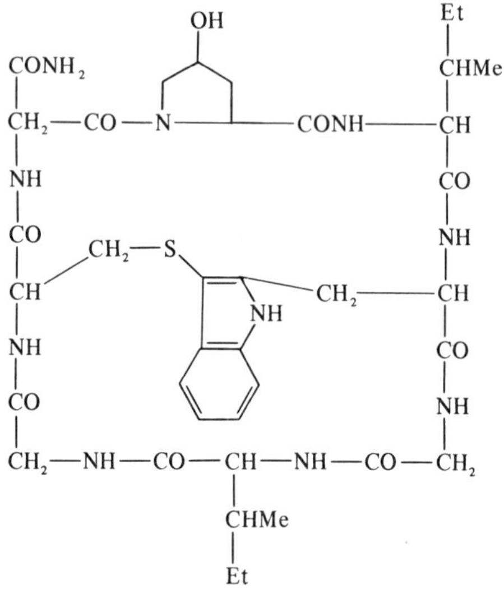

(16)

Four cyclo-octapeptides cyclo(-D- or -L-Ala-Gly-Pro-D- or -L-Phe-)$_2$ have been prepared. In DMSO, n.m.r. studies indicate that all four stereoisomers have *trans*-Gly-Pro peptide bonds and C_2 symmetry. A conformation is proposed for the L-L compound, which has two type I Pro-Phe β-turns and is similar in some ways to the backbone of β-amanitin, although the latter also contains type II turns. It did not prove possible to define closely a conformation for the D-L diastereoisomer.[37] The crystal structure of *S*-deoxo[Ile^3]amaninamide (16), a non-toxic synthetic derivative of amatoxin, also shows a compact conformation similar to that of β-amanatin (with six intramolecular hydrogen bonds), indicating that the 30-fold reduction in binding affinity to RNA polymerase B is not due to changes in the backbone conformation.[38]

When grown in an iron-limiting culture medium, sugar-beet deleterious *Pseudomonas* 7SRl produces extracellularly the yellow-green fluorescent siderophore pseudobactin 7SRl (17), whose three bidentate iron-chelating groups consist of an α-hydroxy acid, an *o*-dihydroxyaromatic residue, and a hydroxamate group. The structure is similar to that of pseudobactin from plant-growth-promoting *Pseudomonas bio*. Only the configuration of the alanine residue is known as yet.[39]

Two minor components, (18) and (19), of the cyclodecapeptide gramicidin S (GS) have been sequenced and found to contain 2-aminobutyric acid (Aba) in place of one or both valine residues of the major component.[40] [(2*R*, 3*R*)-Phe-

[37] K. D. Kopple, K. N. Parameswaran, and J. P. Yonan, *J. Am. Chem. Soc.*, 1984, **106**, 7212.
[38] G. Shohan, D. C. Rees, W. N. Lipscomb, G. Zanotti, and Th. Wieland, *J. Am. Chem. Soc.*, 1984, **106**, 4606.
[39] C.-C. Yang and J. Leong, *Biochemistry*, 1984, **23**, 3534.
[40] S. Nozaki and I. Muramatso, *J. Antibiot.*, 1984, **37**, 689.

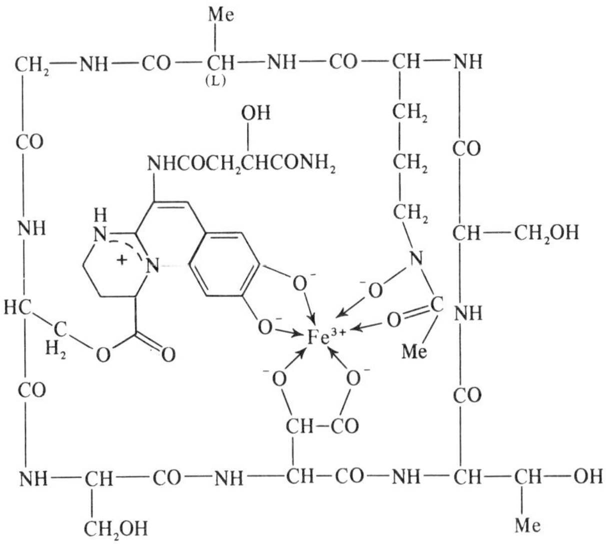

(17)

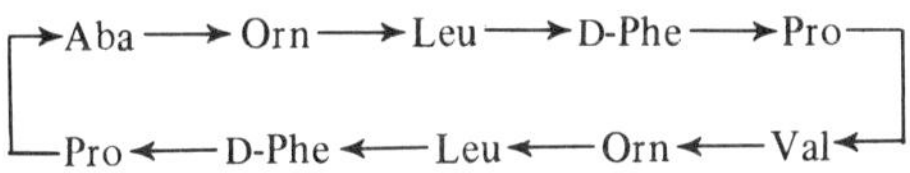

(18)

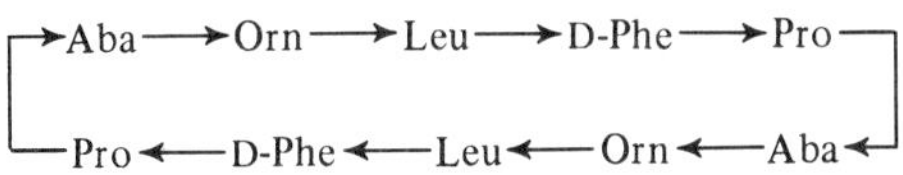

(19)

2,3-$d_2^{4,4'}$]Gramicidin S has also been prepared. Its ^{1}H n.m.r. spectrum in DMSO-d_6 shows a sharp singlet at 2.98 p.p.m. for the (3*S*)-proton of the 4- and 4′-residues, providing support for earlier proposals that among rotamers of the aromatic side chain the one with $\kappa_1 = 180°$ is predominant.[41] A ^{15}N n.m.r. study of GS in water and organic solvents has detected long-range perturbations, the effects being transmitted from one peptide linkage to another *via* intramolecular hydrogen bonds across a total of six bonds.[42] The side-chain and backbone hydrogen-bonding pattern is identical in the computed minimum-energy conformation of GS and in the recently determined *X*-ray structure of the hydrated urea complex of this peptide.[43] Normal-mode calculations have also been carried

[41] K. Tanimura, T. Kato, M. Waki, S. Lee, Y. Kodera, and N. Izumiya, *Bull. Chem. Soc. Jpn.*, 1984, **57**, 2193.
[42] D. H. Live, D. G. Davis, W. C. Agosta, and D. Cowburn, *J. Am. Chem. Soc.*, 1984, **106**, 1939.
[43] G. Nemethy and H. A. Scheraga, *Biochem. Biophys. Res. Commun.*, 1984, **118**, 643.

out on three low-energy structures of GS deduced from conformational-energy calculations. When the results on amide modes are compared with observed bands in the i.r. and Raman spectra of crystalline GS and its *N*-deuteriated derivative, one of these three is clearly disfavoured. Of the other two, the slightly favoured one corresponds to the lowest-energy structure obtained from the energy calculations.[44]

A novel interaction between GS and nucleic acids has been discovered. Complex formation between calf thymus DNA and GS was demonstrated by phase transfer to $CHCl_3$ of ultrasonically irradiated DNA. The stoicheiometry of the interaction is 2:1 (DNA:GS), which is thought to be consistent with a predominantly electrostatic mode of binding.[45] Diphthaloyl derivatives of GS have been subjected to selective *N*-methylation using MeI–Ag_2O in DMF. Alkylation occurred exclusively at Orn and D-Phe residues, giving the tetra-*N*-methyl derivative in high yield. After removal of the phthaloyl groups, the product had essentially the same antimicrobial activity as GS itself.[46] ^{1}H–^{13}C selective NOE studies of GS have given direct experimental proof of an Orn-NH_2 to Phe-CO hydrogen bond.[47]

X-Ray analysis of the crystals of the biologically active analogue [4-*cis*-Br-Pro^7]antamanide (20) shows this cyclodecapeptide to exist as a flat plate with two strong intermolecular hydrogen bonds. The phenyl groups of Phe^5 and Phe^{10} cover one side of the molecule, and on the other four water molecules are hydrogen bonded to NH groups, giving clear hydrophobic and hydrophilic faces to the macrocycle. The only remarkable conformational difference from [Phe^4, Val^6]antamanide lies in the Pro^3 and Phe^6 angles.[48]

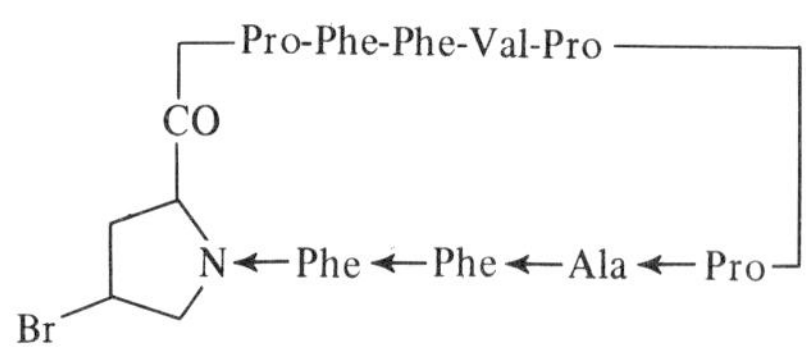

(20)

The total synthesis of cyclosporin, a therapeutically important potent immunosuppressant, has been achieved by cyclization of the linear undecapeptide (21). Although the azide method only gave 15% of cyclosporin, the pentafluorophenol/DCC, mixed-phosphonic anhydride, and BOP-reagent methods all gave 62–65% yields of the desired product. The MeBmt residue was introduced

44 V. M. Naik, S. Krimm, J. B. Denton, G. Nemethy, and H. A. Scheraga, *Int. J. Pept. Protein Res.*, 1984, **24**, 613.
45 E. M. Krauss and S. I. Chan, *Biochemistry*, 1984, **23**, 73.
46 M. Kaiwai, M. Ohya, Y. Butsugan, K. Saito, and T. Higashijima, *Chem. Lett.*, 1984, 1835.
47 N. Niccolai, C. Rossi, P. Mascagni, P. Neri, and W. A. Gibbons, *Biochem. Biophys. Res. Commun.*, 1984, **124**, 739.
48 H. Lotter, G. Rohr, and Th. Wieland, *Naturwissenschaften*, 1984, **71**, 46.

H-D-Ala-MeLeu-MeLeu-MeVal-MeBmt-Abu-Sar-MeLeu-Val-MeLeu-Ala-OH

(21) MeBmt = (4*R*)-4-[(*E*)-2-butenyl]-4,*N*-dimethyl-L-threonine

at the last possible stage in the synthesis of (21) for economic reasons. [D-MeVal^{11}]Cyclosporin was also prepared from the linear undecapeptide obtained by coupling Boc-D-Ala-MeLeu-MeLeu-MeVal-OH with the C-terminal heptapeptide ester using the pivalic anhydride method. Selective inversion of the MeVal configuration occurred during this coupling, probably by enolization.[49] Using a similar synthetic strategy to that used for cyclosporin itself, [MeThr^{1}]-, [Ser^{2}]-, [D-Pro^{3}]-, and [MeLeu^{11}]-cyclosporins have also been synthesized. Like [D-MeVal^{11}]cyclosporin, they were all less potent immunosuppressants than the parent, the most active of them being the [Ser^{2}] analogue.[50]

The synthetic cyclododecapeptide cyclo(-Val-Orn-Leu-D-Phe-Pro-D-Tyr-)$_2$ has been found to have a similar antibiotic activity to GS. This suggests that it may be identical to gratisin, isolated from *Bacillus brevis* in 1973, which has the same sequence but whose constituent amino acid configurations are unknown. Analogue cyclopeptides in which D-Phe and D-Tyr are interchanged show lower activity, but compounds containing D-Phe-D-Tyr-Pro or Pro-D-Phe-D-Phe partial sequences exhibit high activity. A c.d. spectroscopic comparison of these compounds shows that inactive analogues have more labile conformations than those with antibiotic properties.[51]

Cyclic Depsipeptides. – The chlamydocin model cyclo[-(*S*)-3-phenyl-lactoyl-D-Pro-Ala-Aib-] has been prepared in 72% yield by cyclizing (*S*)-3-phenyl-lactoyl-D-Pro-Ala-Aib-NMe$_2$ using hydrogen chloride in toluene at 100 °C. The intermediate Z-Ala-Aib-NH$_2$ was prepared by the reaction of Z-Ala-OH with 3-dimethylamino-2,2-dimethyl-2*H*-azirine.[52] Two cyclotetradepsipeptides, [L-2-amino-4-(*p*-methoxyphenyl)butanoic acid^{3}]AM toxin I and [L-2-amino-6-(*p*-methoxyphenyl) hexanoic acid^{3}]AM toxin I, have been synthesized. They only showed weak toxic activity towards apple leaves, indicating that the length of the side chain in residue three is an important factor for biological activity.[53] Cyclization of H-App-ΔAla-Ala-Hmb-ONSu at 0.3mM concentration gives a mixture of cyclic monomer (AM toxin II, 5% yield) and cyclic dimer, suggesting that the dehydroalanine residue is probably best inserted as a saturated precursor rather than preformed.[54]

Two-dimensional NOESY and COSY ^{1}H n.m.r. has been used to determine the conformation of the complex formed between actinomycin D and the hexanucleoside pentaphosphate dATGCAT. The chromophore of the drug intercalates between the GC base pairs, the pentapeptide lactone nestling in the

[49] R. M. Wenger, *Helv. Chim. Acta*, 1984, **67**, 502.
[50] R. M. Wenger, *Chimia*, 1984, **38**, 11.
[51] M. Tamaki, *Bull. Chem. Soc. Jpn.*, 1984, **57**, 3210.
[52] D. Obrecht and H. Heimgartner, *Helv. Chim. Acta*, 1984, **67**, 526.
[53] H. Miyara, H. Aoyagi, S. Lee, M. Waki, T. Kato, and N. Izumiya, *Int. J. Pept. Protein Res.*, 1984, **23**, 447.
[54] T. Kozono, H. Mihara, H. Aoyagi, T. Kato, and N. Izumiya, *Int. J. Pept. Protein Res.*, 1984, **24**, 402.

HCO-D-Ala-Phe-D-Pro-D-Leu^{t}-Leu^{t}-D-Trp-Arg-D-Cys(O_3H)-Thr-MeGln-D-Leu

Sar ← Thr ← Asn

(22)

minor groove. No significant conformational change of the pentapeptide lactone seems to take place on binding.[55]

The structure of an antimicrobial peptide, discodermin A, from the marine sponge *Discodermia kiiensis* has been elucidated. The cyclohexadepsipeptide ring (22) has an octapeptide tail attached, and contains two residues of the unusual t-leucine. Oxidation of discodermin A with DCC/DMSO followed by 6M HCl hydrolysis and amino acid analysis established that the hydroxyl group of one threonine residue was esterified but that the other one was free. This is the first bioactive peptide isolated from a sponge.[56] The toxic metabolite roseotoxin B from *Trichothecium roseum*, a fungus frequently found growing in improperly stored foodstuffs, has been finally assigned the cyclohexadepsipeptide structure (23), which corresponds to destruxin A except that the valine residue of the latter is *N*-methylated.[57] *X*-Ray-diffraction data support the proposed sequence and show the two β-turns and the two cross-ring hydrogen bonds depicted in (23). Molecular-mechanics calculations indicate that the transannular hydrogen bonds do not significantly rigidify the backbone conformations but are instead a consequence of them.[58]

The acid chloride method of depsipeptide bond synthesis has been used to prepare the hexadepsipeptide derivative (24), which was cyclized to give cyclo(-Val-Lac-)$_3$ (25) in 21% yield (Scheme 3). The 400 MHz ^{1}H n.m.r. spectrum of this product exhibits one signal only each for the amide protons, the methine protons of the lactic acid residues, those of the valine residue, and the side-chain protons. It is, therefore, concluded that the conformation has C_3 symmetry and

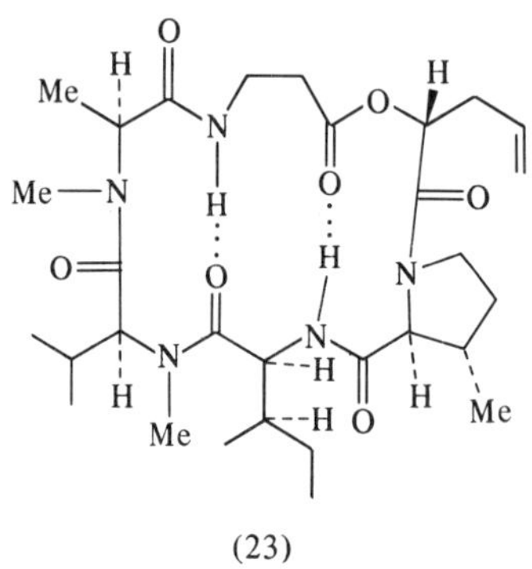

(23)

[55] S. C. Brown, K. Mullis, C. Levenson, and R. H. Shafer, *Biochemistry*, 1984, **23**, 403.

[56] S. Matsunaga, N. Fusetani, and S. Konosu, *Tetrahedron Lett.*, 1984, **25**, 5165.

[57] J. P. Springer, R. J. Cole, J. W. Dorner, R. H. Cox, J. L. Richard, C. L. Barnes, and D. van der Helm, *J. Am. Chem. Soc.*, 1984, **106**, 2388.

[58] J. P. Snyder, *J. Am. Chem. Soc.*, 1984, **106**, 2393.

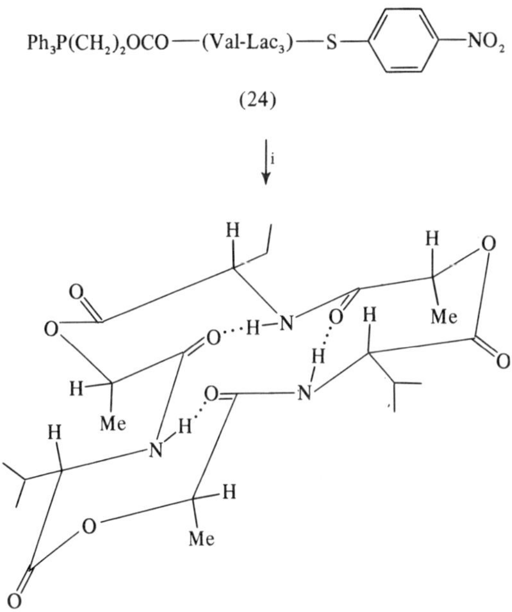

(24) → (25)

Reagent: i, NEt_3, PhSH

Scheme 3

is as shown in (25). In this conformation the side chain of each of the lactic acid residues occupies an 'axial' position, which would account for the difficulties generally experienced in cyclizing depsipeptides containing residues all of the same configuration.[59]

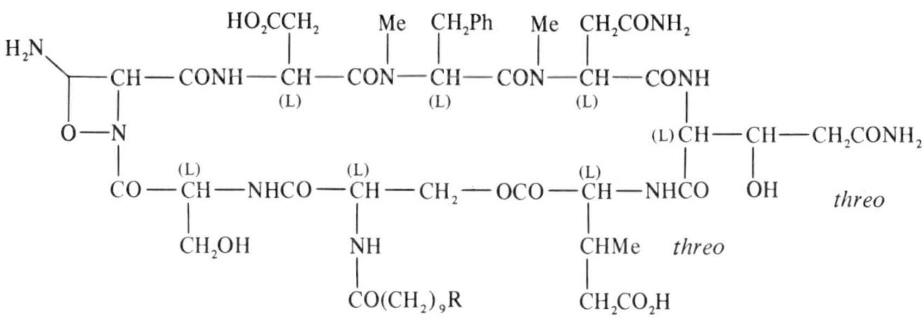

(26) neopeptin A; R = CH_2CHMe_2
neopeptin B; R = $CH_2CHMeEt$
neopeptin C; R = $CHMe_2$

[59] H. Kunz and H.-G. Lerchen, *Angew. Chem., Int. Ed. Engl.*, 1984, **23**, 808.

The structures of three closely related antifungal antibiotics neopeptins A, B, and C (26) from cultures of a *Streptomyces* species resemble the lipopeptins, which were reported in 1980. Treatment of the three components with polymyxin acylase gives the same peptidic product in each case, showing that they only differ in the fatty-acid side chain. A labile basic amino acid derived from the 3-amino-2-oxazetidine-4-carboxylic acid residue could not be isolated after acid hydrolysis; the presence of this residue was, therefore, established by m.s. and ^{1}H n.m.r. evidence.[60] The previously assumed molecular mass of stenothricin has been shown by f.a.b.-m.s. to be too low. Investigation of a partial hydrolysate revealed the presence of α-ketobutyrylvaline, indicating the presence of an α,β-dehydrobutyric acid (Δ-Abu) residue. Fragments obtained by partial hydrolysis after catalytic hydrogenation reveal Δ-Abu to link valine and α,β-diaminopropionic acid (Dpr) residues; the structure must, therefore, be revised to (27).[61]

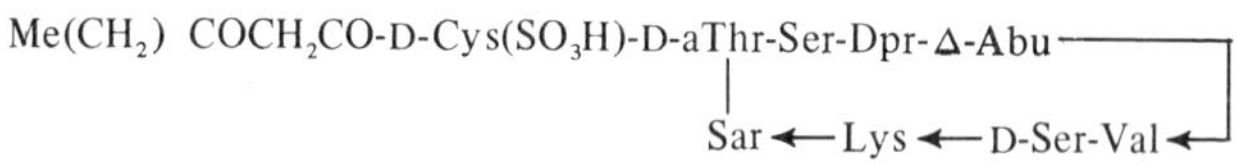

(27)

A second synthesis of triostin A (29) has been reported (Scheme 4), which avoids the racemization of an alanine residue experienced in the original synthesis. N.m.r. spectroscopy shows that *S,S′*-dibenzyldihydrotriostin A (28), like triostin A itself, exists as a mixture of two conformers in solution; this excludes the possibility that the equilibrium in triostin A is a consequence of reversed chirality of the disulphide bond.[62] Nortriostin A, which lacks the *N*-methyl groups present on the cysteine and valine residues of triostin A, has been prepared by a similar route. It lacks antimicrobial activity and exists as a single structure in solution, suggesting that the conformer equilibrium of triostin A is a consequence of the presence of *N*-methyl peptide bonds.[63] *X*-Ray work on triostin A shows a conformation with approximately two-fold symmetry which accounts for the higher binding constants and capacity to bind at GC-rich DNA sequences compared to the des-*N*-methyl analogue.[64] The *X*-ray structure of a triostin A–DNA fragment complex (a double-helical CTGACG sequence) shows the two planar quinoxaline rings to intercalate on the minor groove of the double helix surrounding the CG base pairs at either end. The alanine residues form hydrogen bonds to the guanines, and base stacking in the DNA is perturbed. Overall, the DNA changes shape to accommodate the interaction rather than the peptide. No water molecules intervene in the bonding interactions.[65]

60 M. Ubukata, M. Uramoto, and K. Isono, *Tetrahedron Lett.*, 1984, **25**, 423.
61 M. Rinken, W. D. Lehmann, and W. A. Konig, *Liebigs Ann. Chem.*, 1984, 1672.
62 M. Shin, K. Inouye, and H. Otsuka, *Bull. Chem. Soc. Jpn.*, 1984, **57**, 2203.
63 M. Shin, K. Inouye, N. Higuchi, and Y. Kyogoku, *Bull. Chem. Soc. Jpn.*, 1984, **57**, 2211.
64 G. M. Sheldrick, J. J. Guy, O. Kennard, V. Rivera, and M. J. Waring, *J. Chem. Soc., Perkin Trans. 2*, 1984, 1601.
65 A. H.-J. Wang, G. Ughetto, G. J. Quigley, T. Hakoshima, G. A. van der March, J. H. van Boom, and A. Rich, *Science*, 1984, **225**, 1115.

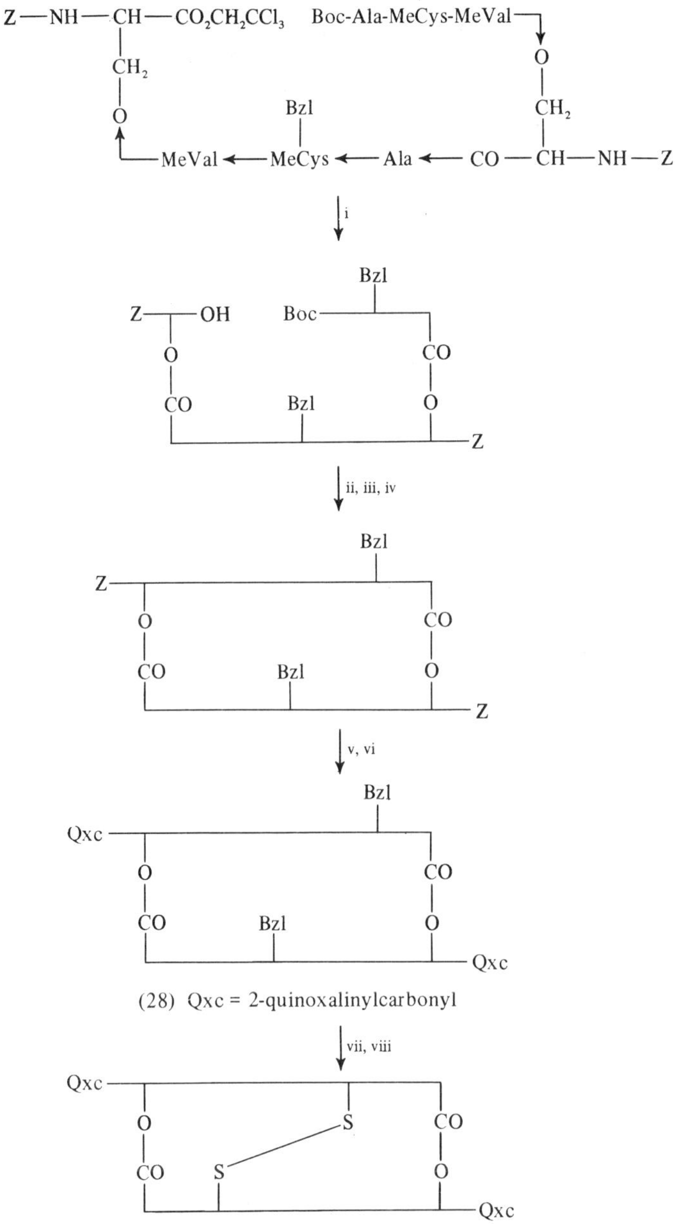

(28) Qxc = 2-quinoxalinylcarbonyl

(29)

Reagents: i, Zn/HOAc; ii, HOSu/DCC; iii, CF_3CO_2H; iv, DIEA; v, CF_3CO_2H/anisole; vi, Qxc-Cl/DIEA; vii, HF; viii, I_2

Scheme 4

Empedopeptin, a new amphoteric water-soluble antibiotic from *Empedobacter holvabium*, has proved to be a cyclononadepsipeptide (30) containing three unusual amino acids. Its water solubility can be related to its hydroxyl content, and although structurally unrelated it has a similar spectrum of antimicrobial activity to vancomycin.[66] The deep-water variety of the blue-green alga *Lyngba majuscula* from the Marshall Islands in the Pacific produces the decadepsipeptide majusculamide C (31); the stereochemistry of the novel constituents 3-amino-2-methylpentanoic acid and 4-amino-2,2-dimethyl-3-oxopentanoic acid has not yet been established. Majusculamide C controls the growth of a number of fungal plant pathogens.[67] Another new cyclodecadepsipeptide is BMY-28260 (32), a broad-spectrum antibiotic from *Bacillus circulans*.

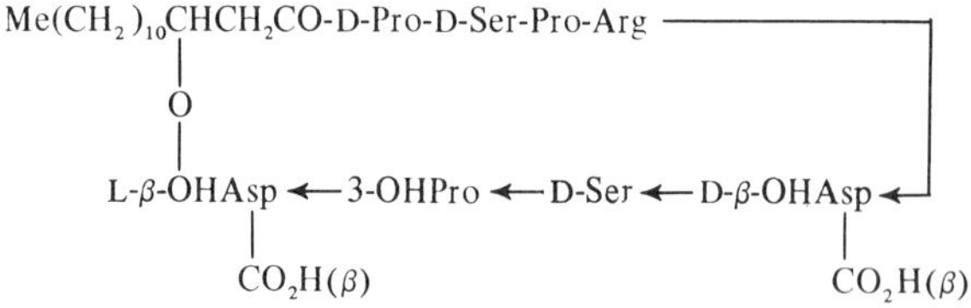

(30)

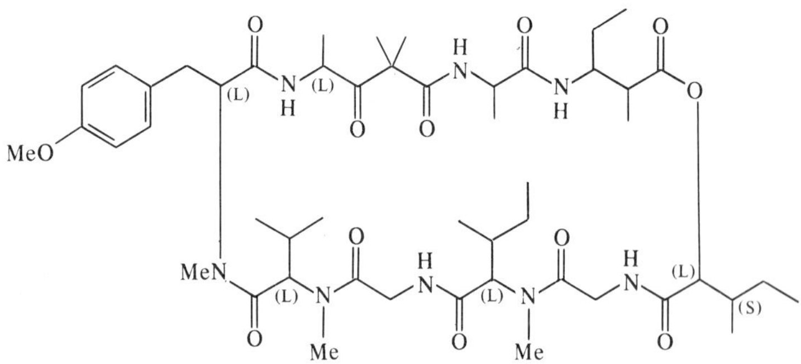

(31)

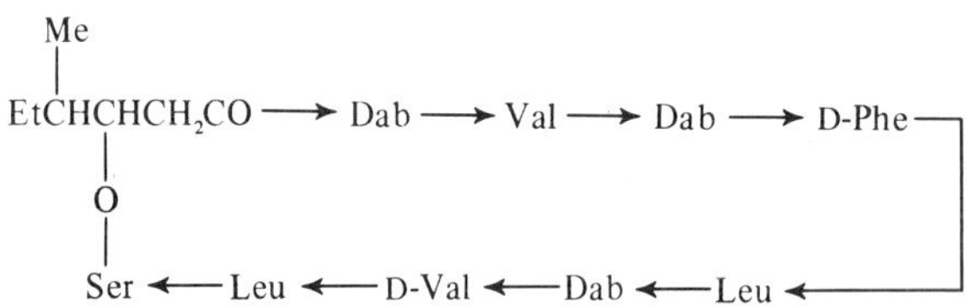

(32)

[66] M. Konishi, K. Sugawara, M. Hanada, K. Tomita, K. Tomatsu, T. Miyaki, and H. Kawaguchi, *J. Antibiot.*, 1984, **37**, 949; K. Sugawara, K. Numata, M. Konishi, and H. Kawaguchi, *ibid.*, p. 958.

[67] D. C. Carter, R. E. Moore, J. S. Mynderse, W. P. Niemczwa, and J. S. Todd, *J. Org. Chem.*, 1984, **49**, 236.

It is closely related to permetin A in structure, differing only in the replacement of L-isoleucine by L-valine.[68]

X-Ray analysis and n.m.r. solution studies of the barium perchlorate complex of valinomycin indicate that in both phases the ionophore is a flat open structure with two barium atoms per molecule and no internal hydrogen bonds. The crystal contains infinite layers of valinomycin molecules interconnected by perchlorate groups, successive layers also sandwiching three solvent molecules.[69]

Other Cyclic Peptides with Modified Backbones. – Cyclization of β-Ala-Phe-Pro-ONp gives 30% of the cyclic tripeptide, but Oab-Phe-Pro-ONp (Oab = *O*-aminobenzoyl) gives 33% of the acylamidine (33). Z-Oab-Phe-Pro-ONp gives as the major product *N*-Z-Oab-cyclo(-Phe-Pro-), while *N*-Me-Oab-Phe-Pro-ONp gives a mixture of the azacyclol (34) (33%) and cyclic tripeptide (17%).[70] Deprotection with tri-t-butylphosphine in an aqueous medium of 2-t-butyldithiopropionyl-Phe-Pro-ONp gives the stable thiacyclol (35), for which an *X*-ray crystallographic analysis is reported. The thiol protective group was chosen to be selectively removable under mild non-oxidative conditions so as to avoid any involvement of the activated carboxy function and of the cyclization products.[71]

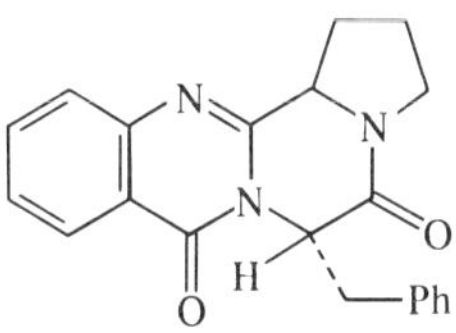

(33)

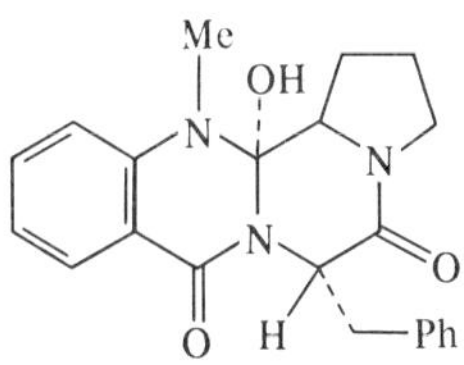

(34)

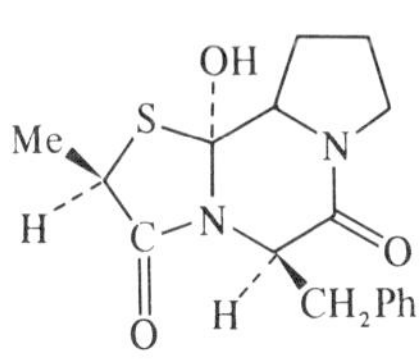

(35)

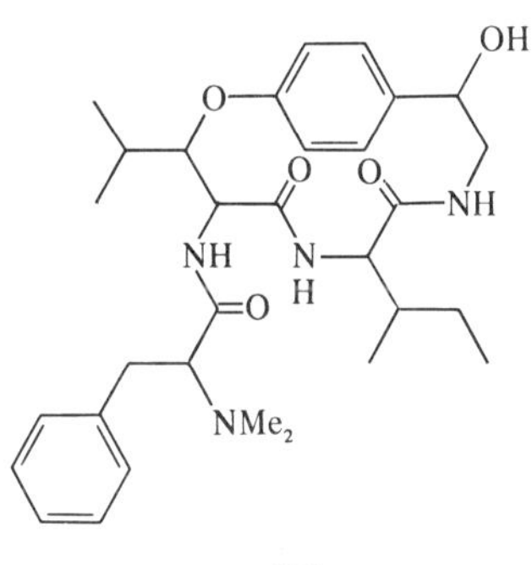

(36)

[68] K. Sugawara, M. Konishi, and H. Kawaguchi, *J. Antibiot.*, 1984, **37**, 1257.

[69] S. Devarajan, M. Vijayan, and K. R. K. Easwaran, *Int. J. Pept. Protein Res.*, 1984, **23**, 324.

[70] F. Pinnen, G. Zanotti, and G. Lucente, *Tetrahedron Lett.*, 1984, **25**, 5201.

[71] G. Zanotti, F. Pinnen, G. Lucente, S. Cerrini, W. Fedeli, and F. Mazza, *J. Chem. Soc., Perkin Trans. 1*, 1984, 1153.

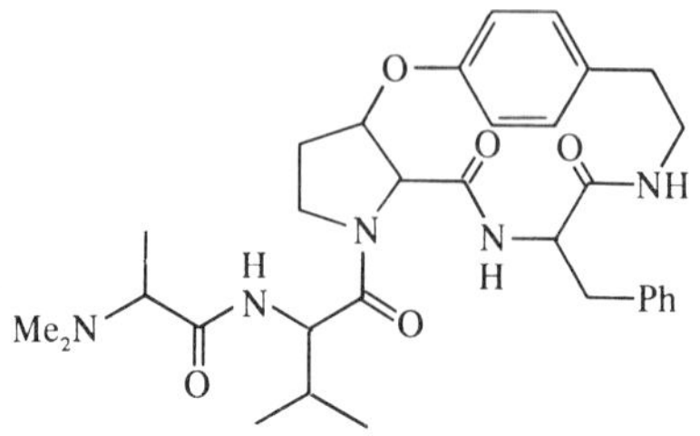

(37)

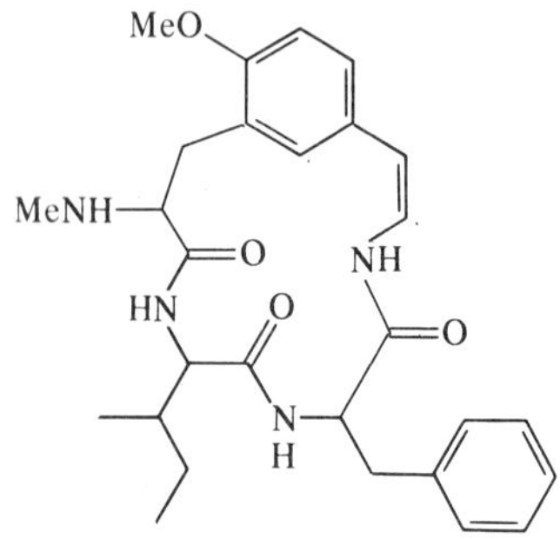

(38)

A new peptide alkaloid, discarine G (36), has been isolated from the bark extract of *Discaria febrifuga*,[72] and the synthesis of dihydromauritine A (37) has been achieved. In the ring-closure step (Scheme 5), use of diphenylphosphoro-

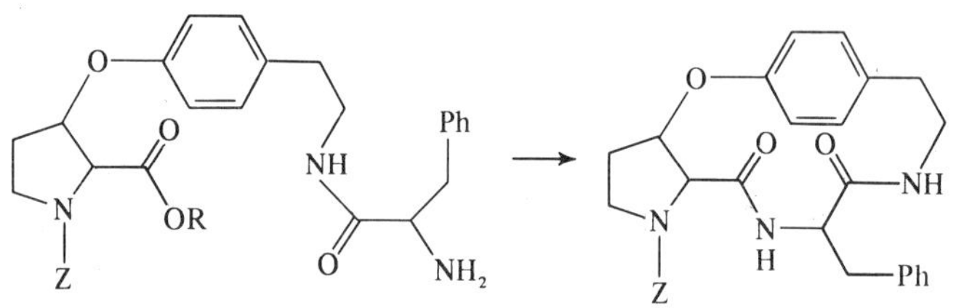

Scheme 5

azidate or DCC/HOBt, even in dilute solution, gave only polymeric products. However, a *p*-nitrophenyl ester cyclization of the trifluoroacetate salt in pyridine at 90 °C gave 8% of cyclic monomer.[73] The total synthesis of mucronin B (38) has also been reported. In this case the cyclization was effected using a pentafluorophenyl ester, a method proved effective in several earlier syntheses of peptide alkaloids.[74]

A new minor fusarinine-type siderophore-like compound has been obtained from *Neurospora crassa*. Named ferric neurosporin, this compound is the ferric chelate of the cyclic triester of N^{α}-acetyl-N^{δ}-hydroxy-N^{δ}-[(*R*)-3-hydroxybutyryl]-D-ornithine. *X*-Ray analysis shows the molecule to be flat, with a Λ-*cis* absolute configuration about the central ferric ion. Conformationally, the N^{α}-acetylornithoyl groups are similar to those in ferric *N*,*N'*,*N''*-triacetylfusarinine.[75] Using molecular mechanics, an extensive low-energy conforma-

[72] R. Herzog, A. Marel, J. Biermann, and W. Voelter, *Hoppe-Seyler's Z. Physiol. Chem.*, 1984, **365**, 1351.
[73] R. F. Nutt, K.-M. Chen, and M. M. Joullié, *J. Org. Chem.*, 1984, **49**, 1013.
[74] U. Schmidt and U. Schanbacher, *Liebigs Ann. Chem.*, 1984, 1205.
[75] D. L. Eng-Wilmot, A. Rahman, J. V. Mendenhall, S. L. Grayson, and D. van der Helm, *J. Am. Chem. Soc.*, 1984, **106**, 1285.

tional region has been calculated for cyclo(-N^{γ}-Mabu-Gly-Ala-Ala-) [Mabu = (R)-α-methyl-γ-aminobutyric acid]. Assuming that enkephalin in its active form shares conformational features with this cyclic peptide, it is deduced that enkephalin contains a Gly^3-Phe^4 type II′ β-bend.[76]

A bicyclic vasopressin analogue (39) has been prepared; it acts as a more selective antagonist of the antidiuretic activity of vasopressin. The Asp-Lys bond was formed as a last step in 22% yield using diphenylphosphoryl azide (DPPA) in dilute DMF solution.[77] Four carba analogues of oxytocin (40) have been synthesized by the solid-phase method. All are more active than oxytocin in *in vitro* uterotonic tests. The cyclic MSH analogue (41) has been prepared by a similar route, and has proved highly active in the frog and lizard skin bioassay systems.[78] Solid-phase synthesis and DPPA ring closure have also been used in the preparation of the cyclic pseudo-hexapeptide analogue of somatostatin cyclo(-Proψ[CH_2S]Phe-D-Trp-Lys-Thr-Phe-). The crystalline product had ~6% of the growth-hormone inhibitory activity of the all-amide cyclic hexapeptide parent in spite of the absence of one of the two postulated intramolecular hydrogen bonds.[79]

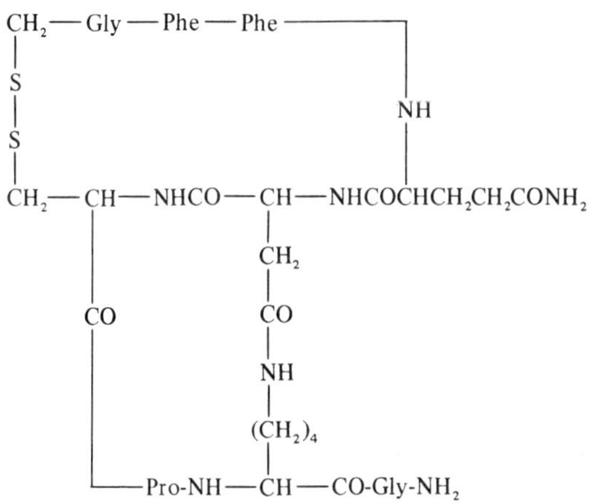

(39)

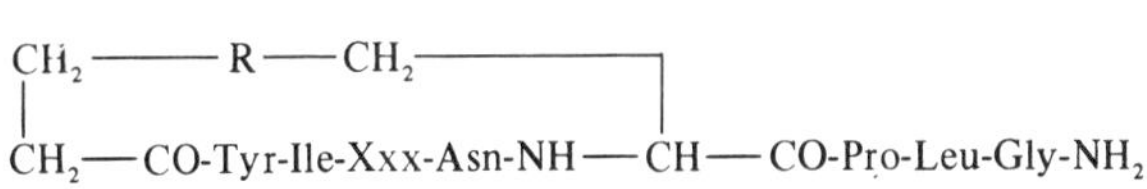

(40) Xxx = Gln or Thr
R = —CH_2S— or —SCH_2—

[76] D. Hall and N. Pavitt, *Biopolymers*, 1984, **23**, 1441.
[77] G. Skala, C. W. Smith, C. J. Taylor, and J. H. Ludens, *Science*, 1984, **226**, 443.
[78] M. Leble and V. J. Hruby, *Tetrahedron Lett.*, 1984, **25**, 2067.
[79] T. W. Gero, A. F. Spatola, I. Terres-Aleman, and A. V. Schally, *Biochem. Biophys. Res. Commun.*, 1984, **120**, 840.

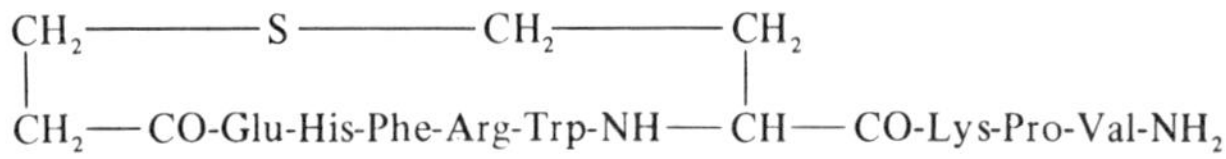

(41)

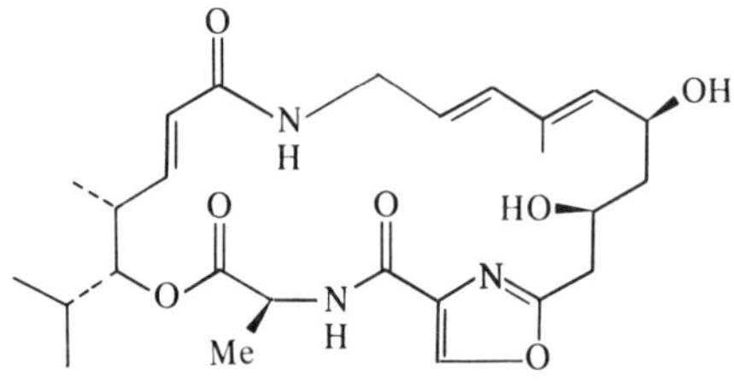

(42)

The biosynthesis of the antibiotic A2315A (42) of the virginiamycin family has been studied by stable-isotope techniques. The basic skeleton proves to be constructed from seven acetate units, valine, glycine, alanine, serine, and a methyl group from methionine. The uncommon D-alanine unit arises from both D- and L-alanine with equal facility.[80] The cyclic nonapeptide N^{α}-Arg-cyclo[-N^{ϵ}-Lys^{1},Gly^{6}-]bradykinin and the cycle decapeptide cyclo(-ϵ-kallidin-) have been prepared; both compounds exhibited a prolonged hypotensive activity.[81]

3 Modified Linear Peptides

Enzyme Inhibitors. – Details of the synthesis of the angiotensin-converting-enzyme (ACE) inhibitor enalapril maleate (43), first reported in 1980, have been published. Work on the diacid corresponding to enalapril, which is clinically effective in the treatment of hypertension and congestive heart failure, has confirmed the *S,S,S*-chirality and has shown that the amide bond is *trans*.[82]

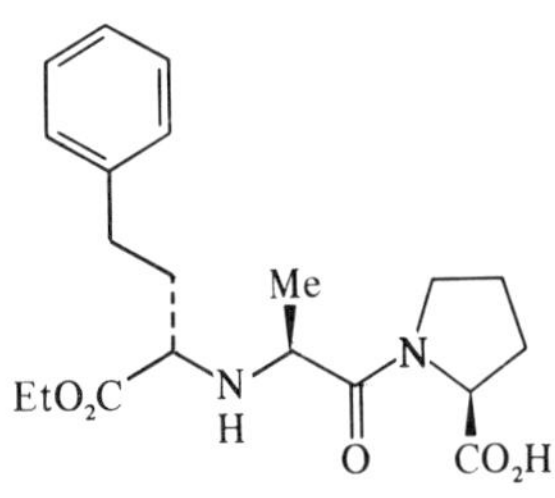

(43)

[80] J. W. Le Fevre and D. G. I. Kingston, *J. Org. Chem.*, 1984, **49**, 2588.

[81] F. K. Mutulis, G. I. Chipens, O. E. Lando, and I. E. Mutule, *Int. J. Pept. Protein Res.*, 1984, **23**, 235.

[82] M. J. Wyvratt, E. W. Tristram, T. J. Ikeler, N. S. Lohr, H. Joshua, J. P. Springer, B. H. Arison, and A. A. Patchett, *J. Org. Chem.*, 1984, **49**, 2816.

Analogues of enalapril that are α-aza substituted show striking alterations in conformational and acid–base properties, but in spite of this they have inhibitor potencies comparable to that of captopril. The most active compound found was (44).[83] Another series of analogues incorporating a ketone substitution into the peptide backbone (45) were most active when R is the side chain of glycine, alanine, or ornithine, although the glycine compound was earlier reported by another group to be of low activity.[84] An extensive series of such ketomethylene ACE inhibitors have been reviewed, and discussed particularly with respect to possible modes of inhibitor binding.[85]

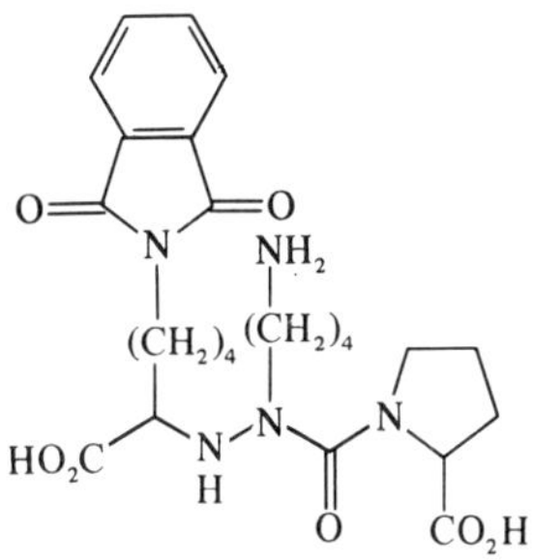

(44)

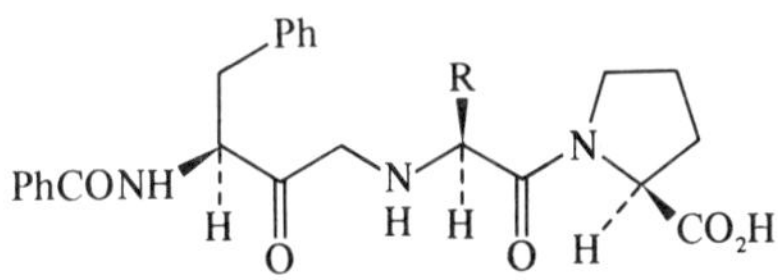

(45)

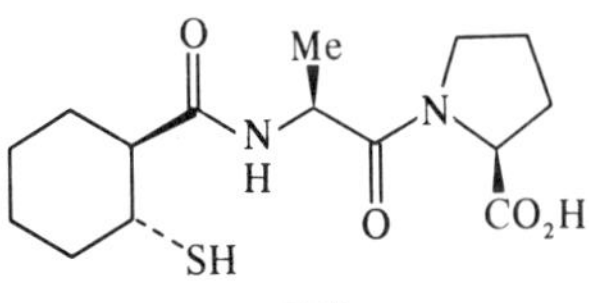

(46)

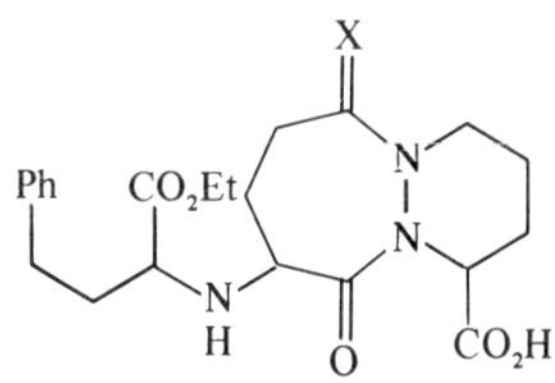

(47) a; X = CH_2
b; X = O

Conformationally constrained ACE inhibitors have been explored by several groups of workers. The mercaptoacyl dipeptide (46) has an *in vitro* potency nearly 10 times that of captopril,[86] and of a series based on a new 7,6-bicyclic system (47) the two most potent compounds (47a and b) have been investigated in animal studies. The former is better absorbed, but both have long-acting

[83] W. J. Greenlee, E. D. Thorsett, J. P. Springer, A. A. Patchett, E. H. Ulm, and T. C. Vassil, *Biochem. Biophys. Res. Commun.*, 1984, **122**, 791.

[84] S. Natarajan, E. M. Gordon, E. F. Sabo, J. D. Godfrey, H. N. Weller, J. Pluščec, M. B. Rom, and D. W. Cushman, *Biochem. Biophys. Res. Commun.*, 1984, **124**, 141.

[85] E. M. Gardan, S. Natarajan, J. Pluščec, H. N. Weller, J. D. Godfrey, M. B. Rom, E. F. Sabo, J. Englebrecht, and D. W. Cushman, *Biochem. Biophys. Res. Commun.*, 1984, **124**, 148.

[86] H. N. Weller, E. M. Gordon, M. B. Rom, and J. Plušček, *Biochem. Biophys. Res. Commun.*, 1984, **125**, 82.

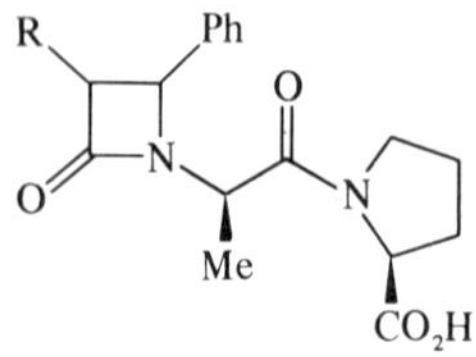

(48) R = PhO, $PhCH_2O$, HO, Cl, or Br

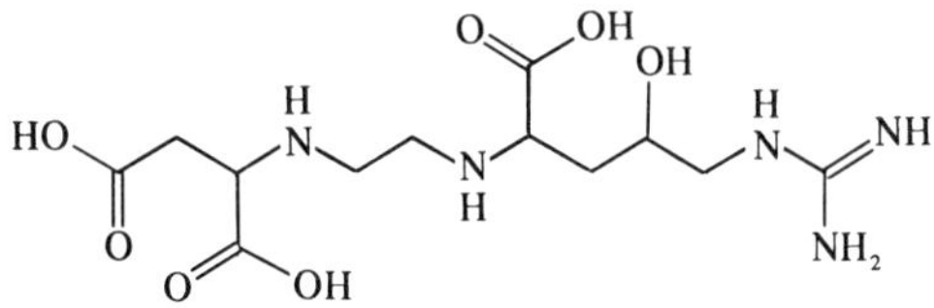

(49)

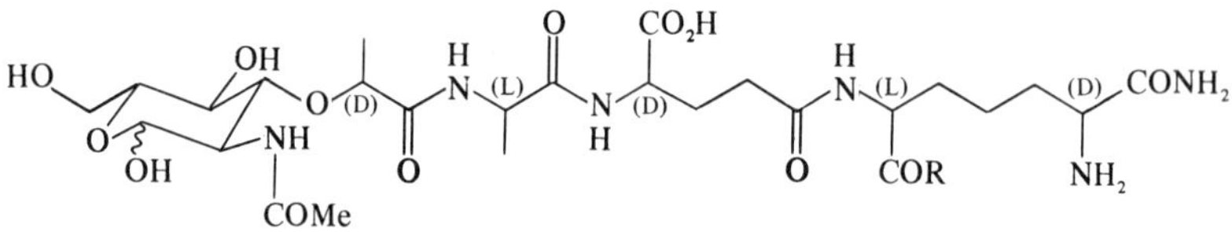

(50) muracein A; R = OH

muracein B; R = 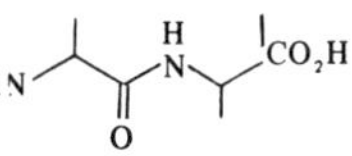

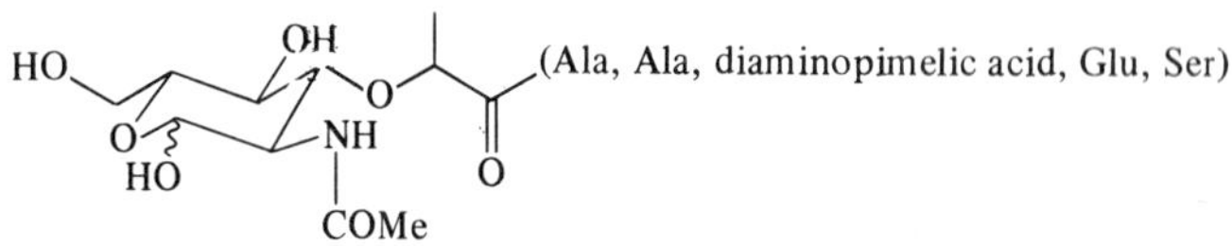

(51)

activity in the rat for inhibition of plasma ACE when administered orally.[87] A series of β-lactam analogues (48), however, have been prepared and found to have little inhibitory potency.[88]

The novel ACE inhibitor (49) has been isolated from *Streptomyces* sp.; the structure is similar to that of marasmine, just lacking one carboxyl group found in the latter.[89] Three new muramyl peptides, muraceins A, B (50), and C, have been extracted from *Nocardia orientalis* and found to be ACE inhibitors also.

[87] M. R. Attwood, R. J. Francis, C. H. Hassall, A. Krohn, G. Lawton, I. L. Natoff, J. S. Nixon, S. Redshaw, and W. A. Thomas, *FEBS Lett.*, 1984, **165**, 201.

[88] C. J. Wharton, R. Wrigglesworth, and M. Rowe, *J. Chem. Soc., Perkin 1*, 1984, 29.

[89] L. Huang, G. Rowin, J. Dunn, R. Sykes, R. Dobna, B. A. Mayles, D. M. Cross, and R. W. Burg, *J. Antibiot.*, 1984, **37**, 462.

The amino acid sequence of muracein C is as yet undetermined, but it contains two alanine residues and one residue each of the other amino acids shown in (51).[90]

Acetyl-Leu-arginal, a specific inhibitor of the enzyme dipeptidyl aminopeptidase III, has been found in the culture filtrate of the bacterium BMG 520-yFz. It is identical in structure with a known leupeptin analogue.[91] A series of proteinase-inhibitor analogues of structure Z-Arg-Xxx-phenylalaninal (Xxx = Leu, Ile, or Val) have been prepared using semicarbazone protection for the aldehyde function. They showed a strong activity towards chymotrypsin; the semicarbazones and dipeptide aldehydes were much less active.[92] The synthetic peptide Suc-Tyr-D-Leu-D-Val-*p*-nitroanilide is an effective and specific inhibitor of human spleen fibrinolytic proteinase (SFP) and human leukocyte elastase-like proteinase (ELP). The tripeptide derivatives Suc-Tyr-Leu-Val-Pipe and Suc-Tyr-D-Leu-D-Val-Pipe (Pipe = 4-methylpiperidine) inhibit SFP very slightly and have no effect on ELP, but both Dan-Tyr-Val-Leu-*p*-nitroanilide (Dan = dansyl) and its LDD-diastereoisomer are SFP and ELP inhibitors.[93]

The mode of binding of the specific thermolysin inhibitor *N*-(1-carboxy-3-phenylpropyl)-Leu-Trp-OH (reported in 1981) has been determined by *X*-ray crystallography. Both oxygens of the carboxymethyl are liganded to the zinc to give overall penta-co-ordination of the metal. The geometry is closer to that observed in the binding of hydroxamates rather than the monodentate binding seen previously for carboxylate–zinc interactions in thermolysin.[94] Twelve tripeptide aldehydes with C-terminal leucinal or phenylalaninal residues have been synthesized and assayed as renin inhibitors. All compounds with an N-terminal aromatic amino acid were active, and Z-Trp-Val-leucinal had an activity towards renin 6.5 times greater than pepstatin. All the active compounds were less inhibitory towards pepsin than was pepstatin.[95]

Dehydropeptides. – Full details of the synthesis of *N*-carboxy α-dehydroamino acid anhydrides and their use in the synthesis of dehydropeptides (Scheme 6) have appeared. Acylation of α-amino acid or dipeptide esters with (53) gives *N*-acetyldehydro di- or tri-peptide esters in almost quantitative yield. By contrast, the free *N*-carboxyanhydride (52) reacts with α-amino acid esters to give *N*-alkylethanedioyl derivatives only.[96] The condensation of *N*-Z- and *N*-Tfa-amino acid amides with pyruvic and phenylpyruvic acids using *p*-toluenesulphonic acid as catalyst has been shown to give reasonably good yields of N-protected dehydropeptides with C-terminal ΔAla and ΔPhe residues, res-

[90] P. D. Singh and J. H. Johnson, *J. Antibiot.*, 1984, **37**, 336.
[91] T. Nishikari, F. Kawahara, H. Naganawa, Y. Muraoka, T. Aoyagi, and H. Umezawa, *J. Antibiot.*, 1984, **37**, 680.
[92] I. J. Galpin, A. H. Wilby, G. A. Place, and R. J. Beynon, *Int. J. Pept. Protein Res.*, 1984, **23**, 477.
[93] Y. Okada, Y. Tsuda, Y. Nagamatsu, and U. Okamoto, *Int. J. Pept. Protein Res.*, 1984, **24**, 347.
[94] A. F. Marzingo and B. W. Matthews, *Biochemistry*, 1984, **23**, 5724.
[95] J.-A. Fehrentz, A. Meitz, B. Castro, C. Cazauban, and D. Nisato, *FEBS Lett.*, 1984, **167**, 273.
[96] C. Shin, Y. Yonezawa, and T. Yamada, *Chem. Pharm. Bull.*, 1984, **32**, 3924.

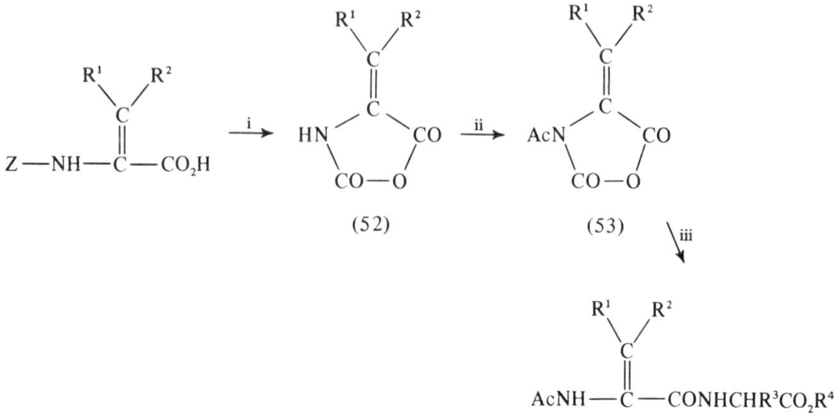

Reagents: i, $SOCl_2$–HOAc; ii, MeCOCl, THF; iii, $NH_2CHR^3CO_2R^4$

Scheme 6

pectively.[97] Z-Dehydroalanine couples with dipeptide esters using DCC as condensing agent in *ca.* 65% yields, but *N*-acetyl-Δ′(*Z*, L)-dehydrodipeptides couple with α-ΔVal-OEt only in 25% yield and give hardly any product with ΔAla esters.[98]

Further examples of the use of the Wittig–Horner reaction in the synthesis of dehydropeptides have been detailed (Scheme 7),[99] while the *X*-ray analysis of (*Z*)-*N*-acetyl-ΔPhe-Ala-OH · $\frac{1}{3}H_2O$ tends to confirm the theory that the conformational flexibility of αβ-unsaturated amino acids is similar to that of the saturated parents.[100] Asymmetric catalytic hydrogenation of tripeptides containing a dehydroamino acid has allowed the successful synthesis of two enkephalin analogues. With the cationic Rh complexes used, the nature of the N-protecting group exerted a significant influence on the asymmetric induction as well as catalyst efficiency.[101] Another group has also reported on the use of various Rh-containing homogeneous catalysts for dehydrodipeptide reduction,[102,103] and the Pd/C-catalysed reduction of chiral tripeptides with a central dehydro residue has been examined.[104]

Peptides Containing α,α-Dialkyl Amino Acids. – Although it was originally thought to be a single compound, hypelcin A has been separated by h.p.l.c.

97 M. Makowski, B. Rzeszotarska, Z. Kubica, and P. Wieczarek, *Liebigs Ann. Chem.*, 1984, 920.

98 C. Shin, Y. Yonezawa, and Z. Tamura, *Chem. Pharm. Bull.*, 1984, **32**, 2825.

99 U. Schmidt, A. Lieberknecht, and J. Wild, *Synthesis*, 1984, 53.

100 V. Busetti, D. Ajo, and M. Casarin, *Acta Crystallogr., Sect. C*, 1984, **40**, 1245.

101 I. Ojima, N. Yoda, M. Yatabe, T. Tanaka, and T. Kogure, *Tetrahedron*, 1984, **40**, 1255.

102 T. Yamagishi, M. Yatagai, M. Hatakeyama, and M. Hida, *Bull. Chem. Soc. Jpn.*, 1984, **57**, 1897.

103 M. Yatagai, T. Yamagishi, and M. Hida, *Bull. Chem. Soc. Jpn.*, 1984, **57**, 823.

104 M. Takasaki and K. Harada, *Chem. Lett.*, 1984, 1745.

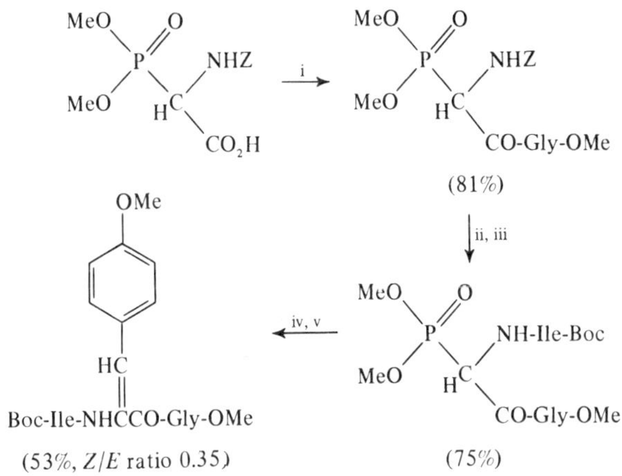

Reagents: i, H-Gly-OMe, DCC; ii, Pd/H_2; iii, Boc-Ile-OH, DCC; iv, $LiNiPr_2$;
v, *p*-$MeOC_6H_4CHO$

Scheme 7

A Ac-Aib-Pro-Aib-Ala-Aib-Aib-Gln-Leu-Aib-Gly-Aib-Aib-Aib-Pro-Val-Aib-Aib-Gln-Gln-Lol
1 2 3 4 5 6 7 8 9 10 11 12 13 14 15 16 17 18 19
A-I ,, ,, ,, ,, ,, ,, ,, ,, -Aib-Leu-Aib-Gly- ,, ,, ,, ,, ,, ,, ,, ,, ,,
A-II ,, ,, ,, ,, ,, ,, -Ala- ,, ,, ,, ,, ,, ,, ,, ,, ,, ,, ,, ,, ,, ,,
A-III ,, ,, ,, ,, ,, ,, -Aib- ,, ,, ,, ,, ,, ,, ,, ,, ,, ,, ,, ,, ,, -C_7H_{16}-NO
A-IV ,, ,, ,, ,, ,, ,, ,, ,, ,, -Ile- ,, ,, ,, ,, ,, ,, ,, ,, ,, ,, -Lol

(54)

into five components (54), which were sequenced by f.a.b.-m.s. The $C_7H_{16}NO$ C-terminus of hypelcin A-III is an amino alcohol residue of unknown structure.[105] Full details have been published of the structure determination of the antibiotic P168 reported earlier (Volume 14 of this title, p. 423); it principally involved in-beam m.s. of the *O,O*-diacetyl derivative.[106]

In the crystal, both Boc-Pro-Aib-Ala-Aib-Ala-OH (alamethicin 2–6) and Boc-Aib-Ala-Aib-Ala-Aib-OMe (suzukacillin A 1–5) form 3_{10}-helices, but they differ in handedness. In contrast to the former, the latter will form a superhelix, which may account for its experimentally observed membrane pore-forming ability. The nonapeptide Boc-Leu-Aib-Pro-Val-Aib_2-Glu(OBzl)-Gln-Pheol (alamethicin 12–20) displays both 3_{10}- and α-helical regions in the solid state.[107] Boc-Ala-Aib-Ala-OMe in the crystal adopts a new type of β-turn with a very wide $4 \rightarrow 1$ hydrogen bond distance of 3.62 Å. This is attributed to strong intermolecular

[105] T. Fujita, Y. Takaishi, K. Matsuura, Y. Takeda, Y. Yoshioka, and H. Bruckner, *Chem. Pharm. Bull.*, 1984, **32**, 2870.
[106] A. Isogai, A. Suzuki, S. Tamura, S. Higashikawa, and S. Kuyama, *J. Chem. Soc., Perkin Trans. 1*, 1984, 1405.
[107] R. Bosch, G. Jung, H. Schmitt, G. M. Shedrick, and W. Winter, *Angew. Chem., Int. Ed. Engl.*, 1984, **23**, 450.

hydrogen bonds forming a two-dimensional network in the *bc* plane.[108] Boc-Aib-Leu-Pro-NHMe in the solid state has no intramolecular hydrogen bond, and each residue occupies a different region of the Ramachandran map. Each molecule is associated with two water molecules.[109] Whereas Piv-Pro-Aib-NHMe and Piv-Pro-D-Ala-NHMe form type II β-turns in both the crystalline and solution states, more variation is seen with Piv-Pro-Val-NHMe and Z-Aib-Pro-Aib-Pro-OMe, with type I β-turns favoured.[110]

The c.d. spectra of Boc-Xxx-(Aib-Xxx)$_n$-OMe (n = 1, 2, or 3) and Boc-(Aib-Xxx)$_5$-OMe (Xxx = Ala or Val) have been examined in several solvents. They behave similarly in all solvents, suggesting that the Aib residues dominate the folding of these peptides. Comparison with n.m.r. studies suggests that estimates of helical content in oligopeptides by c.d. methods may lead to erroneous conclusions.[111] Theoretical calculations on *N*-Ac-Aib-NHMe, *N*-Ac-Iva-NHMe (Iva = isovaline), and *N*-Ac-(Dpg)$_{1\ \mathrm{or}\ 2}$NHMe (Dpg = α,α-di-n-propyl glycine) indicate that as the bulkiness of the C^{α} substituents increases there is a transition from a folded to a fully extended conformation. *X*-Ray analysis of Tfa-(Dpg)$_{1\ \mathrm{or}\ 2}$-DBH (DBH = N',N'-dibenzylhydrazide) confirms these calculations by showing the presence in both cases of the fully extended conformation.[112] The two series Tfa-(Aib)$_{2\ \mathrm{to}\ 5}$-OBu^t and Tfa-(Dpg)$_{2\ \mathrm{to}\ 5}$-OBu^t have also been examined by ^{1}H n.m.r. and i.r. in solution and the solid state. In contrast to the intermolecularly hydrogen-bonded β-structure adopted by the corresponding norvaline compounds, in the series of α,α-dialkyl amino acids intramolecular hydrogen bonding is the dominant factor. There seems to be no conformational transition as the chain length increases, and the Dpg homo-oligomers show exceptional structural stability upon heating.[113]

The analogue [Acc^7]oxytocin (Acc = 1-aminocyclopropane-1-carboxylic acid) has a lower biological activity than oxytocin, but a greater galactogenic selectivity.[114] Of a number of tri- and tetra-peptide kinin analogues, the most potent contained Pro-DL-αMePhe-Arg and Pro-DL-αMePhe-Gly-Pro sequences. Substitution on the N-terminal proline with 4-phenylbutyryl and 4-(4-hydroxyphenyl) butyryl side chains produced enhanced renal vasodilator activity and sometimes selectivity for the renal vasculature.[115] The crystal structure of *N*-acetyl-α-cyclopropylalanyl-Phe-NH$_2$ has been reported.[116]

[108] R. Bosch, G. Jung, K.-P. Voges, and W. Winter, *Liebigs Ann. Chem.*, 1984, 1117.
[109] B. V. Venkataram Prasad, H. Balaram, and P. Balaram, *Int. J. Pept. Protein Res.*, 1984, **24**, 135.
[110] M. Crisma, G. D. Fasman, H. Balaram, and P. Balaram, *Int. J. Pept. Protein Res.*, 1984, **23**, 411.
[111] E. K. S. Vijayakumar, T. S. Sudha, and P. Balaram, *Biopolymers*, 1984, **23**, 411.
[112] E. Benedetti, C. Toniolo, P. M. Hardy, V. Barone, A. Bavoso, B. Di Blasio, P. Grimaldi, F. Lelj, V. Pavone, C. Pedone, G. M. Bonora, and I. Lingham, *J. Am. Chem. Soc.*, 1984, **106**, 8146.
[113] G. M. Bonora, C. Toniolo, D. Di Blasio, V. Pavone, C. Pedone, E. Benedetti, I. Lingham, and P. M. Hardy, *J. Am Chem. Soc.*, 1984, **106**, 8152.
[114] Z. Procházka, M. Lebl, T. Barth, J. Hlaváček, A. Tuka, M. Buděšínsky, and K. Jost, *Collect. Czech. Chem. Commun.*, 1984, **49**, 642.
[115] F. R. Pfeiffer, P. A. Chambers, E. E. Hibbert, P. W. Woodward, and D. M. Ackerman, *J. Med. Chem.*, 1984, **27**, 325.
[116] F. Stierli, R. Prewo, J. H. Bieri, and H. Heimgartner, *Chimia*, 1984, **38**, 435.

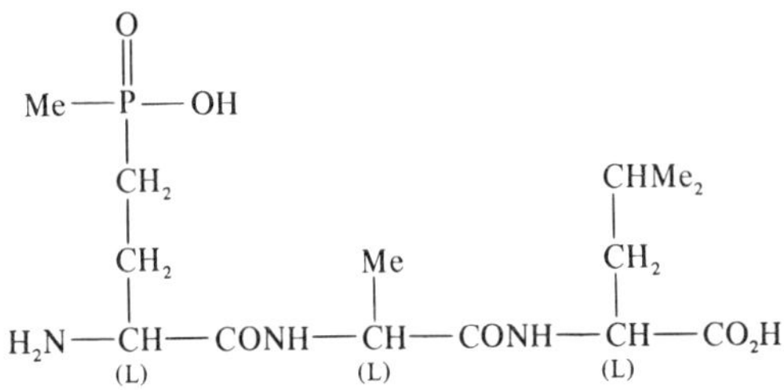

(55)

Phosphopeptides. – A herbicidal antibiotic isolated from *Kitasatosporia phosalacinea* has been characterized as phosphinothricyl-Ala-Leu-OH (55). It differs only from the natural glutamine antimetabolite bialaphos in the replacement of the C-terminal alanine by leucine.[117] A series of *N*-(*P*-substituted phosphinoyl) peptides have been synthesized and their antihypertensive activities tested in spontaneously hypertensive rats. The most potent and long-lasting activity was shown by *N*-(dibenzyloxyphosphinoyl)-Ala-Pro-Pro-OH.[118] The synthesis of some phosphonotripeptides by the Ugi reaction has been explored, and a typical example shown is in Scheme 8.[119] Tripeptides containing central (2-aminoethyl)-phosphonic acid (Aep) residues have also been prepared. Although Z-Aep(OMe)-OH could be coupled to H-Gly-OEt using diphenylphosphoryl azide in 65% yield, *N*,*N*′-dicyclohexylcarbodi-imide, Woodwards reagent K, and a number of standard coupling methods failed to give the desired product. No problem was experienced, however, in coupling to the amino group of H-Aep(OMe)-Gly-

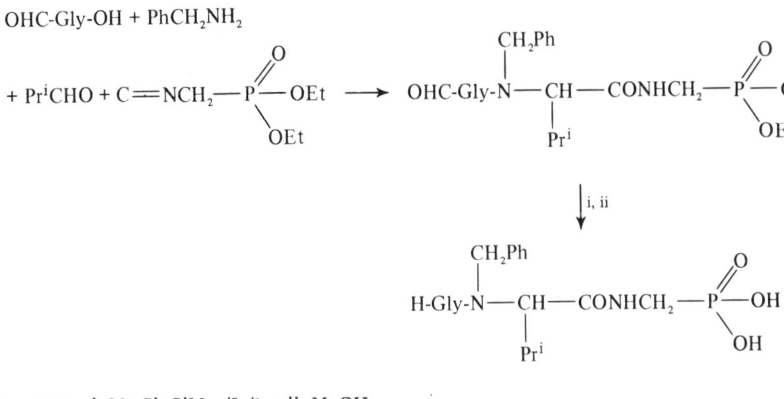

Reagents: i, $Me_3Si–SiMe_3/I_2/h\nu$; ii, MeOH

Scheme 8

[117] S. Omura, K. Hinotozawa, N. Imamura, and M. Murota, *J. Antibiot.*, 1984, **37**, 939.
[118] T. Morikawa, K. Takada, T. Kimura, S. Sakakibara, M. Kurauchi, Y. Ozawa, C. Eguchi, S. Hashimoto, and Y. Yukari, *Biochem. Biophys. Res. Commun.*, 1984, **119**, 1205.
[119] J. Rachon, *Synthesis*, 1984, 219.

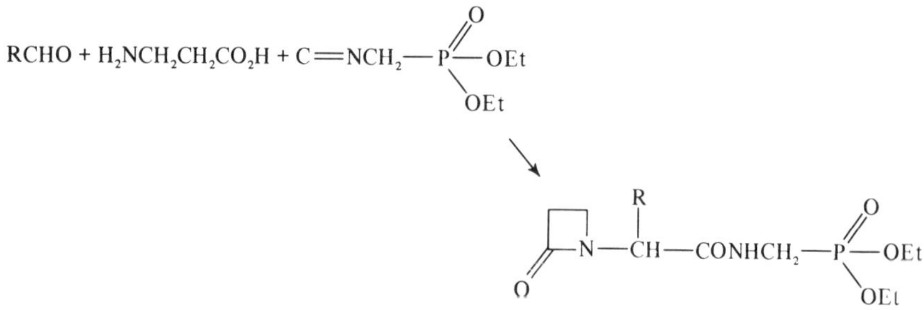

Scheme 9

OEt.[120] The Ugi reaction has also been used to prepare β-lactams containing aminomethylphosphonate groups (Scheme 9).[121]

Amide-bond Analogues. – A number of compounds falling in this category have been discussed in the section on enzyme inhibitors. Two that were not are H-Leuψ[*E*-CH=CH]-Gly-Val-Phe-OMe and His-Leuψ[*E*-CH=CH]-Gly-Val-Phe-OMe. These analogues of known renin inhibitors contain a *trans* carbon–carbon double bond in place of a peptide bond, and their inhibitory activity proved comparable to their fully peptidic parents.[122] An analogue *N*-(tyrosyl-4-aminobutyl)-*N*-methylphenethylamine of enkephalin, which effectively contains a $-CH_2CH_2-$ replacement for the amide bond between two glycine residues, shows analgesic activity in mice on peripheral adminstration.[123] A partially modified retro-inverso analogue of bradykinin-potentiating peptide (56) is active both *in vitro* and *in vivo*, displaying prolonged resistance towards cleavage by ACE.[124]

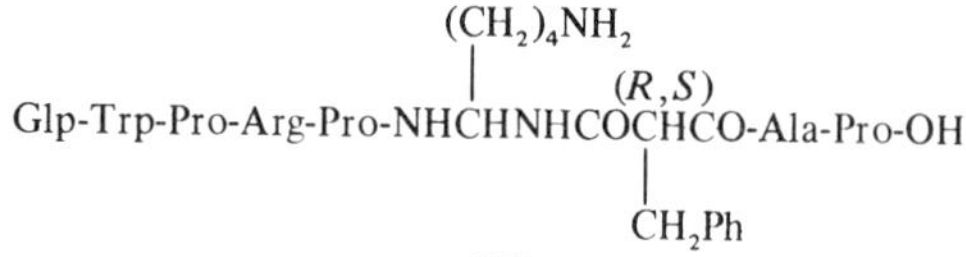

(56)

Two groups have prepared Leu-enkephalin analogues containing thioamide replacements for the peptide bond. They both agree that thionation of the Gly-Gly bond causes potency enhancement with a greater selectivity for δ- over μ-receptors. There is some disparity between the other results, but none of these other reported activities is greater than that of the parent unmodified

[120] K. Yamauchi, S. Ohtsuki, and M. Kinoshita, *J. Org. Chem.*, 1984, **49**, 1158.
[121] J. Rachon, *Chimia*, 1984, **38**, 114.
[122] R. L. Johnson, *J. Med. Chem.*, 1984, **27**, 1351.
[123] M. Maeda, S. Okusada, and K. Kawasaki, *Chem. Pharm. Bull.*, 1984, **32**, 4157.
[124] F. Bonelli, A. Pessi, and A. S. Verdini, *Int. J. Pept. Protein Res.*, 1984, **24**, 553.

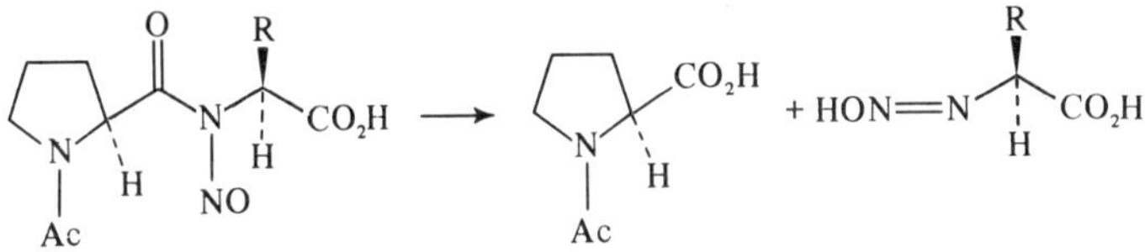

Scheme 10

PhCO-Gly-Gly-OMe	⟶	PhCO-(NO)Gly-(NO)Gly-OMe	100%
Z-Ala-Gly-OEt	⟶	Z-(NO)Ala-(NO)Gly-OEt	40%
		Z-Ala-(NO)Gly-OEt	54%
Z-Ala-Ala-OMe	⟶	Z-(NO)Ala-(NO)Ala-OMe	10%
		Z-Ala-(NO)Ala-OMe	85%
Boc-Ala-Leu-OMe	⟶	Boc-(NO)Ala-Leu-OMe	95%
Z-Val-Leu-OMe	⟶	no reaction	

(57)

peptide.[125,126] Some *N*-(*N*-acetylprolyl)-*N*-nitroso amino acids have been prepared, and their stability in aqueous media has been examined. Rapid decomposition only occurs in strong acid or alkali. Denitrosation is predominant at high acidity, but above pH 1 hydrolytic deamination (Scheme 10) is the major reaction. Overall the results suggest that *N*-nitrosopeptides are sufficiently stable for their absorption through the stomach and the duodenum.[127] The reaction of protected di- and tri-peptides in organic solvents with air-diluted nitrogen oxides (arising from copper wire and conc. HNO_3) has also been examined. In general, the products depend on the steric hindrance of amino acid side chains, as can be seen from the examples given in (57). The nitrosated peptides undergo attack by amines at the carbonyl carbon atom rather than undergoing denitrosation; a typical result is shown in Scheme 11.[128]

PhCO-(NO)Gly-(NO)Phe-OMe ⟶ PhCONHR + N_2=CHCONHR + N_2=C(CH_2Ph)CO_2Me

Reagent: i, $RNH_2/CH_2/0–20\,^{\circ}C$

Scheme 11

[125] G. Lajoie, F. Lepine, S. Lemaire, F. Jolicoeur, C. Aubé, A. Turcotte, and B. Belleau, *Int. J. Pept. Protein Res.*, 1984, **24**, 316.

[126] K. Clausen, A. F. Spatola, C. Lemieux, P. W. Schiller, and S.-O. Lawesson, *Biochem. Biophys. Res. Commun.*, 1984, **120**, 305; K. Clausen, M. Thorsen, S.-O. Lawesson, and A. F. Spatola, *J. Chem. Soc., Perkin Trans. 1*, 1984, 785.

[127] B. C. Challis, J. R. Milligan, and R. C. Mitchell, *J. Chem. Soc., Chem. Commun.*, 1984, 1050.

[128] J. Garcia, J. Gonzalez, R. Segura, and J. Vilorrosa, *Tetrahedron*, 1984, **40**, 3121.

Acylation of a series of N-benzyloxy α-amino acid derivatives with N,N-phthaloylglycine shows that the yield decreases with increasing bulkiness of the substituent even by the acyl chloride or mixed-anhydride procedures. N-Benzyloxy α-amino acid 1-succinimidyl esters are sufficiently stable towards self-condensation to be useful synthetic intermediates. Using a combination of mixed-anhydride and 1-succinimidyl ester couplings, the hexapeptide Ac-[DL-(BzlO)-Ala-Gly]$_3$NHPh has been prepared; each mixed-anhydride acylation was repeated three times. After catalytic hydrogenation this hexapeptide does not appear to form a stable Fe^{III} complex; it seems that an alternating sequence of the N-hydroxy amide units is too dense to allow a stable 1 : 3 complex such as is seen in ferrioxamine B.[129] Examination of CCl_4 or toluene solutions of Boc-[Leu$_3$-Gly-(Dmob)Leu-Leu$_3$]$_n$-OBzl (Dmob = 2,4-dimethoxybenzyl, $n = 1$ or 2) by i.r. indicates predominantly β-sheet structures, as was observed in the solid state; in THF and MeOH random coil predominated. The insertion of proline residues was found to have the same effect on conformation and solubilizing behaviour as protection of the peptide bond.[130]

Peptides Containing Backbone Rings. – The total synthesis of the antibiotic althiomycin (58), found in 1957 in *Streptomyces althioticus*, has been achieved from D-cysteine (Scheme 12). The synthetic material was completely identical with the natural product.[131] An oxytocin analogue in which the Pro^7-Leu^8 dipeptide unit has been replaced by a 2-[1-(2-oxo-3-aminopyrrolidinyl)]-4-methylpentanoic acid unit (59) has been prepared and found to exhibit a higher galactogenic/uterotonic ratio than the parent.[114] X-Ray analysis of Boc-L-aminosuccinyl-Gly-OMe shows a highly extended molecule with the puckered conformation of the ring between that of pure-envelope and pure-twist conformations. Theoretical calculations support these results in suggesting such a sequence adopts a particularly stable type II′ β-bend.[132] Solid-state and solution studies of Boc-L-aminosuccinyl-Ala-Gly-OMe also indicate the adoption of a type II′ β-bend.[133]

Contact of α-protected tripeptides containing a central glutamic acid residue with N,N'-carbonyldi-imidazole generates internal pyroglutamyl residues in *ca.* 80% yield. Usually less than 1% of imide is formed unless there is a hindered N-terminal residue; Boc-Ile-Glu-Gly-NH_2, for example, gave 95% of the imide (60) and only 2% of the lactam. Treatment of the tripeptides containing internal pyroglutamyl residues with phosphate-buffered saline at pH 8.0 at 37 °C predominantly caused peptide-chain cleavage to liberate a pyroglutamyl dipeptide. Ring opening to regenerate the free glutamic acid side chain was less favoured by a factor of 4 to 18, depending on the sequence. It is suggested that some peptide

[129] K. Shimizu, K. Nakayama, and M. Akiyama, *Bull. Chem. Soc. Jpn.*, 1984, **57**, 2456.
[130] M. Narita, K. Ishikawa, H. Nakano, and S. Isokawa, *Int. J. Pept. Protein Res.*, 1984, **24**, 14.
[131] K. Inami and T. Shiba, *Tetrahedron Lett.*, 1984, **25**, 2009.
[132] S. Capasso, C. A. Mattia, L. Mazzorella, and A. Zagari, *Int. J. Pept. Protein Res.*, 1984, **23**, 248.
[133] S. Capasso, L. Mazzarella, F. Sica, and A. Zagari, *Int. J. Pept. Protein Res.*, 1984, **24**, 588.

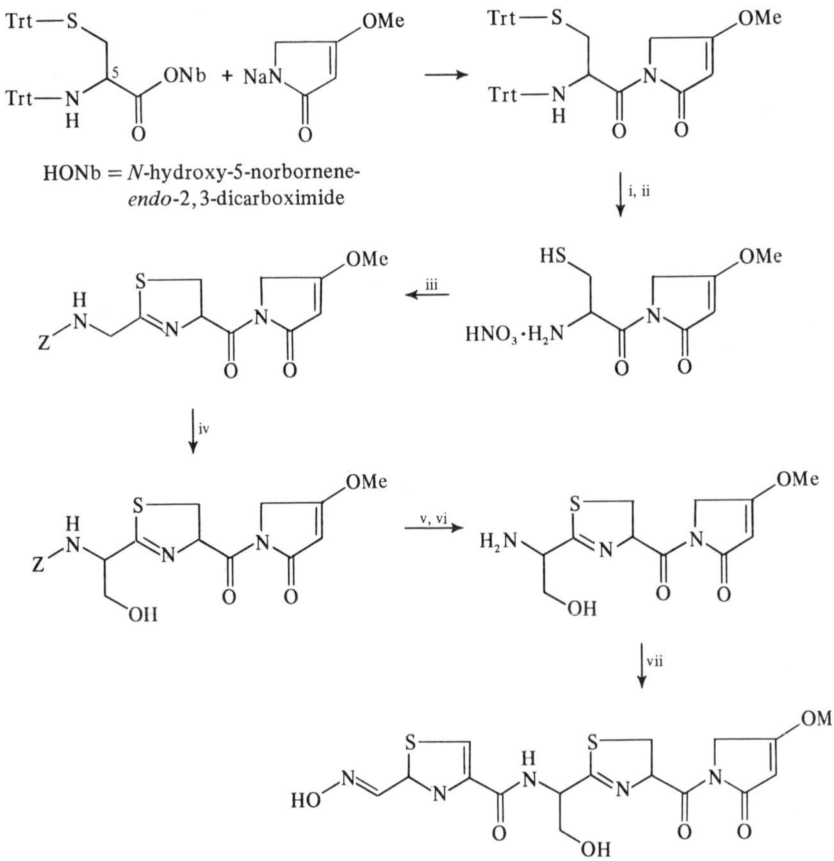

(58)

Reagents: i, $AgNO_3$/pyridine; ii, H_2S; iii, $ZNHCH_2C(=NH)OEt$; iv, HCHO/DMSO; v, HF-anisole; vi, K_2CO_3; vii, HON=CH—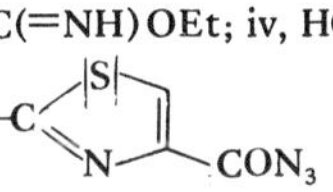

Scheme 12

hormones with N-terminal pyroglutamic acid might arise by regioselective hydrolysis with chain fragmentation of a precursor containing an internal pyroglutamyl residue.[134]

Peptides Containing D-Residues. – Synthetic analogues of hormones and some other well established biologically active peptides that do not involve structural studies are not covered here. Twenty-nine analogues Asp-Xxx-Yyy-OMe (where Xxx and Yyy have only hydrocarbon side chains) of aspartame have been synthesized, 18 of which contain at least one D-residue. It is concluded that for

[134] S. A. Khan and B. W. Erickson, *J. Am. Chem. Soc.*, 1984, **106**, 798.

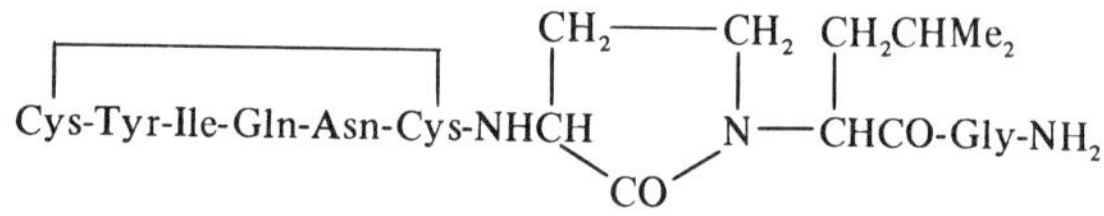

(59)

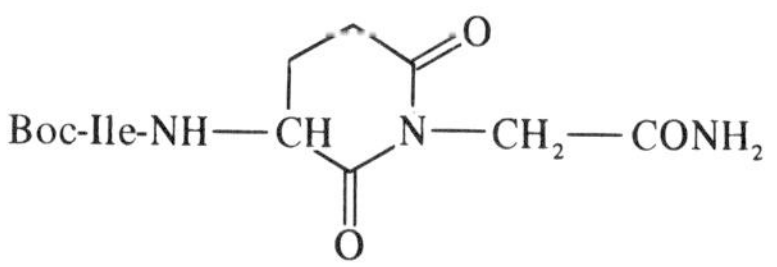

(60)

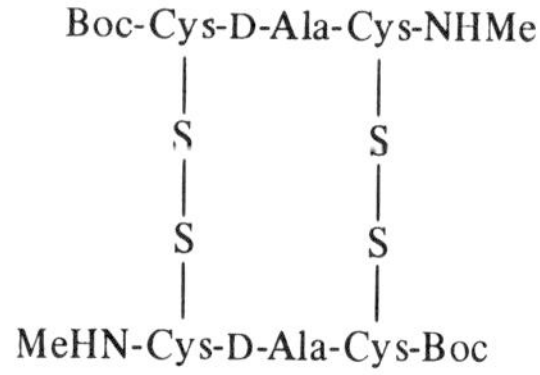

(61)

sweetness in such tripeptides the second residue must have a small alkyl group and be of the D-configuration.[135] Solution-phase Raman spectral differences of H-Ala-Ala-OH and H-D-Ala-Ala-OH have been noted and discussed,[136] while ^{1}H n.m.r. spectroscopy indicates that the cyclic bis-cystine peptide (61) adopts an antiparallel β-sheet conformation in solution. The similarity of the spectra to the L-Ala analogue suggests that the disulphide bridges force the D-Ala into adopting similar ϕ and ψ conformational angles.[137]

In the tetrapeptide Ac-Leu-Xxx-Pro-Val-NHMe, the population of the β-turn conformation decreases as Xxx passes from D-Ala through Gly to L-Ala, agreeing with the magnitudes of the Cotton effects in the c.d. spectra of the *N*-2,4-Dnp *p*-nitroanilide derivatives of these sequences.[138] A polymeric analogue H-(Val-Pro-D-Ala-Val-Gly)$_n$-Val-OMe of elastin has been prepared and found to adopt a similar conformation with a somewhat stabilized β-turn. It coacervates to form a more cohesive viscoelastic material, and when crosslinked by γ-irradiation it undergoes an approximate doubling of the Young's modulus of elasticity.[139] In $CHCl_3$ solutions of HCO-Ile-(D-aIle-Ile)$_4$-OMe, ^{1}H n.m.r. and vapour-pressure osmometry indicate the occurrence of a rapid equilibrium involving the head-to-tail dimerization of $\beta^{4.4}$-helices. This type of equilibrium could be responsible for the formation and breaking down of the ion-conducting channels formed by gramicidin A in lipid bilayers.[140] In $CHCl_3$ solution HCO-Phe-(D-Phe-Phe)$_3$-OMe forms three major species, two dimeric and one tetrameric. In CCl_4 or cyclohexane solution, however, virtually only the tetramer is seen. The n.m.r. data indicate that the dimer in rapid equilibria with the tetramer is a right-handed $\uparrow\uparrow\beta^{5.6}$-helical structure and that the tetramer is formed from this by

[135] Y. Arioshi, *Bull. Chem. Soc. Jpn.*, 1984, **57**, 3197.

[136] M. Diem, M. Reza Oboodi, and C. Alva, *Biopolymers*, 1984, **23**, 1917.

[137] R. Kishore and P. Balaram, *J. Chem. Soc., Chem. Commun.*, 1984, 778.

[138] K. Sato, T. Higoshijima, R. Sugawara, and U. Nagai, *Bull. Chem. Soc. Jpn.*, 1983, **56**, 3699.

[139] D. W. Urry, T. L. Trapane, S. A. Wood, R. D. Harris, J. T. Walker, and K. U. Prasad, *Int. J. Pept. Protein Res.*, 1984, **23**, 425.

[140] G. P. Lorenzi, V. Muri-Valle, and F. Bangerter, *Helv. Chim. Acta*, 1984, **67**, 1588.

a head-to-head association. It is suggested that the linear gramicidins may also form head-to-head dimers of parallel β-helices.[141]

The results of normal-mode calculations on the $\beta^{4.4}$, $\beta^{6.3}$, $\uparrow\downarrow\beta^{5.6}$, and $\uparrow\downarrow\beta^{7.2}$ structures of gramicidin A have been compared with i.r. and Raman spectra of native, crystalline Cs^{I}-bound, and vesicle-bound gramicidin A, leading to the assignment of $\uparrow\downarrow\beta^{5.6}$, $\uparrow\downarrow\beta^{7.2}$, and $\beta^{6.3}$ structures, respectively, to the gramicidin A molecules in the above three systems.[142] The conformation of species 3 of Val-gramicidin A in dioxan has been established by two-dimensional n.m.r. as a left-handed $\uparrow\downarrow\pi\pi_{LD}^{5.6}$-double helix, corroborating earlier proposals.[143] Inclusion of the $-CH_2CH_2OH$ end chain in the computation of the energy profile for single occupancy by Na^+ of the gramicidin A channel modifies the profile obtained without the chain to one in satisfactory agreement with conclusions based on ^{13}C n.m.r. studies.[144] Planar–bilayer studies on the channel activity of des-L-Val^7-D-Val^8-gramicidin A show an increased dispersity of single-channel conductance, supporting the perspective that dispersity derives from different side-chain distributions on the same backbone conformation.[145]

Peptides Containing Other Non-protein Residues. – A new dipeptide lycium-amide isolated from the Chinese crude drug 'Ti-ku'-pi' extracted from the root bark of *Lycium chinense* has been established as *N*-benzoyl-Phe-Pheol *O*-acetate.[146] Another new dipeptide, *S*-(2-carboxy-n-propyl)cysteinyl glycine, has been isolated from *Allium cepa*, a member of the Liliaceae.[147] The unit peptide cadystin, the cadmium-binding peptide of the yeast *Schizosaccharomyces pombe*, has been separated into two main components, cadystin A (62) and cadystin B (63). These structures have been confirmed by synthesis.[148]

The peptide antibiotic galantin I, effective against both Gram-positive and Gram-negative bacteria, found in *Bacillus pulvifaciensis* has been assigned the structure (64). It is a mixture of two congeners containing as the fourth residue either lysine ($n = 4$) or ornithine ($n = 3$) in the ratio of 9:1. The novel cyclic amino acid component galantinic acid has been determined to be (2*S*,4*S*,5*S*)-5-amino-2-carboxymethyl-4-hydroxytetrahydropyran on the basis of n.m.r. and c.d. studies.[149] Galantinic acid has been synthesized in eight steps from a *threo*-3,4-epoxy-2-aminobutanol derivative.[150]

141 G. P. Lorenzi, C. Gerber, and H. Jäckle, *Biopolymers*, 1984, **23**, 1905.
142 V. M. Naik and S. Krimm, *Biochem. Biophys. Res. Commun.*, 1984, **125**, 919.
143 A. S. Arseniev, V. F. Bystrov, V. T. Ivanov, and Yu. A. Ovchinnikov, *FEBS Lett.*, 1984, **165**, 51.
144 C. Etchebest and A. Pullman, *FEBS Lett.*, 1984, **170**, 191.
145 D. W. Urry, S. Alonso-Ramanowski, C. M. Venkatachalan, R. D. Harris, and K. V. Prasad, *Biochem. Biophys. Res. Commun.*, 1984, **118**, 885.
146 M. Naguchi, K. Mochida, T. Shingu, M. Kozuka, and K. Fujitani, *Chem. Pharm. Bull.*, 1984, **32**, 3584.
147 T. Kosai, T. Nishitoba, Y. Shiroshita, and S. Sakamura, *Agric. Biol. Chem.*, 1984, **48**, 2271.
148 N. Kando, K. Imai, M. Isobe, T. Goto, A. Murasugi, C. Wada-Nakagawa, and Y. Hayoshi, *Tetrahedron Lett.*, 1984, **25**, 3869.
149 T. Wakamiya, T. Ando, T. Teshima, and T. Shiba, *Bull. Chem. Soc. Jpn.*, 1984, **57**, 142.
150 Y. Ohfune and N. Kurakawa, *Tetrahedron Lett.*, 1984, **25**, 1587.

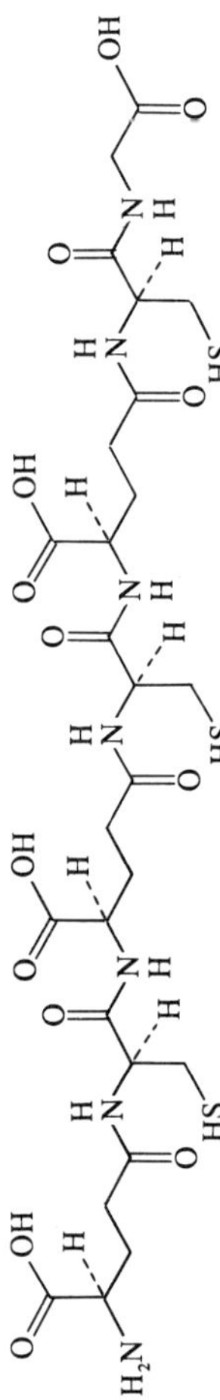

(62)

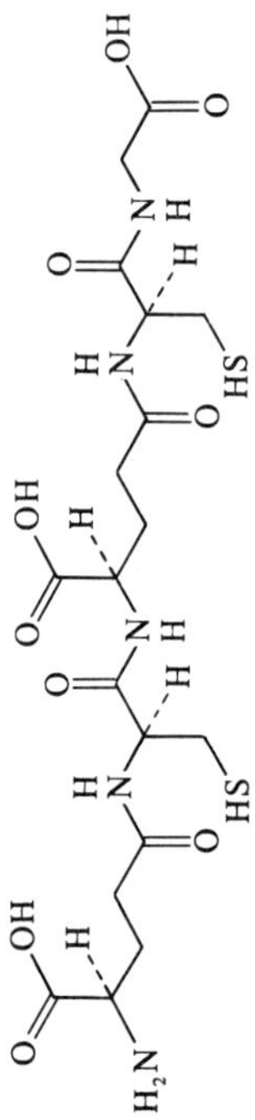

(63)

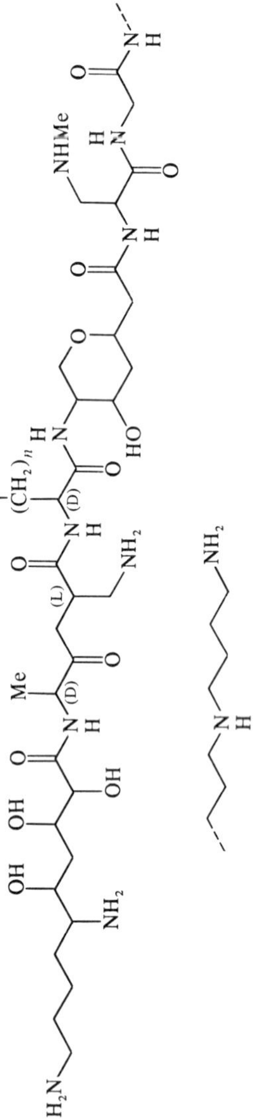

(64)

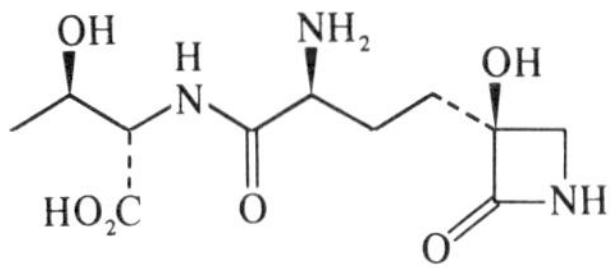

(65)

The stereospecific synthesis of tabtoxin (65), the exotoxin from *Pseudomonas tabaci* responsible for wildfire disease of tobacco plants, has been reported now in full.[151] A [13]-peptide of the sequence H-(Val-Glu-Val-Orn)$_3$-Val-OH has been synthesized as an amphiphilic model of apolipoprotein B. In aqueous solution it exists predominantly as a β-strand and forms monolayers of great stability at the air-water interface. The peptide is able to bind to low-density lipoprotein (LDL) and to associate with synthetic lipids to form a particle of the radius of LDL.[152] Conformational transitions of some fully protected synthetic fragments of a perornithine analogue of the fish protamine thynnin Z1 have been examined. The self-aggregation observed in CH_2Cl_2 is disrupted by addition of DMSO or DMF, and this study has indicated that i.r. is a useful technique to titrate quantitatively the extent of self-aggregation in peptides.[153]

A number of tripeptides of α-aminoadipic acid have been prepared to investigate penicillin biosynthesis. δ-(L-α-Aminoadipoyl)-Cys-Gly-OH inactivates penicillin N synthetase at the same rate as tritium is released from the Cys 3-position, suggesting an inhibition mechanism involving the formation of an enzyme-bound β-lactam.[154] Isopenicillin N synthetase from *Cephalosporium acremonium* converts δ-(L-α-aminoadipoyl)-Cys-D-allylglycine into a mixture of penam and cepham products; two discrete pathways from a common intermediate are suggested.[155] A series of structural variants of δ-(L-α-aminoadipoyl)-Cys-D-Val-OH (ACV) have been prepared, and their effectiveness as substrate for isopenicillin N synthetase has been determined. A minimal structural requirement was an *N*-acyl group having a six-carbon, or equivalent, chain terminating in a carboxyl group.[156] When ACV is incubated with a partially purified *C. acremonium* extract, desacetoxy cephalosporin C was produced. The reaction is sensitive to penicillinase, indicating penicillin N to be an intermediate.[157] Using a cell-free system from *C. acremonium*, δ-(L-α-aminoadipoyl)-Cys-*N*-hydroxy-D-valine is

[151] J. E. Baldwin, P. D. Bailey, G. Gallacher, M. Otsuka, K. A. Singleton, and P. M. Wallace, *Tetrahedron*, 1984, **40**, 3695.

[152] D. Osterman, R. Mara, F. J. Kézdy, E. T. Kaiser, and S. C. Meredity, *J. Am. Chem. Soc.*, 1984, **106**, 6845.

[153] C. Toniolo, G. M. Bonora, F. Marchiori, and G. Barin, *J. Am. Chem. Soc.*, 1984, **106**, 1455.

[154] J. E. Baldwin, E. P. Abraham, C. G. Lovel, and H. H. Toi, *J. Chem. Soc., Chem. Commun.*, 1984, 902.

[155] J. E. Baldwin, R. M. Adlington, A. E. Derome, H.-H. Ting, and N. J. Turner, *J. Chem. Soc., Chem. Commun.*, 1984, 1211.

[156] J. E. Baldwin, E. P. Abraham, R. M. Adlington, G. A. Bahadur, B. Chakravarti, B. P. Domagne-Hayman, L. D. Field, S. L. Flitsch, G. S. Jayatilake, A. Spakovskis, H.-H. Ting, N. J. Turner, R. L. White, and J. J. Usher, *J. Chem. Soc., Chem. Commun.*, 1984, 1225.

[157] Y.-Q. Shen, S. Wolfe, and A. L. Demain, *J. Antibiot.*, 1984, **37**, 1044.

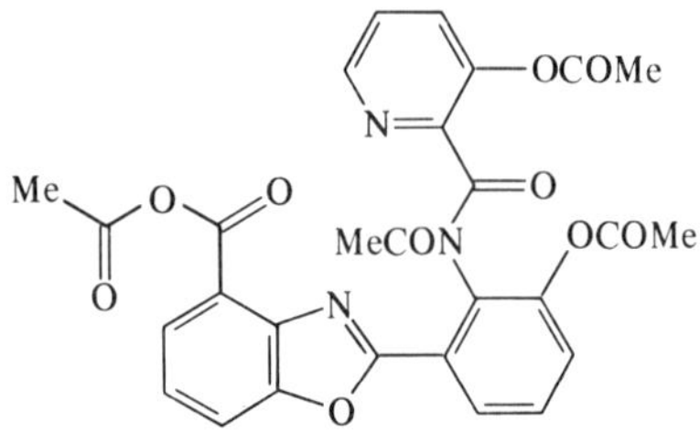

(66)

not converted to isopenicillin N but does inhibit the ability of this system to induce formation of isopenicillin N from ACV.[158] Bis-[(L-cysteinyl-*S*-acetyl)-L-hemicystinyl($S^2 \rightarrow S^{2\prime}$)-D-valine] has also been examined as a substrate of isopenicillin N synthetase.[159]

The TRH analogues Glp-Nva-Pro-NH_2 and Glp-Nle-Pro-NH_2 show a stronger anticataleptic effect than the parent, indicating that the presence of histidine is not essential for CNS activity.[160] *X*-Ray analysis of the tetra-acetyl derivative of an antiviral and antibacterial compound A33853 isolated from a *Streptomyces* sp. found in an Alaskan oil sample shows it to be (66), containing an anhydride group.[161] Acetylenic analogues of tuftsin [Thr-Dah-Pro-Arg (Dah = 2,6-diamino-4-hexynoic acid)] and of a macrophage inhibitory peptide (Thr-Dah-Pro) gave on catalytic tritiation peptides with specific radioactivities of 11.4 and 37.0 Ci $mmol^{-1}$, respectively.[162]

A further selection of peptides containing non-protein residues other than D-amino acids is given in the Table.

Table *Synthetic peptides containing non-protein residues*

	Ref.
α-Isocyanoacetyl peptides as seed-germination inhibitors	163
Ala-Dl-2-(5-fluorouracil)glycine as a 5-fluorouracil-delivery system	164
Pentapeptides containing 4-(5*H*-tetrazolyl)-2-aminobutyric acid as inhibitors of vitamin K-dependent carboxylation	165
[L-mTyr^1]- and [D-mTyr^1]-Leu-enkephalin	166
Boc-Tyr-Gly-Gly-4-BrPhe-Met-OH	167
[Abu^4] and [Cha^4] analogues of [(β-mercapto-β,β-pentamethylenepropionic acid)^1,D-Ile^2,Val^4] arginine vasopressin	168
[3-(2-Benzimidazolyl)Ala^6]-, [3-(2-benzothiazolyl)Ala^6]-, and [3-(2-benzoxazolyl)Ala^6]-LHRH 3-(2-naphthyl)-D-Ala analogues of [*N*-Ac-Pro^1,D-pF-Phe^2]- and [D-pF-Phe^2,D-Trp^3,D-Arg^6]-LHRH	169, 170
[(3-Benzimidazol-2-yl)-D-Ala^6]-, [3-(5,6-dimethylbenzimidazol-2-yl)-D-Ala^6]-, and [3-(2-naphthyl)-D-Ala^6][aza-Gly^{10}]-LHRH	171
[(3-Pyrazolyl)Ala^{119}]ribonuclease (111–124)	172
[L-3,4-DehydroPro^3]tuftsin	213

158 R. L. Baxter, G. A. Thomson, and A. I. Scott, *J. Chem. Soc., Chem. Commun.*, 1984, 32.
159 J. E. Shields, C.-S. Campbell, S. W. Queener, D. C. Duckworth, and N. Neuss, *Helv. Chim. Acta*, 1984, **67**, 870.

4 Conjugate Peptides

Glycopeptide Antibiotics. – *Actinoplanes missouriensis* produces a complex of antibiotics known as actaplanins that have a common aglycon but that differ in the content and/or distribution of attached sugar units, as depicted in (67). The points of substitution have been determined by n.m.r. methods. All species contain either one or two mannose units, and an additional phenolic site in each actaplanin is occupied by either glucose, mannosylglucose, or rhamnosylglucose.[173] Two groups have independently isolated *N*-demethylvancomycin from strains of *Nocardia orientalis*.[174,175] Unlike glycopeptide antibiotics, virtually none of this compound remained bound to the cells of the producing culture.[174] The antimicrobial spectrum is the same as that for vancomycin.[175] Vancomycin and ristocetin are known to act as antibiotics by binding to peptides terminating in D-Ala-D-Ala. Their combinations with model tripeptides Ac_2-L-Lys-Xxx-Yyy-OH (Xxx, Yyy = D-Ala, Gly, or D-Leu) have now been examined and the results discussed.[176] The reaction of demethylbleomycin A_2 with bromoacetyl derivatives leads to the corresponding sulphonium salts, which constitute a new class of biologically active bleomycins.[177]

Other Glycopeptides. – The condensation of *N*-Z-aspartic acid anhydride with β-D-glucopyranosylamine gives a mixture of *N*-(α-aspartyl)- and *N*-(β-aspartyl)-β-glucopyranosylamines that can be separated either by cation-exchange

160 T. Szirtes, L. Kisfaludy, E. Palosi, and L. Szparny, *J. Med Chem.*, 1984, **27**, 741.
161 K. H. Michel, L. D. Boeck, M. M. Hoehn, N. D. Jones, and M. D. Chaney, *J. Antibiot.*, 1984, **37**, 441.
162 G. Auger and D. Blanot, *Int. J. Pept. Protein Res.*, 1984, **24**, 60.
163 K. Nunami, M. Suzuki, K. Matsumoto, N. Yoneda, and K. Takiguchi, *Agric. Biol. Chem.*, 1984, **48**, 1073.
164 W. D. Kingsbury, J. C. Boehm, R. J. Melita, S. F. Grappel, and C. Gilvarg, *J. Med. Chem.*, 1984, **27**, 1447.
165 J. Dubois, S. Gary, M. Gaudy, and A. Marquet, *J. Med. Chem.*, 1984, **27**, 1230.
166 A. M. van den Braken-van Leersum and L. Maat, *Recl. Trav. Chim.*, 1984, **103**, 110.
167 M. Doi, T. Ishida, M. Inoue, T. Fujiwara, K.-I. Tomita, T. Kimura, and S. Sakakibara, *FEBS Lett.*, 1984, **170**, 229.
168 M. Manning, E. Nawrocka, A. Misicka, A. Olma, W. A. Klis, J. Seto, and W. H. Sawyer, *J. Med. Chem.*, 1984, **27**, 423.
169 J. J. Nestor, jun., B. L. Horner, T. L. Ho, G. H. Jones, G. I. McRae, and B. H. Vickery, *J. Med. Chem.*, 1984, **27**, 320.
170 J. J. Nestor, jun., R. Tahilramani, T. L. Ho, G. I. McRae, and B. H. Vickery, *J. Med. Chem.*, 1984, **27**, 1170.
171 T. L. Ho, J. J. Nestor, G. I. McRae, and B. H. Vickery, *Int. J. Pept. Protein Res.*, 1984, **24**, 79.
172 J. Serdijn, W. Bloemhoff, K. E. T. Kerling, and E. Havinga, *Recl. Trav. Chim.*, 1984, **103**, 50.
173 A. H. Hunt, T. K. Elzey, K. E. Merkel, and M. Debono, *J. Org. Chem.*, 1984, **49**, 641.
174 L. V. D. Boeck, F. P. Mertz, R. K. Walter, and C. E. Higgens, *J. Antibiot.*, 1984, **37**, 446.
175 A. H. Hunt, G. G. Marconi, T. K. Elzey, and M. M. Hoehn, *J. Antibiot.*, 1984, **37**, 917.
176 M. P. Williamson, D. H. Williams, and S. J. Hammond, *Tetrahedron*, 1984, **40**, 569.
177 W. J. Vloon, C. Kruk, U. K. Pandit, H. P. Hofs, and J. G. McVie, *Heterocycles*, 1984, **22**, 779.

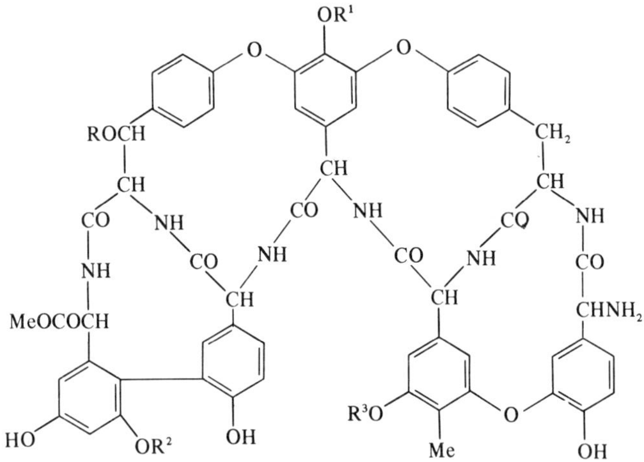

Actaplanin	R^1	R^2	R^3
A	Mannosylglucose	Mannose	Mannose
B_1	Rhamnosylglucose	Mannose	Mannose
B_2	Glucose	Mannose	Mannose
B_3	Mannosylglucose	Mannose	H
C_1	Rhamnosylglucose	Mannose	H
G	Glucose	Mannose	H
C_2 (≡ X)	H	Mannose	Mannose
D_1 (≡ Y)	H	Mannose	H
ψ (≡ Z)	H	H	H
C_3	Glucose	H	Mannose

(67) R = ristosamine

chromatography, through a Cu^{II} complex, or by successive fractional crystallization of the dicyclohexylamine salts.[178] A number of 1-succinimidyl esters of tripeptides have been condensed with glycosylamines having free hydroxyl groups to give *N*-glycopeptides, a typical example being 2-acetamido-1-*N*-[*N*-Boc-aspart-1-oyl-(Phe-Ser-OMe)-4-oyl]-2-deoxy-β-D-glucopyranosylamine.[179] Glycosylated derivatives of Boc-Asn-Gly-NHMe and Boc-Gly-Asn-NHMe have been shown to adopt the folded structure with an intramolecular hydrogen bond involving the NH of the *N*-methylamide group seen in the unglycosylated dipeptide. There appears to be no particular intramolecular interaction between the peptide and carbohydrate moieties, but the latter increases the rigidity of the peptide backbone.[180]

N-Tripeptidyl-D-glucosamines, *e.g.* containing Ala-Leu-Ala, Leu-Phe-Leu, or (β-Bzl-Asp)$_3$ sequences, have been prepared by the stepwise addition of *N*-car-

[178] M. Tamura, H. Nishizaki, C. Mujozaki, and H. Okai, *Bull. Chem. Soc. Jpn.*, 1984, **57**, 3167.

[179] M. Tamura and H. Okai, *Carbohydr. Res.*, 1984, **133**, 207.

[180] H. Ishii, Y. Inouye, and R. Chujo, *Int. J. Pept. Protein Res.*, 1984, **24**, 421.

boxyamino acid anhydrides to D-glucosamine hydrochloride in the presence of equimolar NaOMe in MeOH at −50 °C. At this temperature polymerization does not occur.[181] Carboxyl protection by allyl esters has been found useful in the synthesis of the acid- and base-labile *O*-glycopeptides of Z-Ser-Ala-Ala-Gly-Ala-OH. Deprotection was effected by tetrakis(triphenylphosphane)palladium(0) at 20 °C, the allyl group being transferred to morpholine as an acceptor nucleophile in almost quantitative yield. Other commonly used protecting groups are unaffected by such treatment.[182] A ^{13}C spectral study of the pH behaviour of reductively [^{13}C]methylated glycophorin A glyco-octapeptides and a related glycopentapeptide has been made,[183] and up to 56% of the amino or the carboxyl groups of β-lactoglobulin have been glycosylated with maltose or β-cyclodextrin using the cyclic carbonate method and with glucosamine or glucosamine-octaose using the carbodi-imide method. The electrophoretic mobilities of maltosyl derivatives were essentially unchanged, while those of glucosaminyl derivatives decreased.[184]

Four kinds of disaccharide tri- and tetra-peptide have been isolated from a hydrolysate of *L. plantarum* cell walls by combined catalysis of M-1 *N*-acetylmuramidase and AM_3-endopeptidase. In these MDP-related compounds, the presence of a C-terminal D-amino acid residue appears to play an important role in their pyrogenicity.[185] N^{α}-(*N*-Acetylmuramyl-Ala-D-iGln)-N^{ϵ}-stearyl-Lys-OH (MDP-Lys-L18) is an MDP derivative that has been reported to have an immunomodulating effect in experimental animals. Using immunogens obtained by a water-soluble carbodi-imide-mediated coupling of MDP-Lys-L18 with bovine serum albumin, a sensitive radioimmunoassay has been developed for MDP-Lys-L18.[186] 2-*N*-Octadecanoyl derivatives of 1-*S*-acetyl-, 1-*S*-octadecanoyl-, and 6-*O*-octadecanoyl-1-*S*-octadecanoyl-1-thiomuramyl-Ala-D-iGln-OH have been prepared, but they have only weak immunoadjuvant activity;[187] (*N*-acetyl-6-thiomuramoyl)-Ala-D-iGln-OH and some of its lipophilic alkyl derivatives linked at the C-6 of the sugar skeleton have been made and shown to have no activity.[188]

Two series of MDP derivatives, (68) and (69), have been synthesized and found to possess both adjuvant activity and tumour-suppressive activity. These results confirm that the 5,5-dimethoxy-3-methyl-1,4-benzoquinone ring is essential for tumour-suppressive activity; the most active compound in this respect was *N*-Ac-6-*O*-[10-(5,6-dimethoxy-3-methyl-1,4-benzoquinon-2-yl)decanoyl]muramyl-Val-D-iGln-OMe.[189] A synthesis of MDP capable of producing

[181] M. Oya and T. Negeshi, *Bull. Chem. Soc. Jpn.*, 1984, **57**, 439.

[182] H. Kunz and H. Waldmann, *Angew. Chem., Int. Ed. Engl.*, 1984, **23**, 71.

[183] K. Dill, R. E. Hardy, R. L. Batstone-Cunningham, and M. E. Daman, *Carbohydr. Res.*, 1984, **128**, 183.

[184] R. D. Waniska and J. E. Kinsella, *Int. J. Pept. Protein Res.*, 1984, **23**, 573.

[185] S. Kawata, Y. Yokogawa, E. Takahashi, T. Takamura, Y. Takase, S. Kotani, and M. Tsujimoto, *Agric. Biol. Chem.*, 1984, **48**, 1783.

[186] H. Masayasu, K. Ono, and T. Takegoshi, *Chem. Pharm. Bull.*, 1984, **32**, 4124.

[187] A. Hasegawa, E. Seki, Y. Hioki, M. Kiso, and I. Azuma, *Carbohydr. Res.*, 1984, **131**, 61.

[188] A. Hasegawa, E. Seki, Y. Hioki, M. Kiso, and I. Azuma, *Carbohydr. Res.*, 1984, **129**, 271.

[189] S. Kobayoshi, T. Fukada, H. Yukimasa, M. Fujino, I. Azuma, and Y. Yamamura, *Bull. Chem. Soc. Jpn.*, 1984, **57**, 3182.

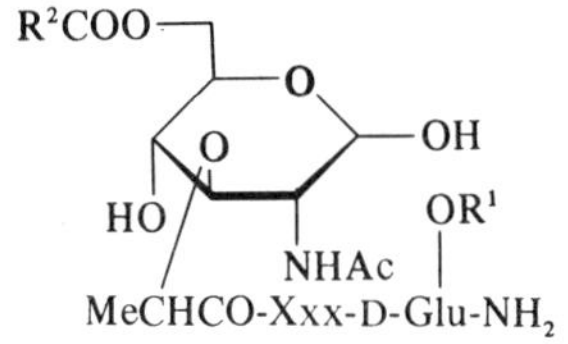

(68) Xxx = Val, Ser(Bzl), Thr(Bzl), Ser, or Thr
R^1 = Me, Et, Pr^i, Bu^t, or n-Hex

R^2 = $Me(CH_2)_{16}$— or

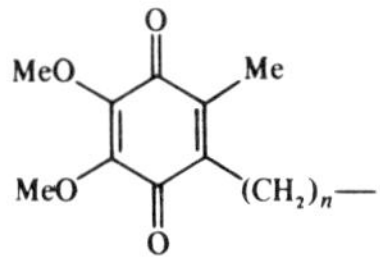

n = 2, 5, 9, 15, or 21

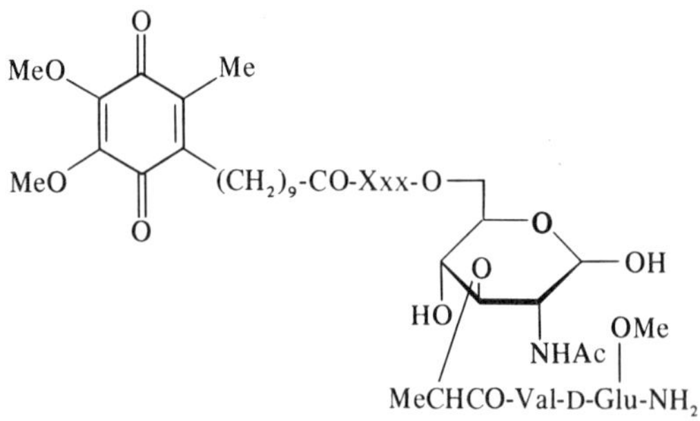

(69) Xxx = Gly or Leu

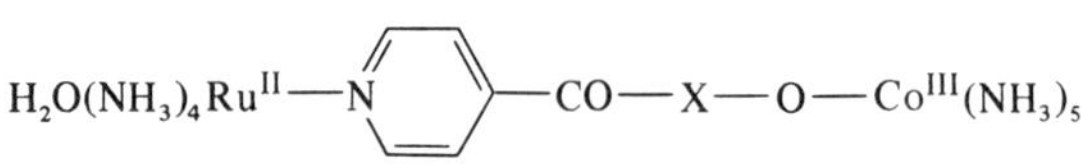

(70) X = Gly-Gly, Gly-Phe, Gly-Leu, or Phe-Phe

gram quantities of highly pure product has been developed; f.a.b.-m.s. was found to be useful for the characterization of the product.[190]

α-Amino Conjugates. – In a series of complexes of the type (70), the effects of the flexible bridging group on the rate of intramolecular electron transfer and its temperature dependence have been studied. Large variations in the latter were observed for the different amino acid bridges.[191] In another series of type (71),

[190] L. R. Phillips, O. Nishimura, and B. A. Fraser, *Carbohydr. Res.*, 1984, **132**, 275.
[191] S S. Isied and A. Vassilian, *J. Am. Chem. Soc.*, 1984, **106**, 1726.

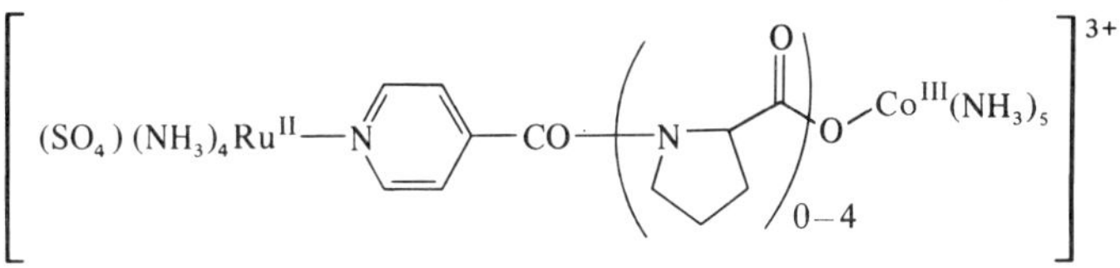

(71)

a decrease in electron-transfer rate occurred on introducing one or two proline bridging groups; the more rapid rate observed with Pro_3 and Pro_4 links indicates a conformation with the metal ions in close proximity.[192]

N-Palmitoyl-Ser-Ser-Asn-Ala-OH, an analogue of the N-terminal part of the lipoprotein from the outer membrane of *E. coli*, has been synthesized. It shows stimulatory activity towards primary B-lymphocytes comparable to the effect of the native lipoprotein. The B-lymphocyte tumour cell line BCl-1 is also activated by the compound.[193] A series of derivatives of the [15]-peptide H-Ser_2-Leu-Lys-Glu-Tyr-Trp-Ser_2-Leu-Lys-Glu-Ser-Phe-Ser-OH with *N*-acyl fatty-acid chains of up to 18 carbon atoms have been prepared. As the chain increased in size, so did the binding to high-density lipoprotein. As judged by c.d. spectra, the association with phospholipid was accompanied by increased α-helical structure.[194] *N*-(2-Chloroethyl)-*N*-nitrosocarbamoyl derivatives of α-melanotropin and gastrin fragment peptides have been prepared by active ester acylation of the peptides. These derivatives significantly increased the life span of L1210 leukaemia-bearing mice.[195]

The biocytin (Bct) derivatives [N^{α}-Bct-Ser^1]- and [12-Bct-N^{α}-dodecanoyl-Ser^1]-[Nle^4, D-Phe^7]-α-MSH have been made. These modifications of the superpotent analogue of α-melanotropin [Nle^4, D-Phe^7]α-MSH proved highly potent, non-biodegradable, and prolonged in their biological activity. Incorporation of biotin was examined because the avidin–biotin system can be used to construct sensitive methods for the visual localization of specific receptors on or within cells.[196] A synthetic [44]-peptide amide with the sequence of human-growth-hormone-releasing factor has been converted to its N^{α}-biotinyl derivative using biotinyl-*N*-1-succinimidyl ester. The product had 60% of the biological activity of the parent.[197]

α-Carboxyl Conjugates. – Z-Gly-Leu-aminohexyl-Sepharose has been used to purify the alkaline protease produced by *Streptomyces violaceorectus* by

[192] S. S. Isied and A. Vassilian, *J. Am. Chem. Soc.*, 1984, **106**, 1732.
[193] W. G. Bessler, M. Cox, K. H. Wiesmuller, and G. Jung, *Biochem. Biophys. Res. Commun.*, 1984, **121**, 55.
[194] G. Ponsin, K. Strong, A. M. Gotto, J. T. Sparrow, and H. J. Pownall, *Biochemistry*, 1984, **23**, 5337.
[195] H. Süli-Vargha and K. Medzihradsky, *Int. J. Pept. Protein Res.*, 1984, **23**, 650.
[196] D. N. Chaturvedi, J. J. Knittel, V. J. Hruby, A. M. de L. Castrucci, and M. E. Hadley, *J. Med. Chem.*, 1984, **27**, 1406.
[197] N. Fujii, M. Shimakura, H. Yajima, F. Shono, M. Tsuda, and A. Yoshitake, *Chem. Pharm. Bull.*, 1984, **32**, 1200.

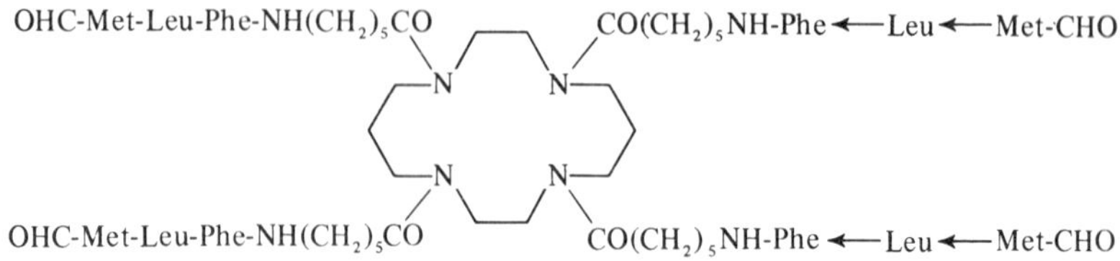

(72)

affinity chromatography, giving material of electrophoretic homogeneity.[198] Cyclic tetrameric clusters of chemotactic peptides linked at their carboxyl ends have been found to be superactive activators of lysozyme release from human neutrophils. The most potent compound tested (72) proved to be 1100 times as active as HCO-Met-Leu-Phe-OMe.[199] A series of tripeptide carbamate esters have been prepared and tested for inhibitory activity towards porcine pancreatic elastase, trypsin, and chymotrypsin. Two of the compounds, methylsuccinyl-Ala_2-Pro-CH_2-*N*-($CHMe_2$)CO_2R (R = phenyl or *p*-nitrophenyl), specifically inhibited elastase but had no effect on the other two serine proteases.[200]

Enzymatic hydrolysis of albomycins δ_1 and δ_2 (73) with several enzymes cleaves off serine-containing nucleosides, but with microsomal leucine amino-peptidases the serine-free nucleosides were obtained instead.[201] Recent work also establishes that in the albomycins the N^5-acetyl-N^5-hydroxyornithines and serine residues are all of the L-configuration, and an *X*-ray analysis of the sulphoxide of the nucleoside moiety confirms that it contains 6-amino-6-deoxy-4-thio-L-glycero-α-L-idoheptanofuranonic acid.[202] N^5-Acetyl-N^5-hydroxy-L-orni-

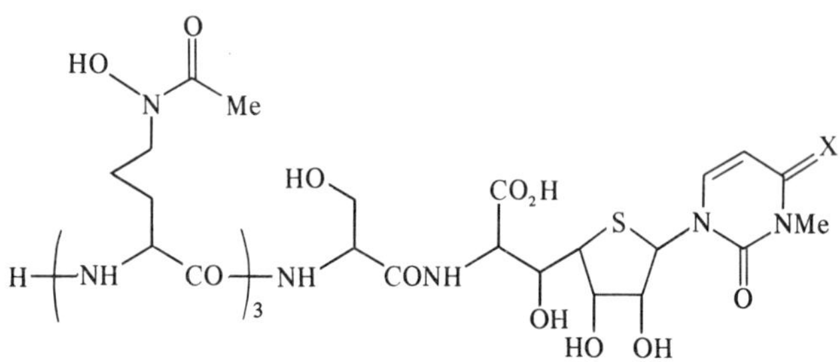

(73) δ_1, X = O; δ_2, X = $NCONH_2$; ϵ, X = NH

[198] Y. Ipoue, S. Nakamura, M. Hamada, K. Aikawa, K. Kitagaki, K. Sato, and E. Yasuda, *Chem. Pharm. Bull.*, 1984, **32**, 4036.

[199] J.-L. Kraus, A. Dipaola, and B. Belleau, *Biochem. Biophys. Res. Commun.*, 1984, **124**, 939.

[200] K. Tsuji, B. L. Aghaa, M. Shinogi, and G. A. Digenis, *Biochem. Biophys. Res. Commun.*, 1984, **122**, 571.

[201] G. Benz, *Liebigs Ann. Chem.*, 1984, 1399.

[202] G. Benz, L. Barn, M. Brieden, R. Grosser, J. Kurz, H. Paulsen, V. Sinnwell, and B. Weber, *Liebigs Ann. Chem.*, 1984, 1408.

thine has been prepared from L-glutamic acid[203] and used in a total synthesis of the tripeptide hydroxamic acid amino terminus of the albomycins.[204]

Two fluorescent peptides H-Leu-Arg$_2$-Ala-Ser-Leu-X [X = 2-(dansylamino)-ethylamino (DAE) or 4-methyl-2-oxo-7-chromenylamino] have been synthesized as substrates of the cyclic AMP-dependent protein kinase (CAPK) reaction. Both peptides were phosphorylated by ATP and the catalytic subunit of CAPK from bovine heart stoicheiometrically. Phosphorylation of the DAE-peptide is accompanied by a 10% increase in the fluorescence intensity at 550 nm, enabling peptide phosphorylation to be assayed by fluorescence spectroscopy.[205] The lymphocyte-differentiating hormone Glp-Ala-Lys-Ser-Gln-Gly$_2$-Ser-Asn-OH, a thymic nonapeptide, has been coupled to both spermidine and spermine. Both conjugates inhibited rather than stimulated incorporation of [^{3}H]thymidine into DNA at levels 20-fold that of spermine.[206] *p*-Oxymethylbenzylcholestan-3β-yl succinates of H$\{$Lys(Boc)$\}_n$OH ($n = 5-10$) and its *N*-α-Boc derivative ($n = 10$ or 20) have been prepared and their conformations examined. From pentamer to decamer, intermolecular β-structure is seen both in the solid state and in solvents of low polarity. The [20]-peptide exhibits a high percentage of α-helical structure both in the solid state and in trifluoroethanol. It seems that the bulky end group does not significantly affect the secondary structure adopted by the peptides.[207]

Side-chain Conjugates. – A naturally occurring nucleopeptide fragment of polio genome RNA in which the phenolic hydroxyl of the dipeptide H-Ala-Tyr-NH$_2$ is linked *via* a phosphodiester bond to the 5'-end of the RNA dimer UpU has been prepared. This derivative was completely digested by ribonuclease.[208] The synthesis of H-Glu-*O*-phospho-Ser-Leu-OH from N^{α}-Boc-*O*-dibenzylphosphono-Ser-OH has been reported. This tripeptide sequence occurs in the heavily phosphorylated area of α- and β-casein chains.[209]

A series of alkyl ethers, fatty-acid esters, benzoates, and other derivatives of RA-V (74), the antitumour cyclic hexapeptide from *Rubiae radix* characterized in 1983, have been made and examined for activity against P-388 lymphocytic leukaemia. The most desirable compounds found in terms of both antitumour activity and toxicity were those with C_1 and C_6 chains on the phenol hydroxyl.[210] The heptapeptide H-Leu-Arg$_2$-Ala-*O*-phospho-Tyr-Leu-Gly-OH, which contains the CAPK substrate amino acid sequence except that serine is substituted by

203 G. Benz, *Liebigs Ann. Chem.*, 1984, 1424.
204 G. Benz and D. Schmidt, *Liebigs Ann. Chem.*, 1984, 1434.
205 M. Kondo, K. Takaki, R. Kuroki, A. Tada, K. Fukumoto, and J. Sunamoto, *Bull. Chem. Soc. Jpn.*, 1984, **57**, 2957.
206 H.-M. Shieh, D. Campbell, and K. Folkers, *Biochem. Biophys. Res. Commun.*, 1984, **122**, 21.
207 C. Toniolo, G. M. Bonora, I. F. Lüscher, and C. H. Schneider, *Int. J. Pept. Protein Res.*, 1984, **23**, 47.
208 C. Shattenkerk, C. T. J. Wreesman, M. J. de Graaf, G. A. van der Marel, and J. H. van Boom, *Tetrahedron Lett.*, 1984, **25**, 5197.
209 P. F. Alewood, J. W. Perich, and R. B. Johns, *Tetrahedron Lett.*, 1984, **25**, 987.
210 H. Itokawa, K. Takeya, N. Mori, T. Sonobe, W. Serisawa, T. Hamanaka, and S. Mihoshi, *Chem. Pharm. Bull.*, 1984, **32**, 3216.

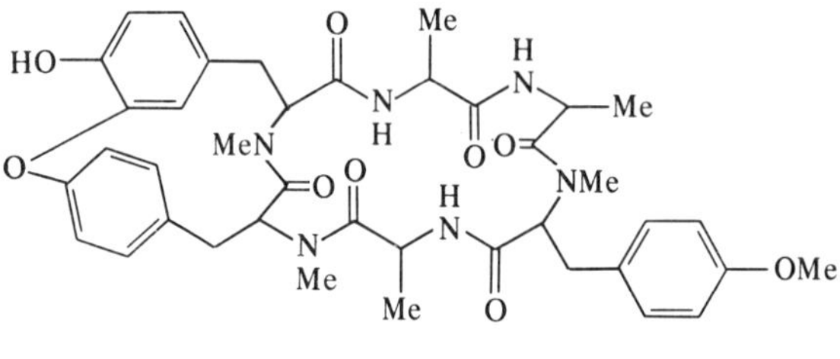

(74)

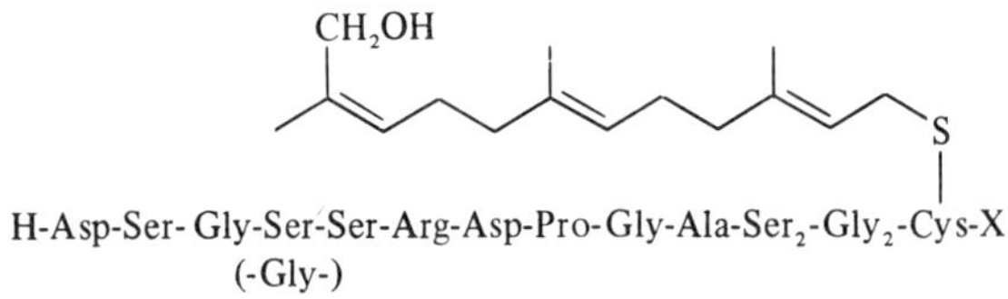

(75) I, X = OMe; VIII, X = OH

tyrosine, has been prepared by solid-phase methodology using N^{α}-Boc-*O*-diMephosphono-Tyr-OH.[211] The structures (75) of two peptidyl sex hormones of the fungus *Tremella brasiliensis* have been determined. Both hormones exhibit microheterogeneity at the fourth residue (Gly or Ser) and have a unique isoprenoidal group blocking the sulphydryl group of the C-terminal Cys residue.[212]

[211] R. M. Valerio, P. F. Alewood, R. B. Johns, and B. E. Kemp, *Tetrahedron Lett.*, 1984, **25**, 2609.
[212] Y. Ishibashi, Y. Sakagami, A. Isogai, and A. Suzuki, *Biochemistry*, 1984, **23**, 1399.
[213] A. A. Amoscato, G. F. Babcock, and K. Nishioka, *Peptides*, 1984, **5**, 489.

5

β-Lactam Antibiotic Chemistry

BY J. BRENNAN

1 Introduction

The high level of activity in this field, which was generated by the discovery of such compounds as clavulanic acid and thienamycin and by the synthesis of the penems in the mid-1970s continues apace, while studies on biosynthetic pathways have produced a number of subtle and intriguing insights. Reviews have included a general overview,[1] a description of methods applied to the stereospecific synthesis of chiral β-lactam rings,[2] and a discussion on the relationship between structure, β-lactamase stability, and antibacterial activity in compounds of potential clinical interest;[3] the proceedings of an international symposium held this year were published in 1985.[4]

2 Nomenclature

Monocyclic compounds will be numbered as azetidine derivatives whereas bicyclic compounds will be named according to the penam–penem (1) and cepham–cephem (2) systems, with numbering always commencing at the position corresponding to sulphur.

3 New Natural Products

Several groups have reported the isolation from bacterial sources of 7α-formamido cephalosporin derivatives (3a–f) bearing highly substituted C-3′ acyloxy side

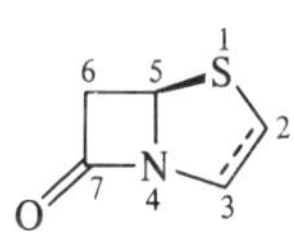

(1)

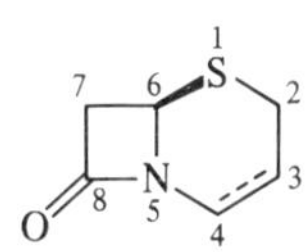

(2)

[1] S. M. Roberts, *Chem. Ind. (London)*, 1984, 162.
[2] R. Labia and C. Morin, *J. Antibiot.*, 1984, **37**, 1103.
[3] C. M. Cimarusti, *J. Med. Chem.*, 1984, **27**, 247.
[4] 'Recent Advances in the Chemistry of β-Lactam Antibiotics', ed. A. G. Brown and S. M. Roberts, Special Publication No. 52, The Royal Society of Chemistry, London, 1985.

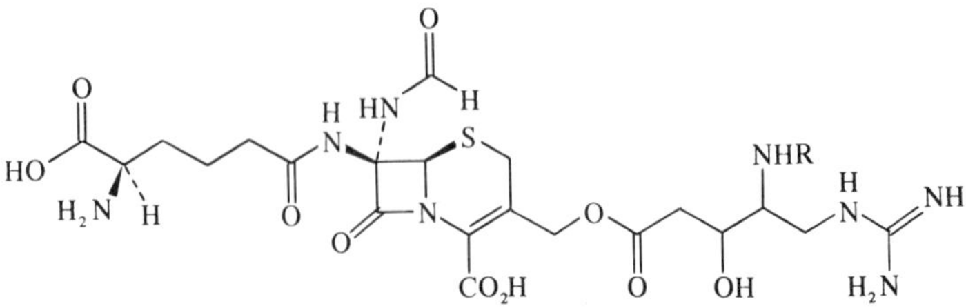

(3) a; R = Ala-H
b; R = Ala-Ala-H
c; R = Ala-Ala-Ala-H
d; R = Ser-H
e; R = Ser-Ser-H
f; R = Ser-Ser-Ala-H

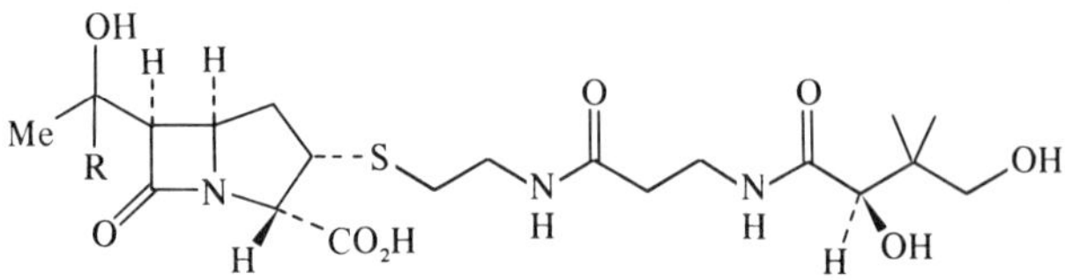

(4) a; R = H
b; R = Me

chains.[5–8] These have been variously named as chitinovorins,[5] cephabacins,[7] and chitinobolins,[8] and they display weak to moderate antibacterial activity.

Carbapenams OA6129D (4a) and OA6129E (4b), which may be of significance in carbapenem biosyntheses, have been isolated from a streptomycete and display extremely weak antibacterial activity.[9] Although strictly beyond the scope of this review, obafluorin (5), a naturally occurring β-lactone, is apparently

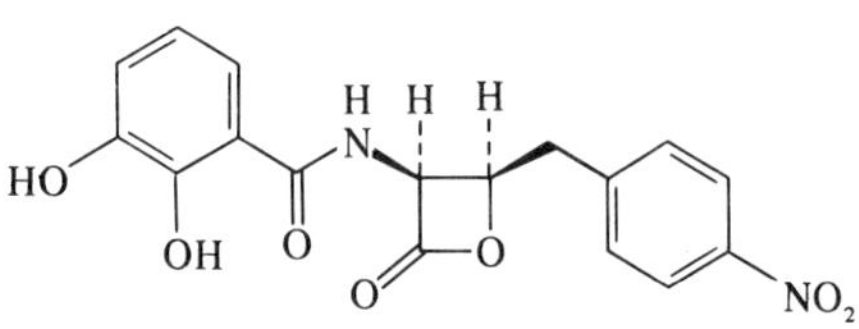

(5)

[5] P. D. Singh, M. G. Young, J. H. Johnson, C. M. Cimarusti, and R. B. Sykes, *J. Antibiot.*, 1984, **37**, 773.

[6] J. Shoji, T. Kato, R. Sakazaki, W. Nagata, Y. Terni, Y. Nakagawa, M. Shiro, K. Matsumoto, T. Hattori, T. Yoshida, and E. Kondo, *J. Antibiot.*, 1984, **37**, 1486.

[7] S. Tsubotani, T. Hida, F. Kasahara, Y. Wada, and S. Harada, *J. Antibiot.*, 1984, **37**, 1546.

[8] Y. Nozaki, K. Okonogi, N. Katayama, H. Ono, S. Harada, M. Kondo, and H. Okazaki, *J. Antibiot.*, 1984, **37**, 1555.

[9] T. Yoshioka, A. Watanabe, I. Kojima, Y. Shimauchi, M. Okabe, Y. Fukagawa, and I. Ishikura, *J. Antibiot.*, 1984, **37**, 211.

produced by a wide variety of *Pseudomonas* strains; the compound exhibits low antibacterial activity and a high susceptibility to β-lactamases.[10]

4 Biosynthesis

Studies on the penicillin and cephalosporin biosynthetic pathways have been the subject of a large number of reports this year, with high-field n.m.r. techniques playing an important role in many of these.

Cyclization of δ-(L-α-amino-δ-adipoyl)-L-cysteinyl-D-valine (LLD-ACV) (6) to the free β-lactam-4-thiol intermediate (7) during the biosynthesis of isopenicillin N (8) appears to be highly unlikely. Attempts to synthesize and isolate (7), which, it is claimed,[11] was isolated from a *Penicillium chrysogenum* system, revealed that while moderately stable at pH 1.5 and 20 °C ($t_{1/2}$ = 25 min) the compound was extremely labile at biological pH values; from these results the earlier report was considered to have been in error and the question of the presence of free (7) in penicillin biosynthesis to be unresolved.[12]

Incubation of the *N*-hydroxy LLD-ACV analogue (9) with a *Cephalosporium acremonium* did not yield isopenicillin N[13] and therefore failed to support an earlier model study by the same group which had suggested that N-activation might be a key step in biosynthetic β-lactam formation.

That β-lactam formation is a rate-determining step which precedes closure of the C–S bond in LLD-ACV (6) was supported by incubation of deuteriated compounds (10a) and (10b).[14] Thus isopenicillin synthetase from a *C. acremonium* gave a slower conversion of (10a), which was deuteriated at the site of β-lactam N–C bond formation, than of the undeuteriated substrate (6), whereas no significant discrimination in rates was observed between (6) and (10b), which was deuteriated at the site of C–S ring closure.

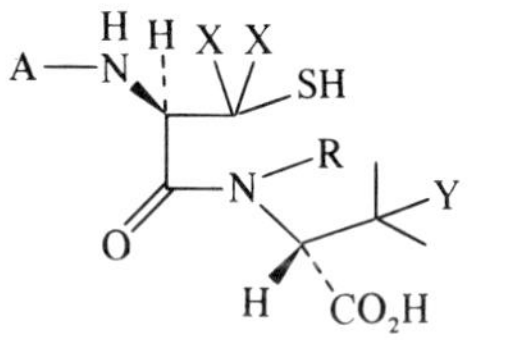

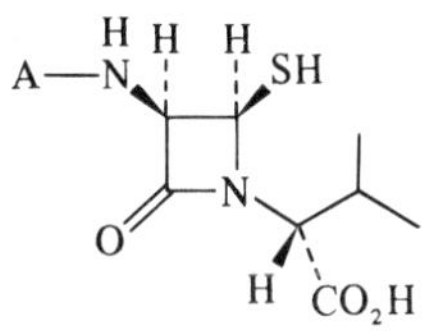

(7)

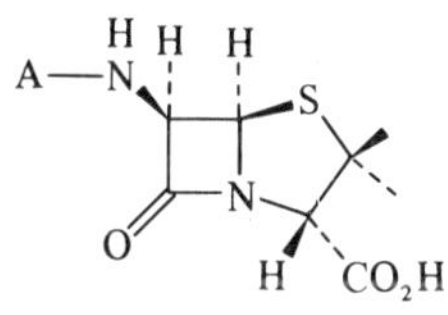

(8)

(6) R = H, X = H, Y = H
(9) R = OH, X = H, Y = H
(10) a; R = H, X = D, Y = H
b; R = H, X = H, Y = D

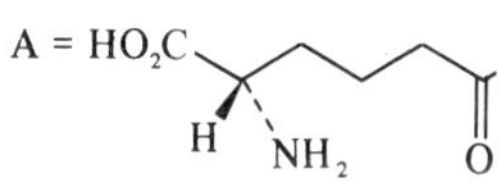

[10] J. Scott Wells, W. H. Trejo, P. A. Principe, and R. B. Sykes, *J. Antibiot.*, 1984, **37**, 802.

[11] B. Meesschaert, P. Adriaens, and H. Eyssen, *J. Antibiot.*, 1980, **33**, 722.

[12] J. E. Baldwin, E. P. Abraham, R. M. Adlington, M. J. Crummin, L. D. Field, G. S. Jayatilake, R. L. White, and J. J. Usher, *Tetrahedron*, 1984, **40**, 1907.

[13] R. L. Baxter, G. A. Thomson, and I. A. Scott, *J. Chem. Soc., Chem. Commun.*, 1984, 32.

[14] J. E. Baldwin, R. M. Adlington, S. E. Moroney, L. D. Field, and H.-H. Ting, *J. Chem. Soc., Chem. Commun.*, 1984, 984.

Extensive studies involving modification of the valine moiety of LLD-ACV (6) have shown that *C. acremonium* will tolerate considerable variation of this residue provided that closure to a five-membered ring, or larger, remains possible. Thus, inhibition of the synthetase system was observed with the glycine-substituted analogue (11), and this was consistent with enzymic recognition of the substrate and with cyclization to a monocyclic intermediate such as (13). Absence of a suitable site on (13) for C–S bond formation then frustrated expulsion of a bicyclic product and therefore blocked the enzyme.[15] In the labelled system (12) ^{3}H release increased with enzyme concentration and parallelled the rate of decrease in enzymic activity. Reductive degradation of the blocked enzyme with NaB[^{3}H]$_4$ yielded (12), which was considered to support the presence of a monocyclic intermediate such as (13), although the alternative of an enzyme-bound thioaldehyde could not be discounted.

Introduction of substrate analogues of LLD-ACV in which sulphur–carbon bond formation was possible yielded a number of interesting bicyclic structures.[16] Compound (14) yielded the known penam (15) and novel cephams (16a and b);

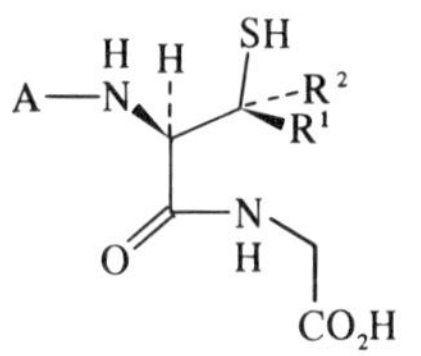

(11) $R^1 = R^2 = H$
(12) $R^1 = H,\ R^2 = {}^3H$
$R^1 = {}^3H,\ R^2 = H$

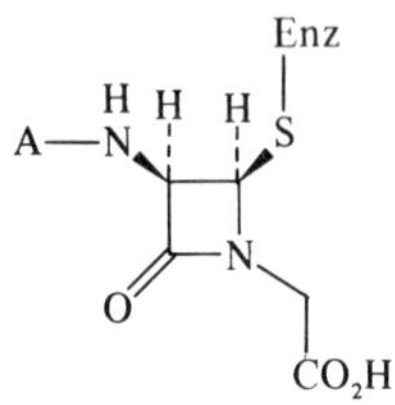

(13)

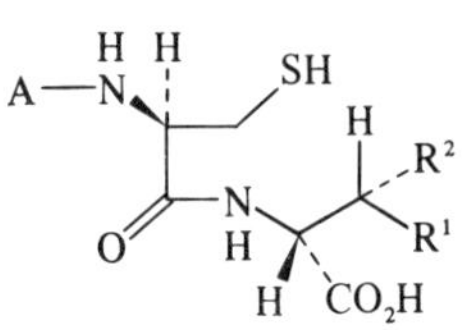

(14) $R^1 = Et,\ R^2 = Me$
(17) $R^1 = H,\ R^2 = Et$

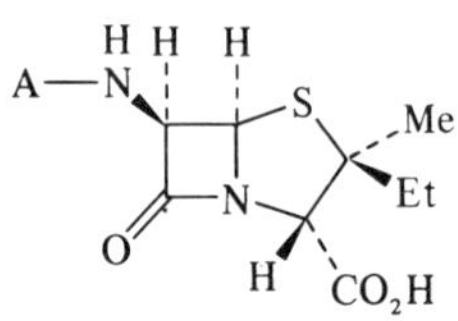

(15)

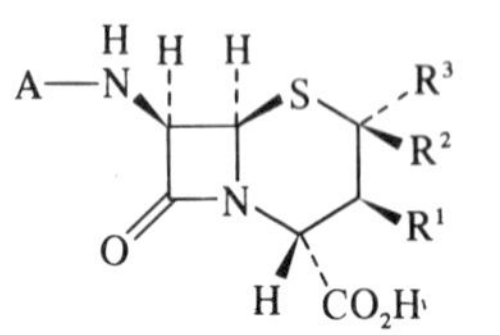

(16) a; $R^1 = R^2 = Me,\ R^3 = H$
b; $R^1 = R^3 = Me,\ R^2 = H$
(18) a; $R^1 = R^2 = H,\ R^3 = Me$
b; $R^1 = R^3 = H,\ R^2 = Me$

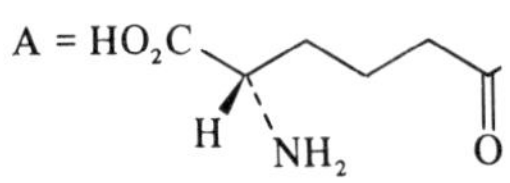

[15] J. E. Baldwin, E. P. Abraham, C. G. Lovel, and H.-H. Ting, *J. Chem. Soc., Chem. Commun.*, 1984, 902.
[16] J. E. Baldwin, R. M. Adlington, N. J. Turner, B. P. Domayne-Hayman, H.-H. Ting, A. E. Derome, and J. A. Murphy, *J. Chem. Soc., Chem. Commun.*, 1984, 1167.

compound (17) yielded only the cephams (18). These results were presented as further support for the dual-pathway mechanisms in which the types of products are determined by C–H bond-dissociation energies and radical stabilities. Thus with (14) formation of a tertiary radical leads to a penam product whereas formation of a less stable secondary radical leads to the less strained cephams; with (17), where both penam and cepham products could arise *via* a secondary radical, only the latter products (18) were in fact observed.

Extension of this approach to the introduction of allylic substrate (19) led to the possibility of generating the allylic radical (20), which could undergo cyclization to yield five- or seven-membered rings.[17] Experiment showed that both of these processes were effective and gave rise to (21) and (22), with a predominance of the latter suggesting a lower energy barrier for this cyclization. Hydroxylated bicyclic [4.2.0] and [5.2.0] systems were also shown to be produced, but these were thought to have arisen by a different pathway.

Studies on the interaction of isopenicillin N synthetase with analogues of LLD-ACV (6) in which the α-aminoadipoyl residue had been modified have been reported by a number of groups.[18,19,20] The results indicate that to be processed by the enzyme the minimum requirements are that the substituting residue must be a six-membered chain bearing a terminal carboxylic acid group and if an

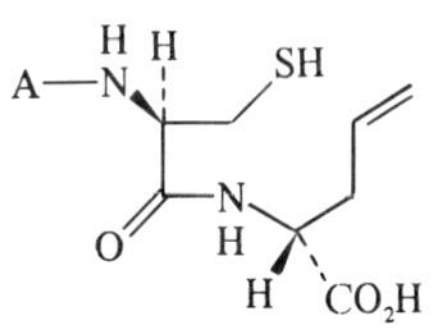

(19)

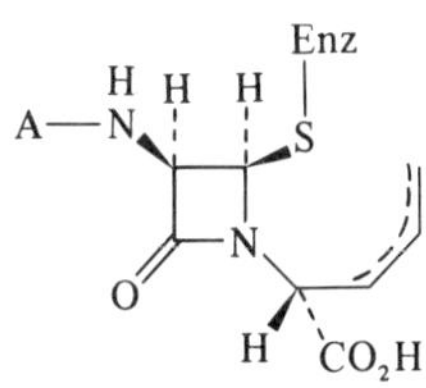

(20)

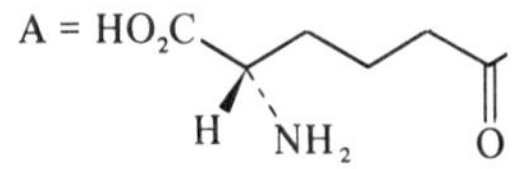

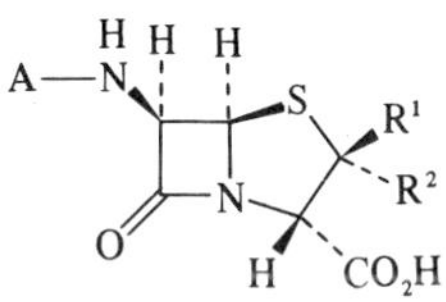

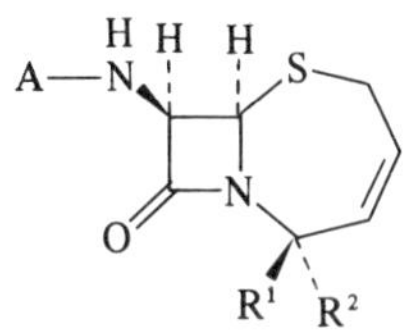

(21) a; R^1 = H, R^2 = 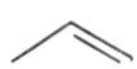

b; R^1 = 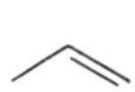, R^2 = H

(22) a; R^1 = H, R^2 = CO_2H
b; R^1 = CO_2H, R^2 = H

[17] J. E. Baldwin, R. M. Adlington, A. E. Derome, H.-H. Ting, and N. J. Turner, *J. Chem. Soc., Chem. Commun.*, 1984, 1211.

[18] J. E. Baldwin, E. P. Abraham, R. M. Adlington, G. A. Bahadur, B. Chakravarti, B. P. Domayne-Hayman, L. D. Field, S. L. Flitsch, G. S. Jayatilake, A. Spakovskis, H.-H. Ting, N. J. Turner, R. L. White, and J. J. Usher, *J. Chem. Soc., Chem. Commun.*, 1984, 1225.

[19] S. E. Jensen, D. W. S. Westlake, R. J. Bowers, C. F. Ingold, M. Jouany, L. Lyubechansky, and S. Wolfe, *Can. J. Chem.*, 1984, **62**, 2713.

[20] J. E. Shields, C. S. Campbell, S. W. Queener, D. C. Duckworth, and N. Neuss, *Helv. Chim. Acta*, 1984, **67**, 870.

amino group is present in the α-position it must be free of *N*-acyl substitution. Thus, whereas (23) and (24) were substrates, (25) and (26) were not, and such observations are consistent with the presence of a basic anchoring site on the synthetase enzyme that interacts with the adipoyl terminal acid group and locates the elements of the bicyclic β-lactam system at the correct sites on the enzyme for the ring-forming reactions to take place. A report that DLD-ACV (27), in which the adipoyl side-chain chirality is reversed, was a substrate for an isopenicillin N synthetase from *C. acremonium* and that this gave rise to penicillin N (28)[18] conflicts with another report that no penicillin N could be isolated from a similar enzyme system,[19] although the latter group identified deacetoxycephalosporin C (29) derived from penicillin N when they combined synthetase and expandase enzyme systems and introduced (27) as a precursor.[21]

Whereas (23) was utilized with only 1–2% of the efficiency of LLD-ACV (6) by *Streptomyces clavuligerus*,[19] the thia isostere (24) was processed by the *C. acremonium* system with similar efficiency to (6), although slightly more slowly.[20] Further studies with (24) showed the synthetase product (30) to be a substrate for the isomerase and expandase enzyme systems from *S. clavuligerus*, giving rise to cephem (32); the existence of a 1:1 equilibrium between (30) and (31) on this pathway could be detected by h.p.l.c., whereas in the natural system the equilibrium (8)⇌(28) has so far proved impossible to observe (Scheme 1). In common with the synthetase, the isomerase and expandase enzymes processed the thia analogues with similar efficiency to the natural substrates but more slowly.[22]

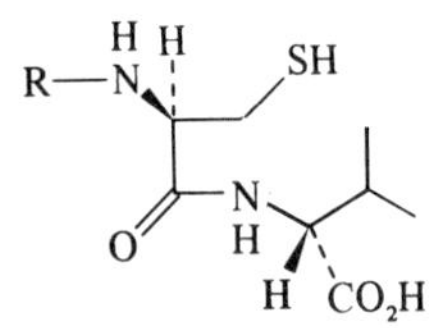

(6) R = 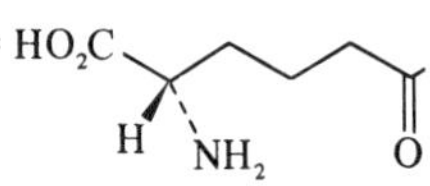

(23) R = 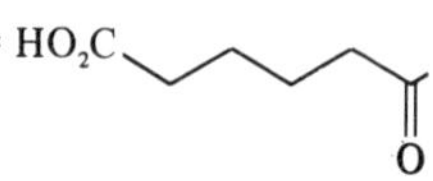

(24) R = 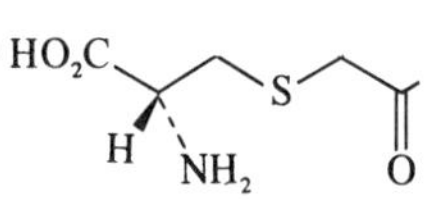

(25) R = 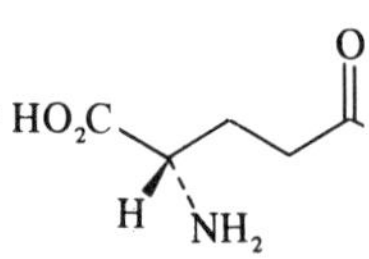

(26) R = 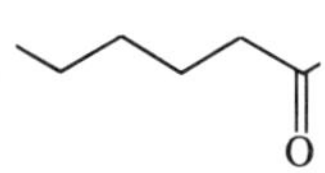

(27) R = 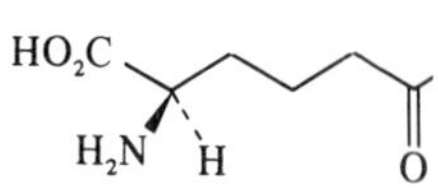

[21] Y.-Q. Shen, S. Wolfe, and A. L. Demain, *J. Antibiot.*, 1984, **37**, 1044.
[22] R. J. Bowers, S. E. Jensen, L. Lyubechansky, D. W. S. Westlake, and S. Wolfe, *Biochem. Biophys. Res. Commun.*, 1984, **120**, 607.

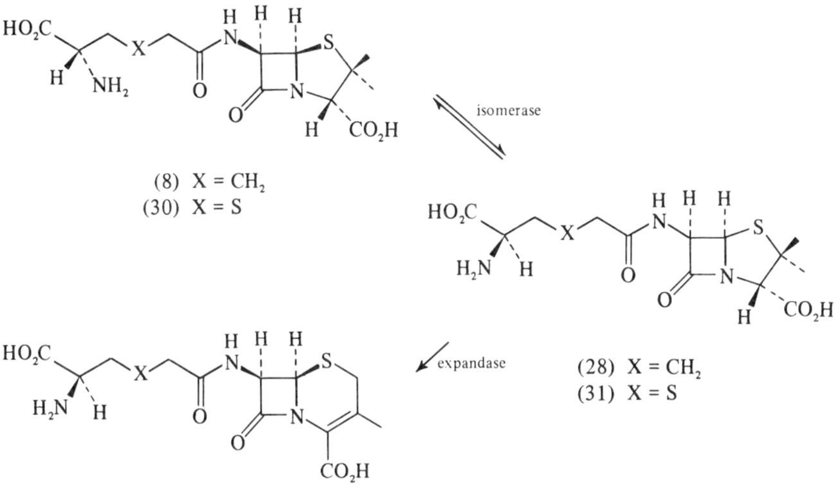

(8) X = CH_2
(30) X = S

(28) X = CH_2
(31) X = S

(29) X = CH_2
(32) X = S

Scheme 1

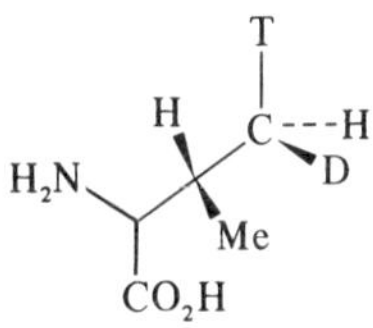

(33)

A partially purified enzyme from an industrial strain of *C. acremonium* was shown to be active as both a penicillin N expandase and a deacetoxycephalosporin C C-3′ hydroxylase;[23] the two functions required the same cofactors and could not be separated by SDS polyacrylamide electrophoresis, from which the authors suggested the presence of a single bifunctional enzyme. Feeding of the isotopically labelled valine precursor (33) to *C. acremonium* CW19 and elegant chemical and enzymic degradation of the labelled cephalosporin C product demonstrated that hydroxylation at C-3′ of deacetoxycephalosporin C (29) had taken place with retention of stereochemistry on the C-3′ methyl group.[24]

In an attempt to characterize 2-hydroxymethyl penam and 3-hydroxy cepham, biosynthetic intermediates in cephem biosynthesis broths were subjected to oxidation conditions designed to stabilize the intermediates as their sulphoxides, but only sulphoxides of penicillin N and cephalosporin C (34) were isolated.[25] In

[23] A. Scheidegger, M. T. Künzi, and J. Nusch, *J. Antibiot.*, 1984, **37**, 522.
[24] C. A. Townsend and E. B. Barrabee, *J. Chem. Soc., Chem. Commun.*, 1984, 1587.
[25] S. Kukolja, S. W. Queener, R. D. Miller, D. C. Duckworth, D. E. Dorman, L. L. Huckstep, and N. Neuss, *Helv. Chim. Acta*, 1984, **67**, 876.

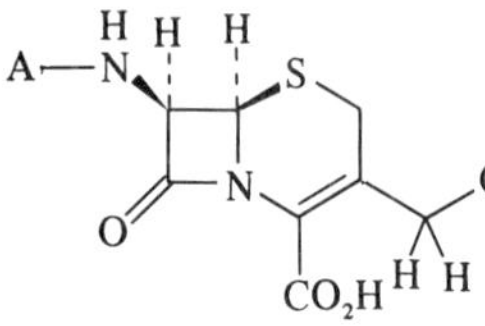

(34) R = Ac
(36) R = H

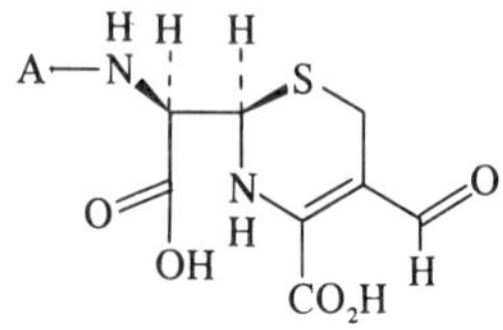

(35)

A = 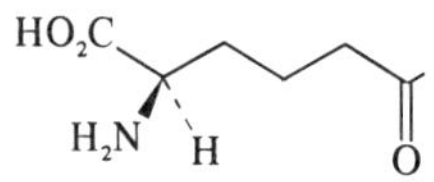

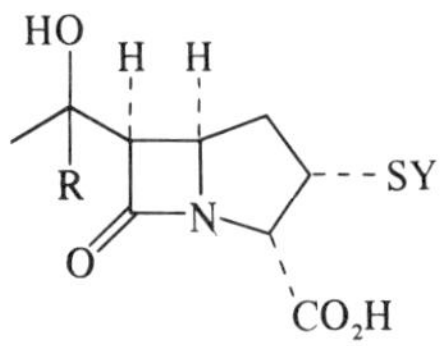

(4) a; R = H
b; R = Me

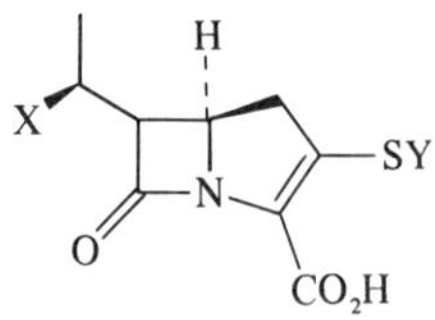

(37) X = H, C-6—R
(38) X = OH, C-6—R
(39) X = OH, C-6—R
(40) X = OSO_3H, C-6—R

Y = 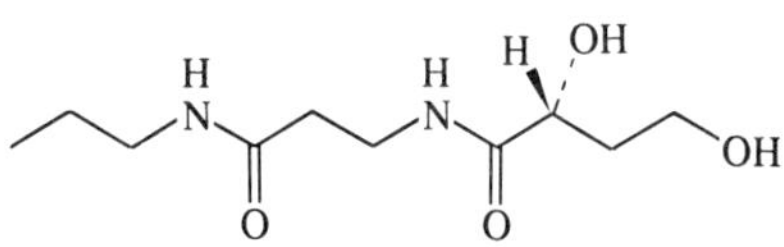

accord with a previous failure to incorporate penicillin N β-sulphoxide into cephem biosynthesis,[26] neither this nor the α-isomer was a substrate for deacetoxycephalosporin C synthetase. Isolation of (35) from a mutant cephamycin producer was presented as support for the hypothesis that deacetylcephalosporin C (36) is an intermediate in cephamycin C biosynthesis and that C-7 methoxylation is a late event on the biosynthetic pathway.[27]

Some studies on the less well established carbapenem biosynthetic sequence have also been reported. Isolation of (4a) and (4b) from *Streptomyces* OA6129, a producer of the carbapenems OA6121A (37), OA6121B1 (38), OA6121B2 (39), and OA6121C (40), was in accord with the proposal that a carbapenam bearing the pantotheinyl substituent would be a carbapenem precursor. Whereas *Streptomyces fulvoviridis* Normally produces *C*-2-substituted carbapenems such as (41) and (42), a blocked mutant from this strain produced only the OA6129

[26] J. E. Baldwin, M. Jung, P. Singh, T. Wan, S. Haber, S. Herchen, J. Kitchin, A. L. Demain, N. A. Hunt, M. Kokshake, T. Konowi, and M. Yoshida, *Philos. Trans. R. Soc. London, B*, 1980, **289**, 169.

[27] A. Rein, B. Arison, T. W. Miller, B. Lago, and C. L. Ginther, *J. Antibiot.*, 1984, **37**, 664.

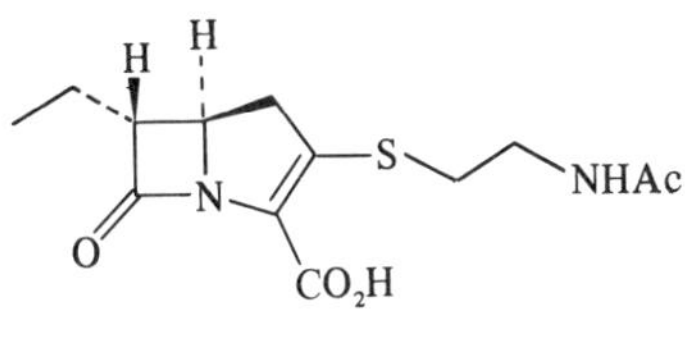

(41)

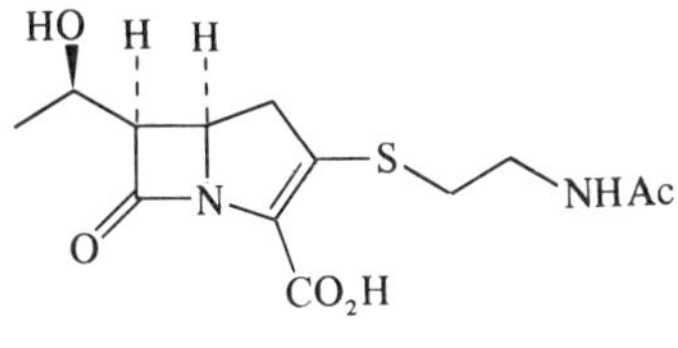

(42)

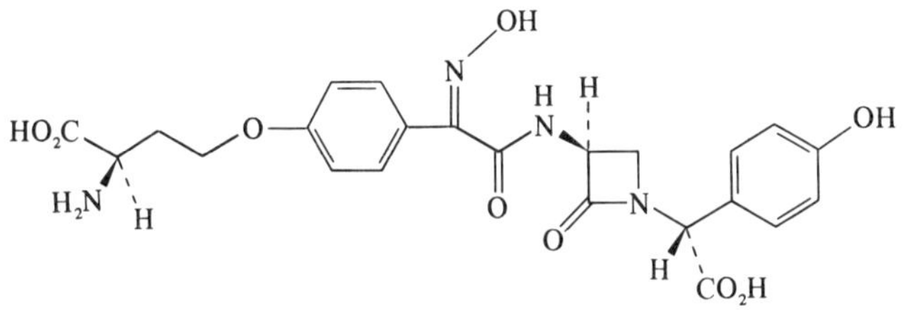

(43)

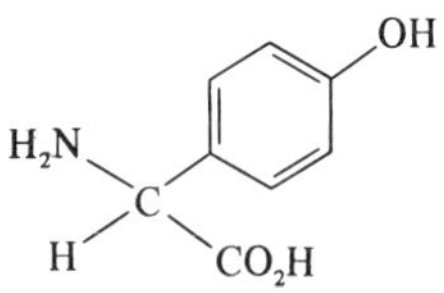

(44)

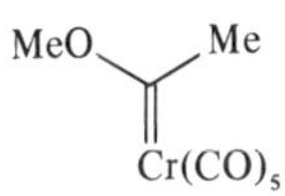

(45)

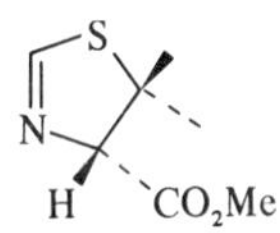

(46)

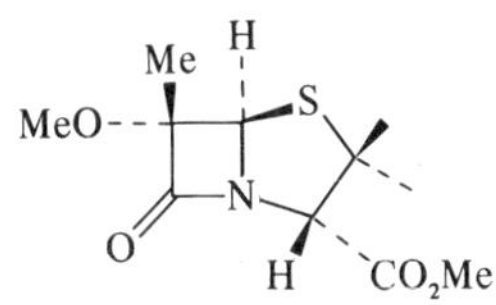

(47)

group, which could then be converted into the normal products by incubation with the parent strain;[28] isolation of an acylase from *S. fulvoviridis* and reaction of this with (37) in the presence of acetyl CoA resulted in the formation of (41), and similar results were observed for a number of OA6129 carbapenems.[29]

In the biosynthesis of the monocyclic β-lactam compound nocardicin A (43) it has been shown with D,L-[2-^{13}C,^{15}N](*p*-hydroxyphenyl) glycine (44) that this amino acid is a precursor to both the β-lactam ring and the side chain, and in the latter case it presumably undergoes oxidation during biosynthesis.[30]

5 The Chemistry of Penicillins and Cephalosporins

The ready availability of the penicillin nucleus from natural sources makes it an attractive system for chemical modification to novel antibacterials, and most reports relate to this. An approach to the chemical synthesis of penam systems has appeared, however, and this utilized the photolytic reaction of chromium–carbene complexes such as (45) with imine (46) to yield penam (47).[31]

[28] Y. Fukagawa, M. Okabe, S. Azuma, I. Kojima, T. Ishikura, and K. Kubo, *J. Antibiot.*, 1984, **37**, 1388.
[29] K. Kubo, T. Ishikura, and Y. Fukagawa, *J. Antibiot.*, 1984, **37**, 1394.
[30] C. A. Townsend and G. M. Salituro, *J. Chem. Soc., Chem. Commun.*, 1984, 1631.
[31] L. S. Hegedus, M. A. McGuire, L. M. Schultze, C. Yijun, and O. P. Anderson, *J. Am. Chem. Soc.*, 1984, **106**, 2680.

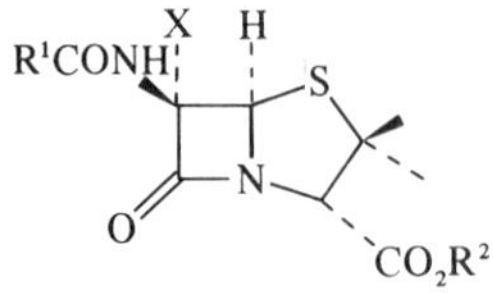

(48) X = CHO
(49) X = NHCHO

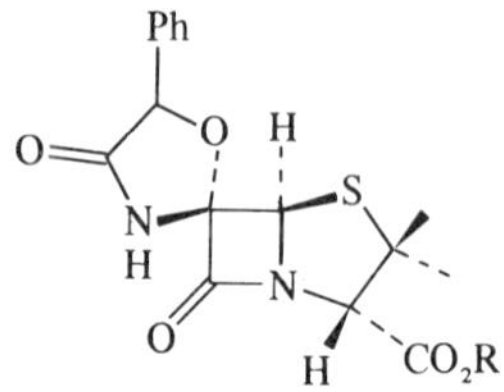

(50)

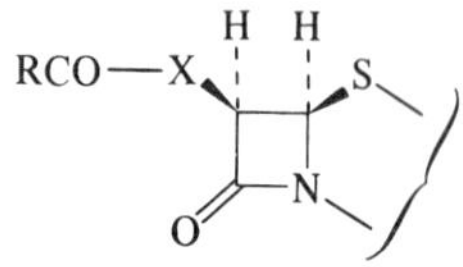

(51) X = O, S, or CH_2

Oxidation of 6α-hydroxymethyl penicillins with SO_3-pyridine yielded 6α-formyl derivatives (48) that could be further converted into acetyl and Wittig-derived products; 6α-substitution on the penicillin nucleus has normally been associated with a diminution of biological activity, and although the free acids of some 6α-formyl derivatives showed antibacterial activity none was β-lactamase stable.[32] In contrast, many 6α-formamido acids (49) showed good Gram-negative activity and β-lactamase stability;[33] they were synthesized by reaction of the appropriate 6α-thiomethyl derivatives with *N,N*-bis(trimethylsilyl)formamide in the presence of mercury(II) acetate. Similar 7α-cephem derivatives were synthesized but were of little biological interest, although it is worth noting that such compounds were also isolated from natural sources this year.[5-8] Spirocyclic penicillanates (50) were synthesized by oxidation (Bu^tOCl) at C-6 of the 6β-(α-hydroxyphenylamino) penicillanates, whereas the corresponding spiro cephalosporanates were prepared by reaction of 7α-methylthiacephems bearing the same 7β side chain. Although the free acids were active they were less so than analogous compounds in which spiro substitution was absent.[34] *In vitro* activities for penicillanic and cephalosporanic acid derivatives of the type (51) have been reported; these were much less antibacterially active than the corresponding acylamino derivatives, the general order of activity being $X = NH \gg O > CH_2 > S$.[35]

Decomposition of 6-diazopenicillanate (52) by $BF_3 \cdot Et_2O$ in the presence of substituted benzaldehydes gave rise to one, two, or three products (53), (54),

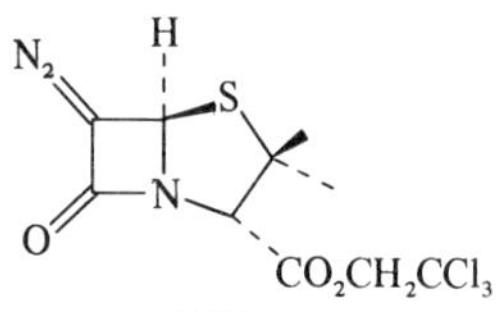

(52)

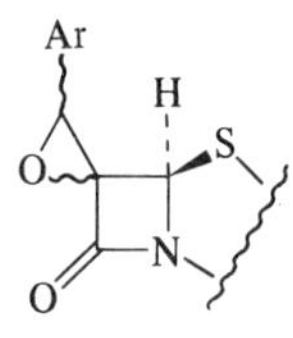

(53)

[32] A. W. Guest and P. H. Milner, *Tetrahedron Lett.*, 1984, **25**, 4845.
[33] P. H. Milner, A. W. Guest, F. P. Harrington, R. J. Ponsford, T. C. Smale, and A. V. Stachulski, *J. Chem. Soc., Chem. Commun.*, 1984, 1335.
[34] P. G. Sammes, S. Smith, and B. C. Ross, *J. Chem. Soc., Perkin Trans. 1*, 1984, 2117.
[35] J. C. Sheehan, E. Chacko, T. J. Commons, Y. S. Lo, D. R. Ponzi, A. Schwabacher, N. Solomon, and A. L. Demain, *J. Antibiot.*, 1984, **37**, 1441.

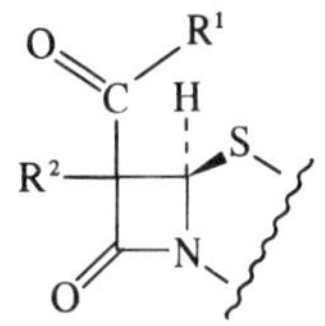

(54) $R^1 = H$
$R^2 = Ar$

(56) R^1 - alkyl or aryl
R^2 = alkyl or aryl

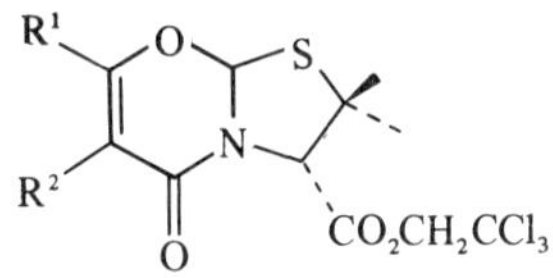

(55) $R^1 = H$
$R^2 = Ar$

(57) R^1 = alkyl or aryl
R^2 = alkyl or aryl

and (55) depending upon the nature of the aryl substituent. It was proposed that (55) resulted from heterolysis of the C-5–C-6 bond of (54), followed by cyclization on to the enolate oxygen.[36] Analogous reactions of (52) with ketones yielded only compounds of the types (56) and (57), whereas failure to observe the conversion of 6β-acyl compounds (56) into (57) led to the suggestion that interconversion only took place with the 6α isomers.[37]

Reaction of (52) with allylic sulphides or selenides gave products that could be oxidized to yield unstable allylidene derivatives, and these could be trapped with thiols to form stable 6β-vinyl derivatives (Scheme 2). Related products were synthesized by Michael addition of the C-6 magnesium derivative of benzyl 6,6-dibromopenicillanate to methyl α-(phenylsulphinyl)acrylate; the free acids of these 6β-vinylpenicillanate derivatives showed very low antibacterial activity.[38]

Unlike a number of previously reported bisnorpenicillin derivatives, the derivative (58), synthesized *via* the reaction of a 4-mercaptoazetidin-2-one with

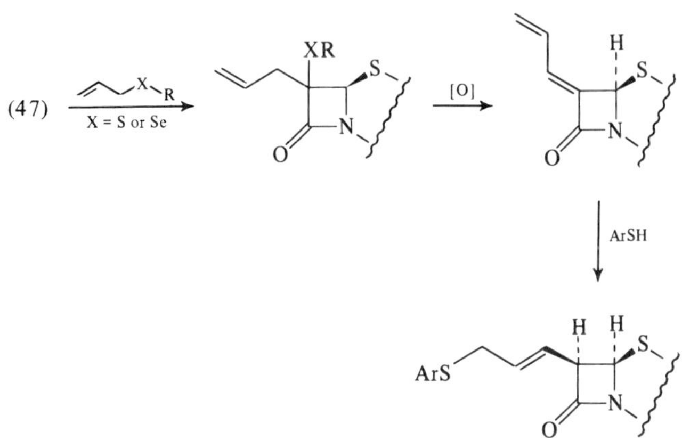

Scheme 2

[36] V. J. Jephcote, D. I. John, P. D. Edwards, K. Luk, and D. J. Williams, *Tetrahedron Lett.*, 1984, **25**, 2915.
[37] V. J. Jephcote and D. I. John, *Tetrahedron Lett.*, 1984, **25**, 2919.
[38] C. D. Foulds, A. A. Jaxa-Chamiec, A. C. O'Sullivan, and P. G. Sammes, *J. Chem. Soc., Perkin Trans. 1*, 1984, 21.

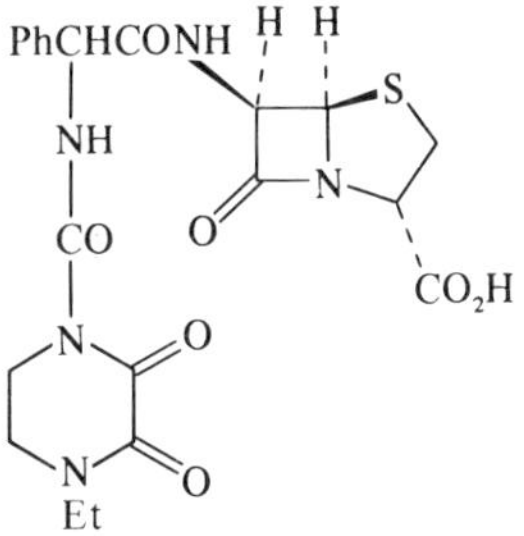

(58)

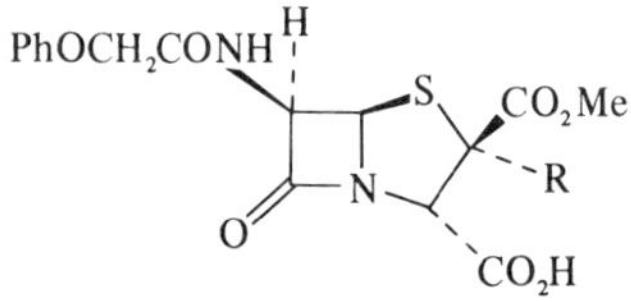

(59) R = H
(60) R = CO_2Me

benzyl α-bromoacrylate, exhibited superior antibacterial activity and β-lactamase resistance to the parent penicillanate.[39]

The esters (59) and (60), which were synthesized *via* stabilized carbanion attack upon an azetidinone disulphide, were only weakly active against Gram-positive bacteria.[40]

Chlorinolysis of disulphide (61) with SO_2Cl_2 took place initially at the S–S linkage to yield upon subsequent reaction 2α-chloromethylpenicillanates; further chlorinolysis of these with the same reagent then took place at the C-5 bond.[41] Free-radical bromination of trichloroethyl 6β-phenylacetamidopenicillanate with *N,N′*-dibromo-5,5-dimethylhydantoin resulted in monobromination of the side-chain methylene group in an *R*:*S* ratio of 3:2; cephalosporanates bearing the same side chain could be reacted similarly.[42] Further chemistry of penicillin-derived thietan-2-ones has been reported, including the conversion of (62) into D-2-methylpenicillamine (63).[43]

A full paper describing the diborane reduction of benzhydryl 6β-phenylacetamidopenicillanate to (64) includes details of attempts to cyclize the latter to

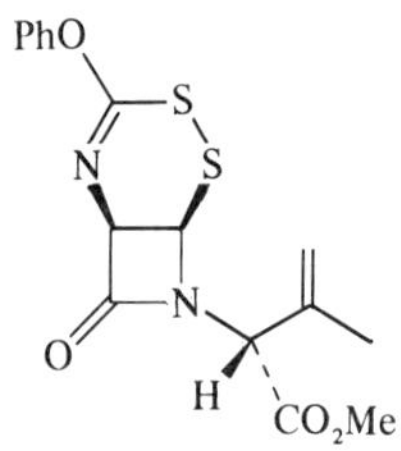

(61)

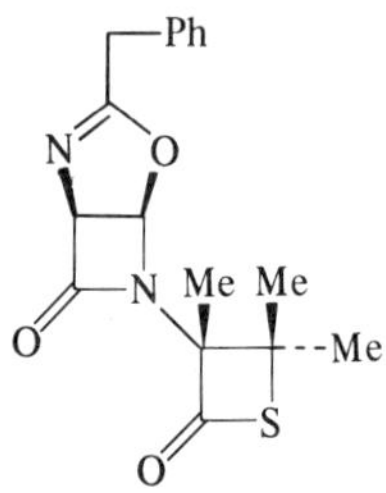

(62)

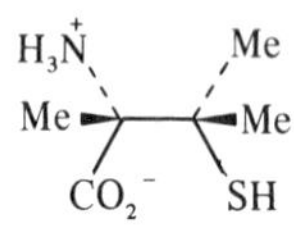

(63)

[39] M. J. Driver, P. H. Bentley, R. A. Dixon, R. A. Edmonson, A. C. Kaura, and A. W. Taylor, *J. Antibiot.*, 1984, **37**, 297.
[40] C. U. Kim, P. F. Misco, U. J. Haynes, and D. N. McGregor, *Tetrahedron Lett.*, 1984, **25**, 5593.
[41] R. E. Micetich, R. Singh, W. O. Merlo, D. M. Tetteh, C.-C. Shaw, and R. B. Morin, *Heterocycles*, 1984, **22**, 2757.
[42] J. Cooper, D. C. Humber, B. Laundon, A. G. Long, and A. L. Lynd, *Tetrahedron*, 1984, **40**, 4153.
[43] M. M. L. Crilley and R. J. Stoodley, *J. Chem. Soc., Perkin Trans. 1*, 1984, 1127.

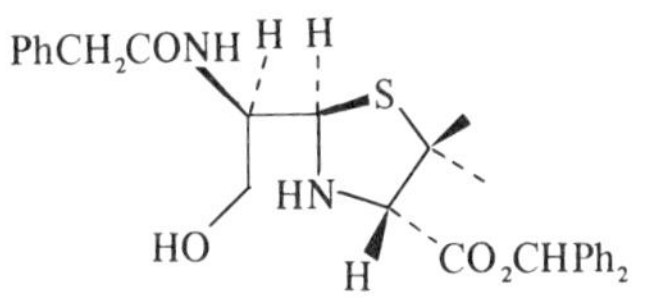

(64)

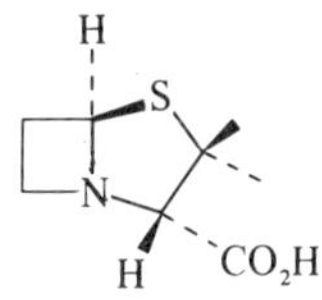

(65)

an azetidine analogue of penicillin G. Whereas these attempts were unsuccessful, apparently because of preferential reaction of the activated hydroxyl with the acylamino side chain, cyclization of a simpler derivative, lacking the side chain, yielded bicyclic analogue (65); this was, however, only a weak β-lactamase inhibitor.[44]

High yields of penam and cephem oxides have been reported by the use of H_2O_2–CH_2Cl_2 and acetic acid for the synthesis of sulphoxides, whereas substituting formic for acetic acid and extending reaction times gave the sulphone derivatives.[45]

A further variation on the commercially important conversion of penicillin to cephalosporin has been described in which thiazolidine-azetidinones (66) were converted into benzothiazolylazetidinone disulphides (67), and treatment of these with ammonia induced cyclization to the corresponding cephalosporins (68).[46] Reaction of compounds (69) with ammonium cerium(IV) nitrate led to oxidation at the benzylic position to give keto derivatives (70), which upon functionalization, ring opening, and cyclization gave cephalosporin derivatives (71), on which further elaboration of the C-7 side chain could be carried out; thus the usual C-7 side-chain cleavage and reacylation steps were circumvented.[47] The ratio of penam to cephem products from the Ag^+–$ClCH_2CO_2H$-catalysed closure of disulphide (72) was altered when the ester function was reduced; it was proposed that this resulted from the change in electronic densities in the episulphonium-ion intermediate.[48]

Reaction of Δ^2 or Δ^3 cephem 3-chlorides or 3-tosylates with azide yields the corresponding 3-azido compounds, which exist preferentially in the Δ^3 configuration, resulting in addition reactions of the azido group characteristic of an electron-poor azide; a free-acid derivative showed diminished biological activity compared to the corresponding 3-acetoxymethyl compound.[49] Thermolysis of ester (73) in methanol yielded the ring-expanded product (74) as the major product, presumably *via* an azirine intermediate.[50] Tricyclic cephems (75) have been synthesized by Wittig reactions of cephem sulphoxide C-3′ phosphorane derivatives with acrylaldehyde, whereas the parent (unoxidized) systems reacted

[44] P. G. Sammes, S. Smith, and G. Woolley, *J. Chem. Soc., Perkin Trans. 1*, 1984, 2603.
[45] R. G. Micetich, R. Singh, and S. N. Maiti, *Heterocycles*, 1984, **22**, 531.
[46] T. Torii, H. Tanaka, T. Siroi, and M. Sasaoka, *Tetrahedron Lett.*, 1984, **25**, 2017.
[47] S. Torii, H. Tanaka, T. Siroi, and M. Sasaoka, *Tetrahedron Lett.*, 1984, **25**, 1801.
[48] A. Balsamo, P. M. Benedini, B. Macchia, F. Macchia, and A. Rosello, *J. Chem. Soc., Perkin Trans. 1*, 1984, 413.
[49] D. O. Spry and A. O. Bhala, *Heterocycles*, 1984, **22**, 2487.
[50] D. O. Spry, A. O. Bhala, W. A. Spitzer, N. D. Jones, and J. K. Schwartzendruber, *Tetrahedron Lett.*, 1984, **25**, 2531.

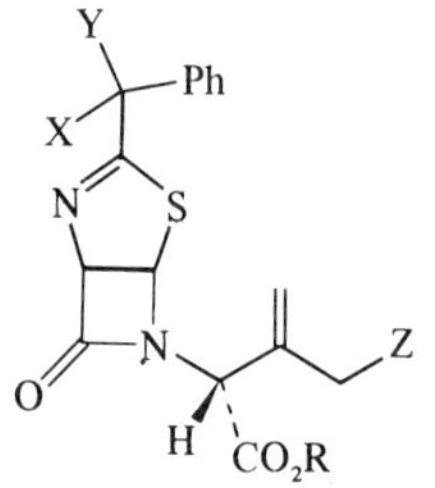

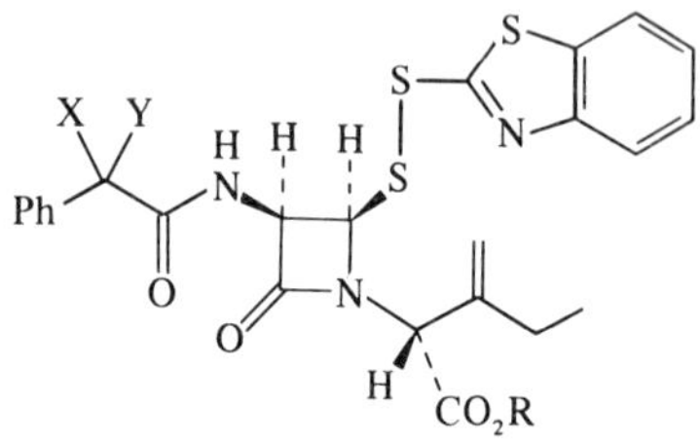

(66) X = Y = H, Z = 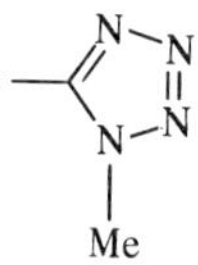

(67) X = Y = H, Z = 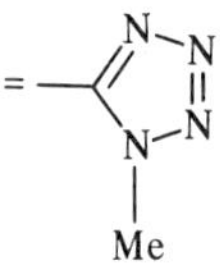

(69) X = Y = Z = H
(70) X, Y = =O, Z = H

(72) X = Y = Z = H

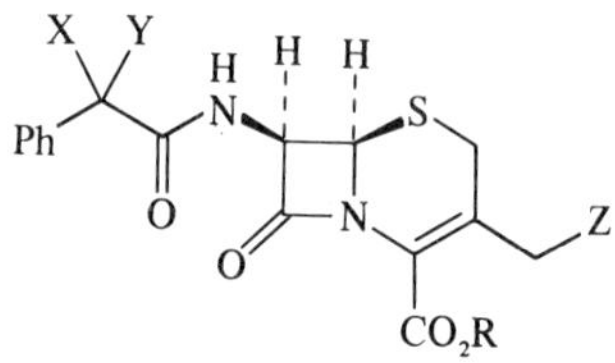

(68) X = Y = H, Z = 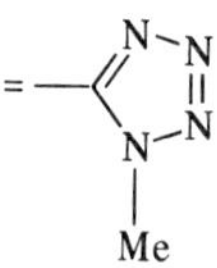

(71) X, Y = =O, Z = 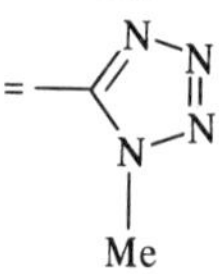

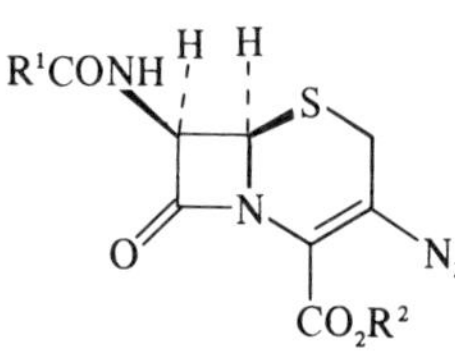

(73)

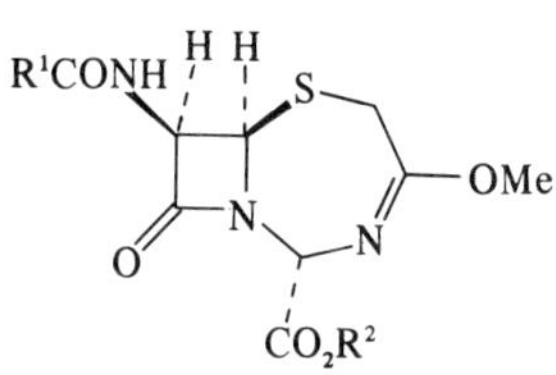

(74)

with cyclization on to C-4 and a concomitant Δ^3-Δ^2 migration; the free acids showed poor biological activity whereas aromatization of the carbocyclic ring and Raney nickel desulphurization yielded nocardicin analogues.[51] A synthesis

[51] M. Hatanaka, Y. Yamomoto, and T. Ishimaru, *J. Chem. Soc., Chem. Commun.*, 1984, 1705.

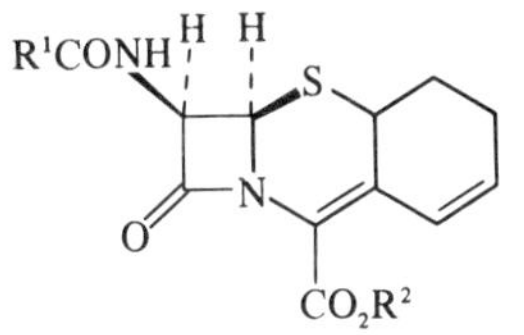

(75)

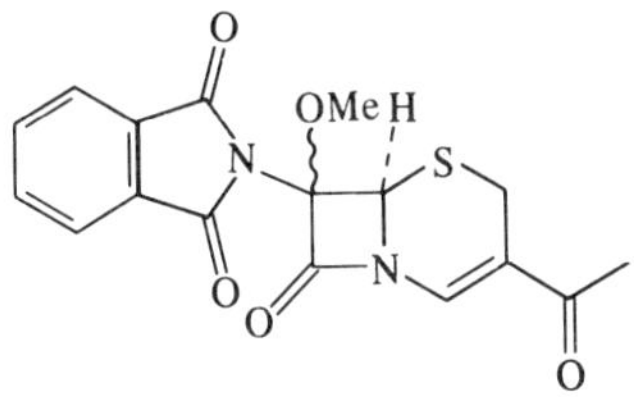

(76)

of the cephamycin system (76) has been reported utilizing an acyclic precursor that bears all of the ultimate cephem C-7 substituents.[52]

6 Clavulanic Acid

A full description of the general chemistry of clavulanic acid (77), including esterification, oxidation, reduction, isomerization, and elimination reactions, has appeared.[53] Hydrolysis of (77) under neutral or alkaline conditions ultimately yields 1-amino-4-hydroxybutan-2-one, which dimerizes to form pyrazine (78); under acid conditions, however, the ketoamine salt is stable.[54] The utility of (77) in the synthesis of novel β-lactam systems was illustrated by the reaction of ester (79) to yield its 3-hydroxy derivative; this was in equilibrium with the keto form (80) and allowed the synthesis of 1-oxacephem (81) *via* a Wittig

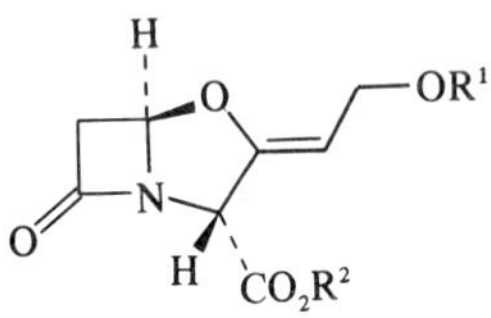

(77) $R^1 = R^2 = H$
(79) $R^1 = Me$, $R^2 = CH_2Ph$

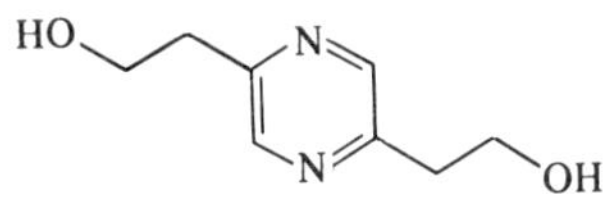

(78)

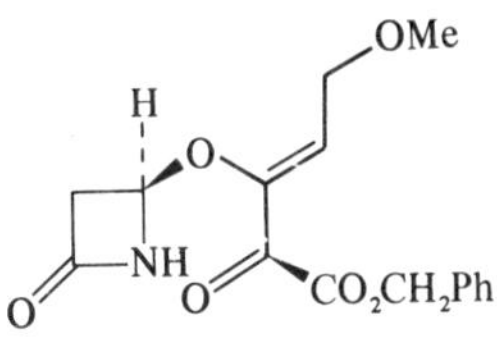

(80)

[52] M. Lees, M. Chehna, M. A. Riahi, D. Duguay, and H. Quinion, *J. Chem. Soc., Chem. Commun.*, 1984, 157.
[53] A. G. Brown, D. F. Corbett, J. Goodacre, J. B. Harbridge, T. T. Howarth, R. J. Ponsford, I. Stirling, and T. J. King, *J. Chem. Soc., Perkin Trans. 1*, 1984, 635.
[54] M. J. Finn, M. A. Harris, E. Hunt, and I. I. Zomaya, *J. Chem. Soc., Perkin Trans. 1*, 1984, 1345.

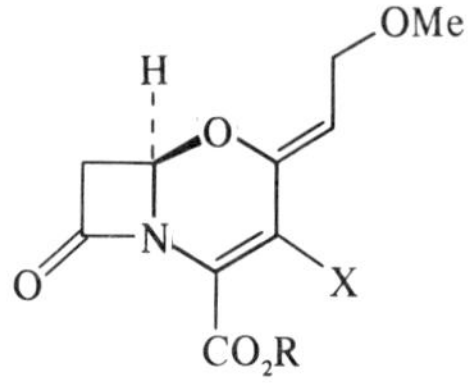

(81) $X = CO_2CH_2Ph$
(83) $X = OH$

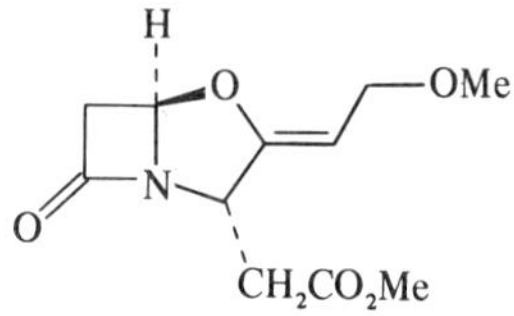

(82)

cyclization. The homologous ester (82) was similarly reacted with selenium dioxide, and the β-keto ester product was converted into 1-oxacephem (83) by diazo transfer and carbenoid insertion into the lactam N–H bond.[55]

7 Penem Chemistry

These highly active penam–cephem hybrids have continued to attract the attention of a number of groups. Trialkylphosphite-mediated cyclization of thiol esters on to oxalimides has been shown to be a useful approach to the bicyclic system[56] (Scheme 3). An analysis of the probable pathways and intermediates led to the proposal that such reactions proceed *via* a carbenoid intermediate such as (84) followed by formation of its phosphite derivative (85) and cyclization on to a thiol ester with the loss of trialkyl phosphate.[57] The main attraction of this

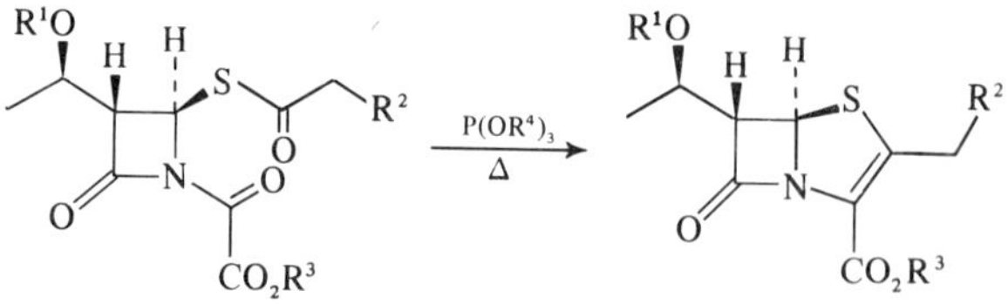

Scheme 3

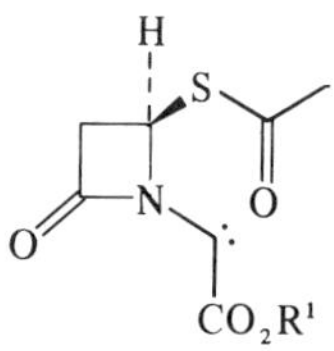

(84)

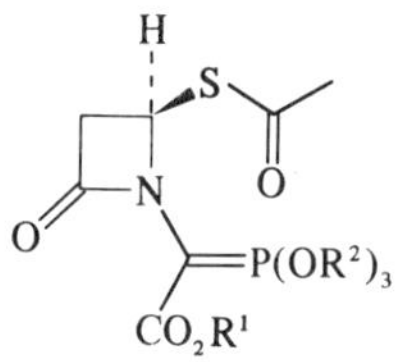

(85)

[55] G. Brooks, B. C. Gasson, T. T. Howarth, E. Hunt, and K. Luk, *J. Chem. Soc., Perkin Trans. 1*, 1984, 1599.
[56] C. Battistini, C. Scarafile, M. Foglio, and G. Franceschi, *Tetrahedron Lett.*, 1984, **25**, 2395.
[57] E. Perrone, M. Alpegiani, A. Bedeschi, F. Guidici, and G. Franceschi, *Tetrahedron Lett.*, 1984, **25**, 2399.

method is that modified penicillins frequently bear this nitrogen-substitution pattern, which normally requires further degradation and reactivation prior to cyclization to a penem. A further example of the utility of this route was illustrated in a report of the chiral synthesis of a 6-(1-hydroxyethyl)penem ester from a 6-(1-hydroxyethyl)penam *S*-oxide *via* the corresponding cephem with retention of the C-5 stereochemistry.[58] Extension of a previously described triphenylphosphine-mediated contraction of 2-thiacephems yielded penem derivatives but as the normally inactive C-5 *S*-enantiomers.[59] 6-Substituted 2-(heteroarylthiomethyl)penems have been synthesized as racemates from 4-acetoxyazetidin-2-ones by routine methods.[60] Penems have been isomerized to 2-exopenems (86), which were oxidized at sulphur and at the double bond to yield (87) and (88), respectively; none of the free acids showed antibacterial or β-lactamase inhibitory activity.[61] A related 2-thioxopenam ester (89) was synthesized from (90) by conversion of the exocyclic sulphur into its sulphoxide, thermolysis, and triphenylphosphine deoxygenation of the sulphenic acid intermediate.[62]

The 6-acetonylidene penem ester (91), an analogue of the powerful β-lactamase inhibitor 6-acetonylidenepenicillanic acid (92), has been synthesized from an

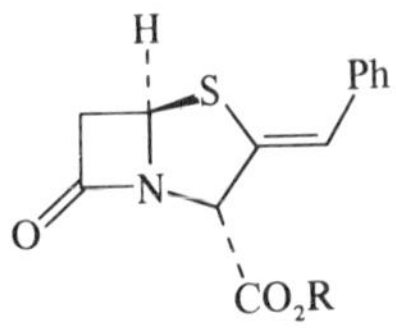

(86)

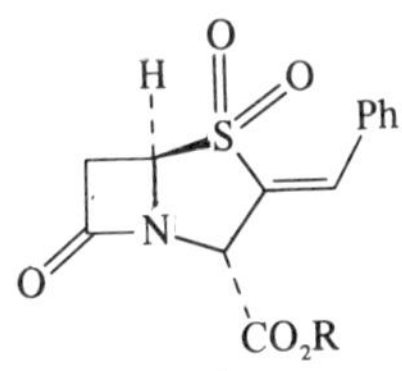

(87)

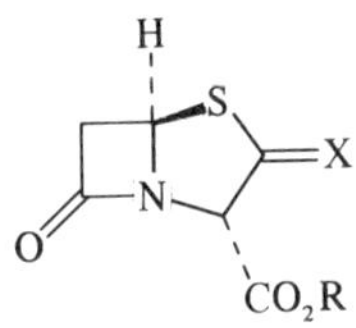

(88) X = O
(89) X = S

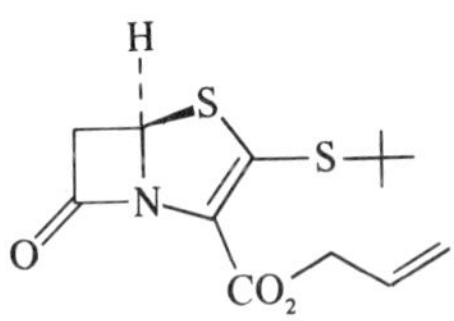

(90)

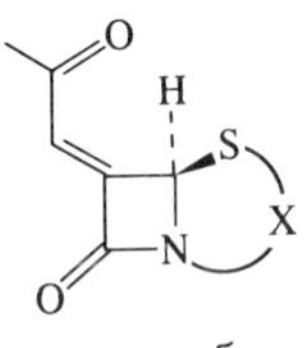

(91) X =

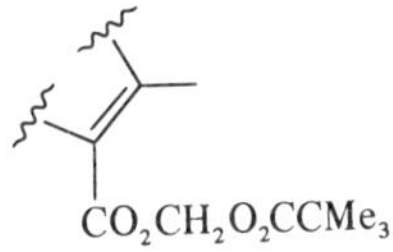

(92) X =

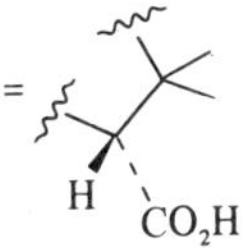

[58] M. Alpegiani, A. Bedeschi, E. Perrone, and G. Franceschi, *Tetrahedron Lett.*, 1984, **25**, 4171.

[59] M. Alpegiani, A. Bedeschi, E. Perrone, and G. Franceschi, *Tetrahedron Lett.*, 1984, **25**, 4167.

[60] M. Alpegiani, A. Bedeschi, G. Franceschi, F. Guidici, G. Nannini, and E. Perrone, *Gazz. Chim. Ital.*, 1984, **114**, 319.

[61] J. L. Douglas, A. Martel, G. Caron, M. Menard, L. Silveira, and J. Clardy, *Can. J. Chem.*, 1984, **62**, 2282.

[62] U. Krahmer-Seifert and G. Emmer, *Heterocycles*, 1984, **22**, 375.

ester of the latter. The exocyclic double bond was masked by Michael addition of benzenethiol, and the penem system was constructed by conventional methods, after which the exocyclic double bond was reintroduced by an oxidation–elimination sequence on the phenylthioether. The ester (91) inhibited β-lactamases in cell-free systems but was not antibacterially active.[63]

8 Carbapenem and Carbapenam Chemistry

In this section, only the chemistry of bicyclic molecules will be described; the synthesis of known and potential precursors to these systems will be included in the section on azetidinones, since the relationship between such precursors and total-synthetic schemes can vary from obvious to tenuous.

Thienamycin (94) and modified thienamycin derivatives continue to attract attention, partly as a result of the need to identify compounds that retain thienamycin's potency whilst possessing a high resistance to renal dehydropeptidase. A chiral synthesis of 6-epithienamycin (93) from D-glucose has been described; this compound displayed significant biological activity.[64] Several highly modified systems have been reported; they include (95) and (96), which were both synthesized by variations on the keto–carbenoid insertion route in which protected phosphonyl[65] and tetrazolyl[66] groups replaced the customary ester function. Compounds (96) were more biologically active than (95) and were stable to dehydropeptidase but possessed less antibacterial activity than the natural 3-carboxylates. Cyclopropyl derivative (97) was synthesized *via* a

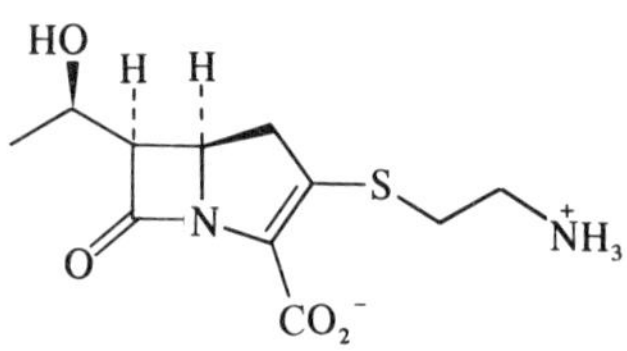

(93)

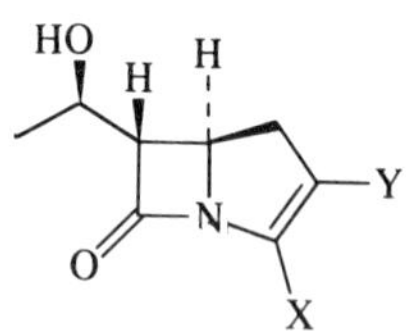

(94) X = CO_2^-, Y = 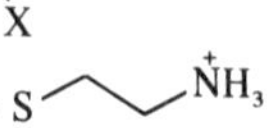

(95) X = 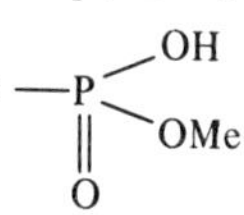 , Y = SR

(96) X = 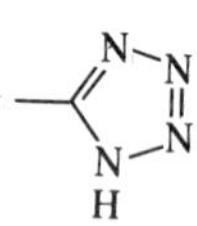 , Y = SR

(97) X = CO_2H, Y = 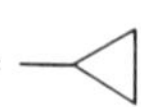

63 S. Adam, *Heterocycles*, 1984, **22**, 1509.
64 A. Knierzinger and A. Vasella, *J. Chem. Soc., Chem. Commun.*, 1984, 9.
65 A. Andrus, B. G. Christensen, and J. V. Heck, *Tetrahedron Lett.*, 1984, **25**, 595.
66 A. Andrus, J. V. Heck, B. G. Christensen, and B. Partridge, *J. Am. Chem. Soc.*, 1984, **106**, 1808.

Wittig cyclization on to the appropriate cyclopropyl ketone but showed poor chemical stability.[67] Application of the trialkylphosphite-mediated penem synthesis[56, 57, 58] to carbapenems has also been reported, with the synthesis by this route of (98) from (99)[68] and of the 2-unsubstituted system (100).[56]

The structurally related 6α-ethyl carbapenem PS-5 (41) has been synthesized in racemic form *via* 2–3 bond formation utilizing a selective Dieckmann cyclization of (101) by lithium hexamethyldisilazide to yield the intermediate 2-oxo system (102);[69] the 6β-ethyl system (6-*epi*-PS-5) was also synthesized by this route. The application of a novel 3–4 cyclization based upon singlet-oxygen cleavage of (103) to α,β-diketoester (104) has been described; after cleavage of the silyl N-protection the system spontaneously cyclized to the 2-oxo-3-hydroxy-

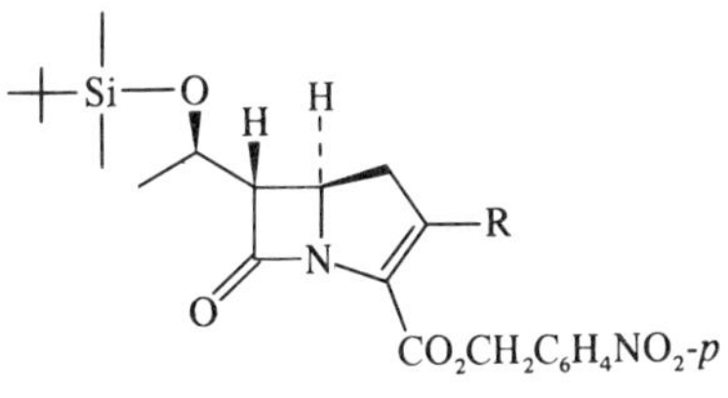

(98) R = S~~NHAc

(100) R = H

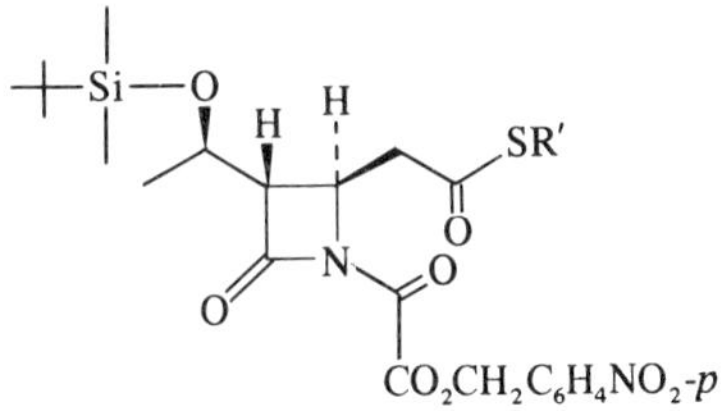

(99)

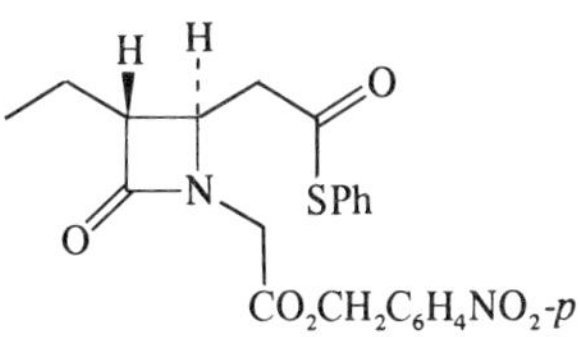

(101)

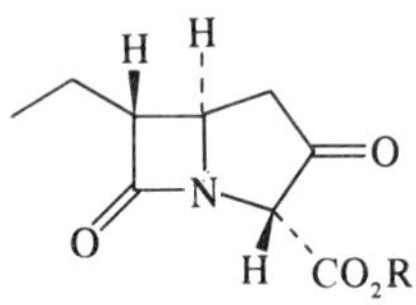

(102)

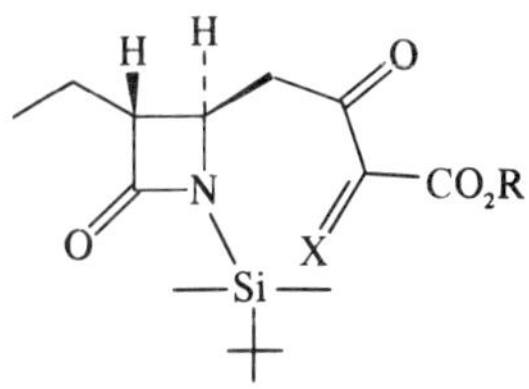

(103) X = $CHNMe_2$
(104) X = O

3-carboxy carbapenam, which upon reaction with thionyl chloride and zinc-acetic acid reduction yielded the 2-oxo system (102).[70] A more conventional phosphorane–thiol ester 2–3 closure has also been utilized in the synthesis of racemic 6-*epi*-PS-5.[71]

A general approach to the synthesis of asparenomycins (105) has been described in which the unsaturation at C-6 was introduced at a late stage by

[67] D. H. Shih, J. A. Fayter, and B. G. Christensen, *Tetrahedron Lett.*, 1984, **25**, 1639.
[68] A. Yoshida, Y. Tajima, N. Takeda, and S. Oida, *Tetrahedron Lett.*, 1984, **25**, 2793.
[69] M. Hatanaka, H. Nitta, and T. Ishimaru, *Tetrahedron Lett.*, 1984, **25**, 2387.
[70] H. H. Wasserman and W. T. Han, *Tetrahedron Lett.*, 1984, **25**, 3747.
[71] R. Cecchi, D. Favara, D. Omodei-Sale, A. Depaoli, and P. Consonni, *Gazz. Chim. Ital.*, 1984, **114**, 225.

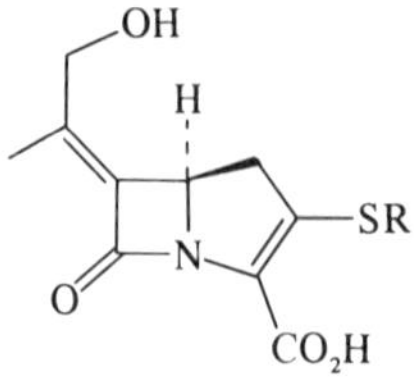

(105)

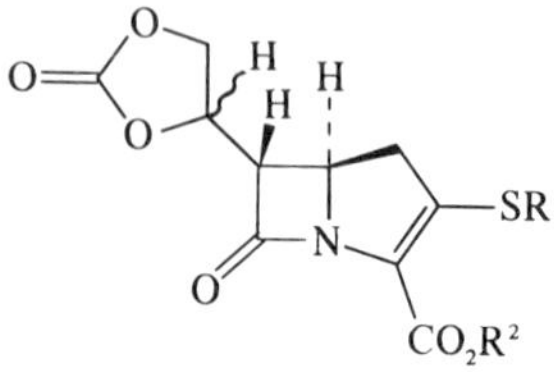

(106)

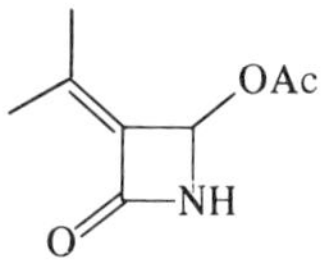

(107)

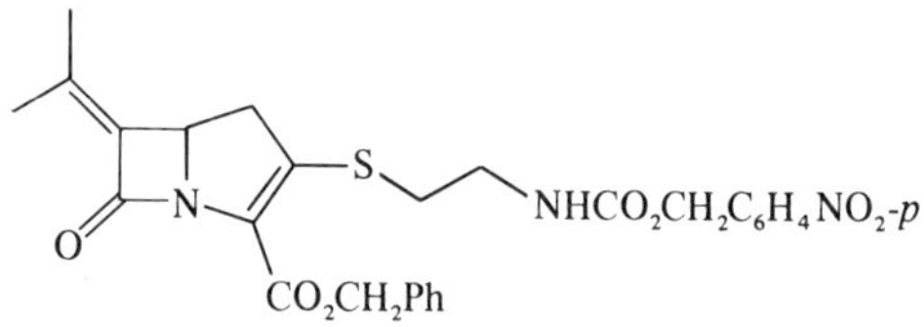

(108)

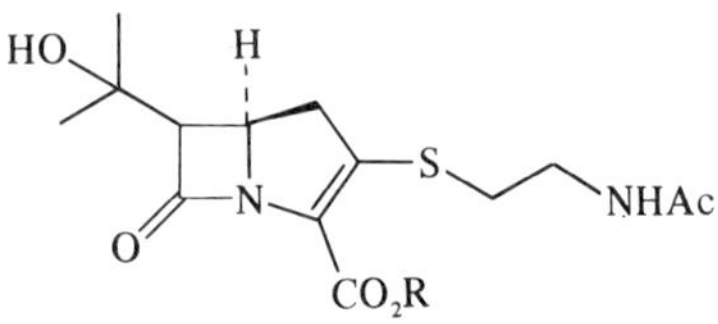

(109) C-6 *S*
(110) C-6 *R*

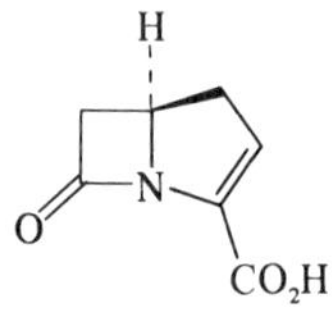

(111)

base-catalysed carbon dioxide elimination from (106).[72] Alternatively (107), synthesized from the appropriate allene and chlorosulphonyl isocyanate, has been utilized in the synthesis of the asparenomycin (108).[73] Carpetimycins (109) and 6-*epi*-carpetimycins (110) have been synthesized in a regio- and stereo-selective fashion from a C-3 unsubstituted azetidinone derived from penicillin.[74] The absolute configuration of SQ27860 (111), the first carbapenem isolated from a bacterium, has been confirmed by a chiral total synthesis of its *p*-nitrobenzyl ester from D-glucosamine.[75]

Reaction of carbapenem 2-sulphoxides with the carbanions derived from nitromethane and methyl cyanoacetate led to substitution of the sulphoxides by carbon to yield 2-nitromethyl carbapenems in the former case and 2-*exo*-methylene carbapenams in the latter.[76] Fluorinated 2-oxocarbapenams (112) reacted with many C, N, O, and S nucleophiles with ring cleavage at C-2, whereas

[72] H. Ona and S. Uyeo, *Tetrahedron Lett.*, 1984, **25**, 2237.
[73] J. D. Buynak, H. Pajouhesh, D. L. Lively, and Y. Ramalakshmi, *J. Chem. Soc., Chem. Commun.*, 1984, 948.
[74] H. Hirai, K. Sawada, M. Aratani, and M. Hashimoto, *Tetrahedron Lett.*, 1984, **25**, 5075.
[75] M. Miyashita, N. Chida, and A. Yoshikoshi, *J. Chem. Soc., Chem. Commun.*, 1984, 195.
[76] T. Yoshioka, K. Yamamoto, Y. Shimauchi, Y. Fukagawa, and T. Ishikura, *J. Chem. Soc., Chem. Commun.*, 1984, 1513.

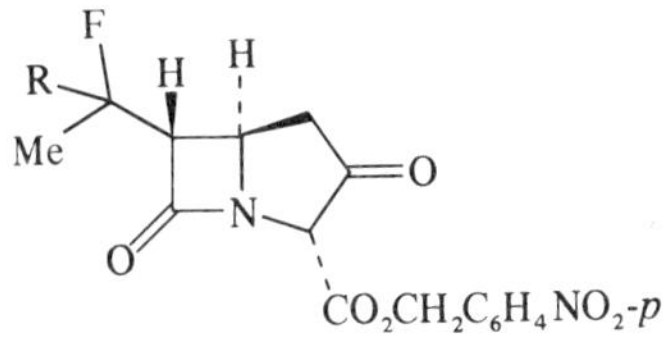

(112) R = H or Me

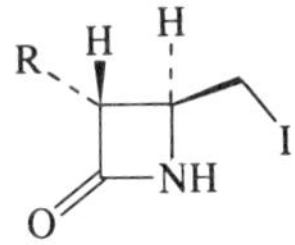

(113) R = H or 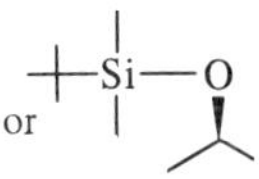

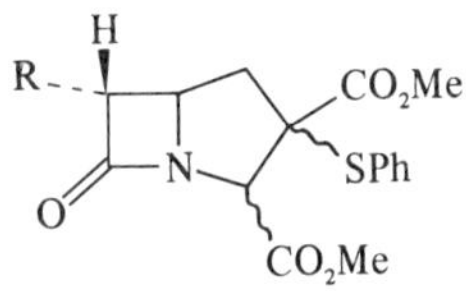

(114) R = H or 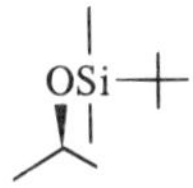

stabilized phosphoranes reacted to give mixtures of carbapenems and 2-*exo*-methylene carbapenams; attempts to deprotect these by hydrogenolysis led to partial or complete reduction of the double bonds.[77] Iodomethyl azetidinones (113) formed *N*-anions that reacted with Michael addition to dimethyl (phenylthio)fumarate to give carbapenams (114) in low yields.[78]

9 Azetidinone Chemistry

A large number of reports have appeared on the synthesis of substituted azetidin-2-ones, some exemplifying new or modified methods, some chiral, some stereoselective, some general, and some directed towards specific known intermediates or products. Rather than attempting to classify syntheses under one or several of such headings they will be grouped according to the ultimate bond-forming reaction in the azetidin-2-one synthesis, *i.e.* 1–2, 2–3, 3–4, 4–1, or concerted and rapid two-step reactions. Monobactams and related monocyclic systems will be included in this section.

1-2 Bond-forming Reactions. – Full synthetic details of the synthesis of tabtoxin (115) have appeared; a key intermediate was the spirocyclic compound (116), the azetidinone ring of which was formed from the β-amino acid using 2,2-dipyridyldisulphide and triphenylphosphine. The ready transacylation of the β-lactam carbonyl on to the side-chain amino group was prevented by maintaining the latter in its masked form until a late stage in the synthesis.[79] A stereocontrolled

[77] J. G. De Vries, G. Hauser, and G. Sigmund, *Tetrahedron Lett.*, 1984, **25**, 5989.
[78] K. Fujimoto, Y. Iwano, and K. Hirai, *Tetrahedron Lett.*, 1984, **25**, 1151.
[79] J. E. Baldwin, P. D. Bailey, G. Gallacher, M. Otsuka, K. A. Singleton, and P. M. Wallace, *Tetrahedron*, 1984, **40**, 3695.

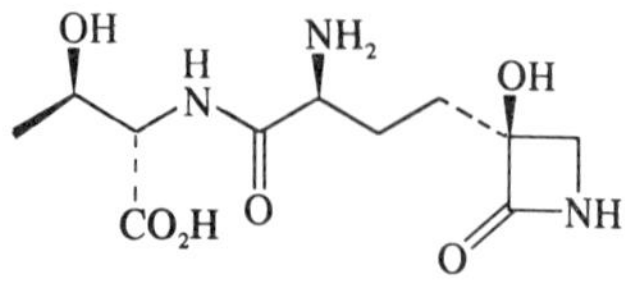

(115)

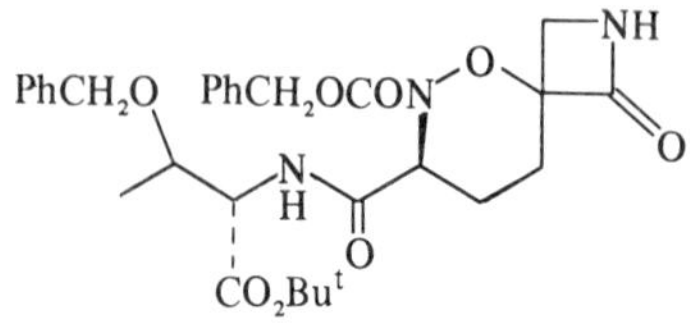

(116)

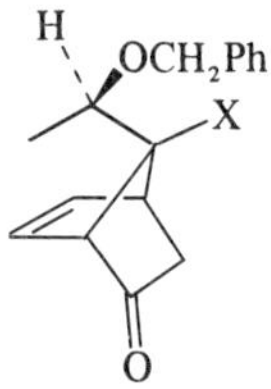

(117) X = H

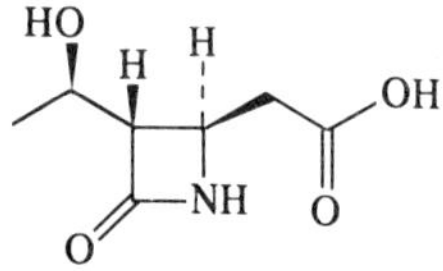

(118)

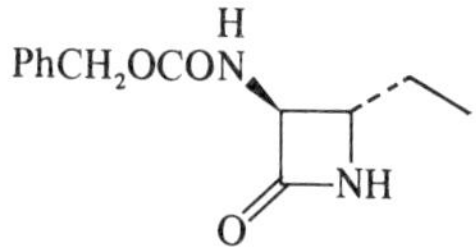

(119)

route from (117) to carbapenem intermediate (118) proceeded *via* β-lactam formation with dicyclohexylcarbodi-imide and has been suggested as a potential route to 6,6-disubstituted carbapenems (X = F, OH, or Me).[80] Chiral azetidinone (119) was synthesized from 2-amino-2-deoxy-D-glucose, utilizing 2,2-dipyridyl-disulphide and triphenylphosphine in the ring-formation step,[81] whereas the monobactam analogue precursor (120) was cyclized with DABCO and tosyl chloride, ultimately yielding compounds showing 'appreciable' biological activity.[82] The *N,O*-bistrimethylsilyl derivative of 4,4,4-trifluoro-3-aminobutanoic acid reacted with methylmagnesium bromide to yield 4-(trifluoromethyl)-azetidin-2-one, which was ultimately converted into a 4-trifluoromethyl sulphazecin analogue; this compound showed poor biological activity.[83] The reaction of ketene methyl trimethylsilyl acetals with α-methoxymethyl amine derivatives (121) or (122) in the presence of trimethylsilyl triflate or titanium tetrachloride yielded *N*-substituted β-amino methyl esters (123) and (124); the former were suggested as β-lactam precursors,[84] and conversion of the latter into β-lactams

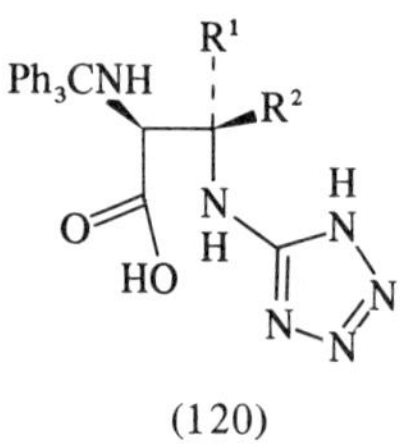

(120)

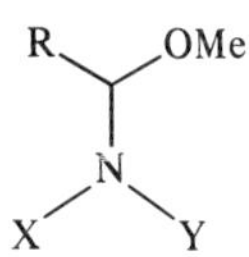

(121) X = Y = $SiMe_3$
(122) X = Me, Y = CO_2Me

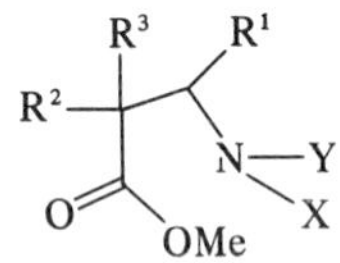

(123) X = Y = $SiMe_3$
(124) X = Me, Y = CO_2Me

80 P. A. Grieco, D. L. Flynn, and R. E. Zelle, *J. Am. Chem. Soc.*, 1984, **106**, 6414.
81 S. Hanessian and S. P. Sahoo, *Can. J. Chem.*, 1984, **62**, 1400.
82 M. Klich and G. Teutsch, *Tetrahedron Lett.*, 1984, **25**, 3849.
83 P. F. Bevilacqua, D. D. Keith, and J. L. Roberts, *J. Org. Chem.*, 1984, **49**, 1434.
84 K. Okano, T. Morimoto, and M. Sekiya, *J. Chem. Soc., Chem. Commun.*, 1984, 883.

by N-deprotection and treatment with methylmagnesium bromide was demonstrated.[85] A stereoselective route to *β*-amino acyl derivatives (125) from the reaction of the lithium enolate of (126) with 1,2-diphenyletheneimine has been described; mild oxidation of (125) led to spontaneous formation of azetidinone (127).[86] Ring contraction of 5-nitroisoxazolidines by thermal or photochemical methods gave rise to *β*-lactams by N–C bond formation *via* an intermediate nitroacyl species.[87]

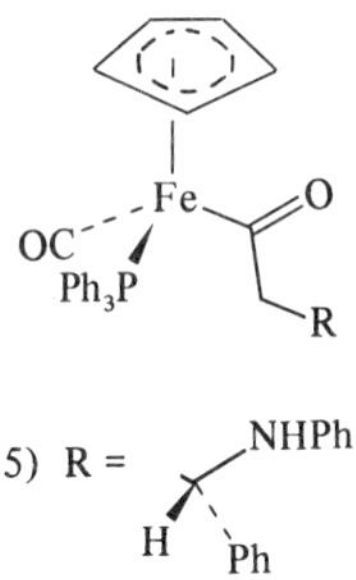

(125) R =

(126) R = H

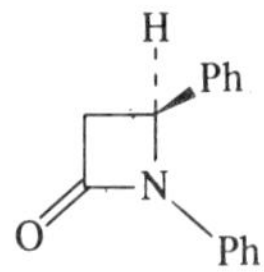

(127)

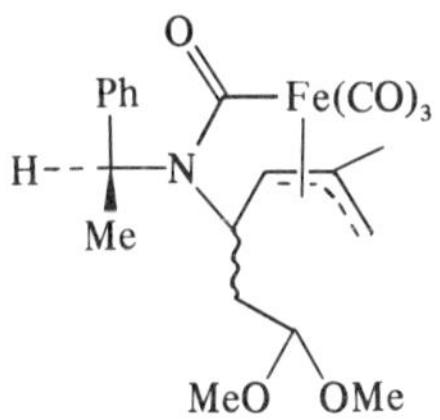

(128)

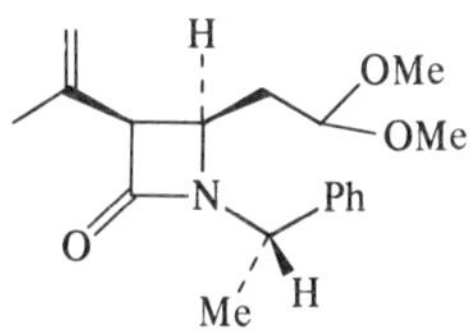

(129)

2–3 Bond-forming Reactions. – The only example of this unusual closure is a chiral adaptation of a previously reported organometallic route: diastereoisomeric lactams (128) were separated and the appropriate isomer was oxidatively cyclized to (129), which was converted into a known thienamycin precursor.[88]

3–4 Bond-forming Reactions. – The application of photolytic methods to this mode of closure has been notable. 2-Pyridone photocylizations yield bicyclic [2.2.0] isomers that incorporate an azetidinone moiety; thus 4-acetoxy derivatives (130) were isomerized to (131), which upon hydrolysis reduction, retro-aldol ring opening, and further reduction yielded 3-unsubstituted or 3-alkyl-4-(2-hydroxyethyl)azetidinones (132).[89] Isomerization of pyridones (133) and (134)

[85] T. Shono, K. Tsubata, and N. Okinaga, *J. Org. Chem.*, 1984, **49**, 1056.
[86] K. Broadley and S. G. Davies, *Tetrahedron Lett.*, 1984, **25**, 1743.
[87] A. Padwa, K. F. Koehler, and A. Rodriguez, *J. Org. Chem.*, 1984, **49**, 282.
[88] S. T. Hodgson, D. M. Hollinshead, and S. V. Ley, *J. Chem. Soc., Chem. Commun.*, 1984, 494.
[89] C. Kaneko, T. Naito, and A. Saito, *Tetrahedron Lett.*, 1984, **25**, 1591.

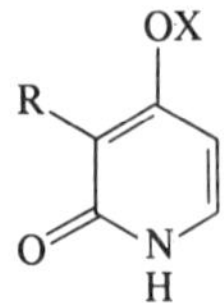

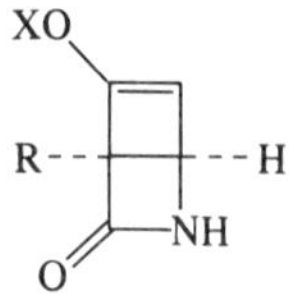

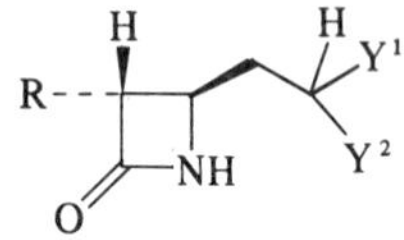

(130) X = Ac
(133) X = CH_2Ph
(134) X = Si+

(131) X = Ac
(135) X = CH_2Ph
(136) X = Si+

(132) Y^1 = H, Y^2 = OH
(137) Y^1, Y^2 = =O

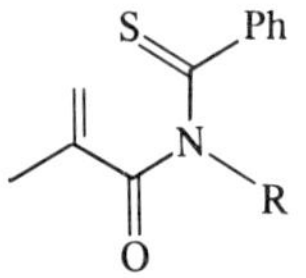

(138)

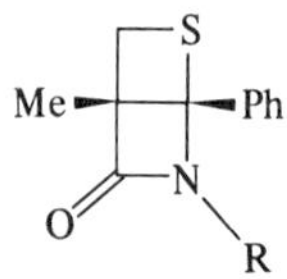

(139)

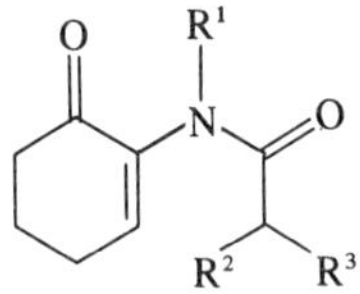

(140)

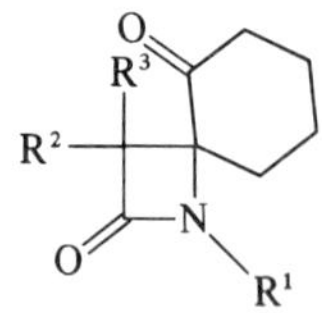

(141)

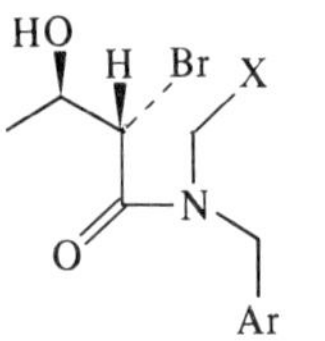

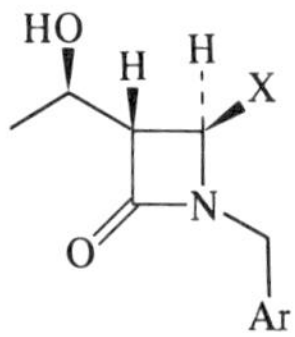

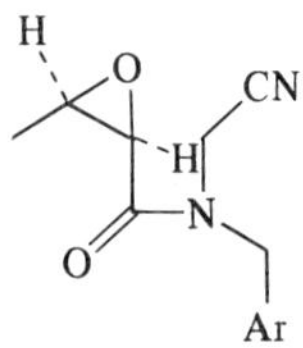

(144)

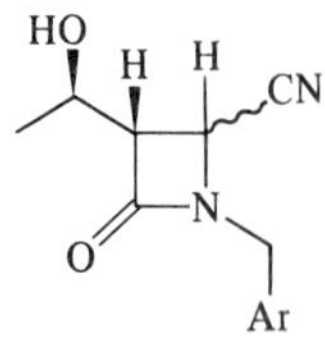

(145)

(142) X = $\mathrm{P(O)(OEt)_2}$
(146) X = CO_2Bu^t

(143) X = $\mathrm{P(O)(OEt)_2}$
(147) X = CO_2Bu^t

to (135) and (136) followed by reduction and O-deprotection yielded a bicyclic hydroxyl compound that was described as a stable synthetic equivalent of the azetidinone acetaldehyde (137).[90] Thioimides (138) underwent photocyclization to the bicyclic [2.2.0] systems (139) in good yields,[91] and photolysis of vinyl amides (140) led to the formation of 4-spiro-substituted azetidinones (141).[92]

Reaction of chiral bromophosphonate (142) with lithium hexamethyldisilazide led to cyclization to the β-lactam (143);[93] in a similar fashion chiral epoxynitrile (144) was cyclized to a mixture of 4α- and 4β-cyanoazetidinones (145) in the ratio 1:2.3.[94] In a further example of the application of this approach (147) was obtained in good yield from (146), which had been synthesized from

[90] N. Katagiri, M. Sato, T. Naito, M. Muto, T. Sakamoto, S. Saikawa, and C. Kaneko, *Tetrahedron Lett.*, 1984, **25**, 5665.

[91] M. Sakamoto, Y. Omote, and H. Aoyama, *J. Org. Chem.*, 1984, **49**, 396.

[92] M. Ikeda, T. Uchino, H. Ishibashi, Y. Tamura, and M. Kido, *J. Chem. Soc., Chem. Commun.*, 1984, 758.

[93] M. Shiozaki and H. Masuko, *Heterocycles*, 1984, **22**, 1727.

[94] M. Shiozaki, H. Maruyama, and N. Ishida, *Heterocycles*, 1984, **22**, 1725.

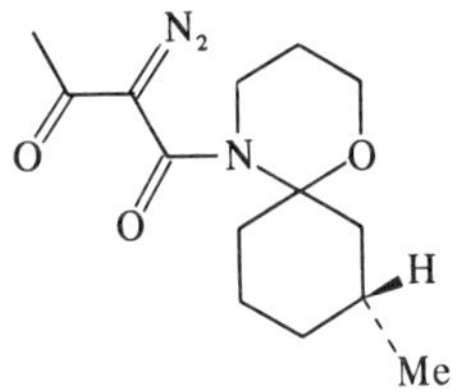

(148)

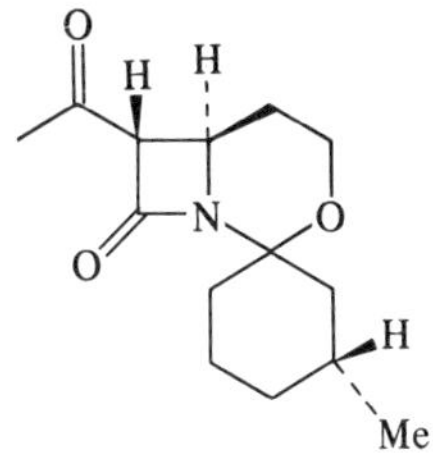

(149)

L-threonine.[95] Modification of a previously reported carbenoid insertion reaction by use of a chiral tetrahydro-1,3-oxazine derivative (148) yielded a diastereoisomeric mixture of products from which (149) was separated and utilized in the synthesis of a chiral carbapenem.[96]

4-1 Bond-forming Reactions. – Oxamazins (150), a new class of monocyclic β-lactam antibacterials, have been synthesized *via* ring closure of (151) by triphenylphosphine and diethyl azodicarboxylate,[97] but attempts to synthesize similarly nocardicin analogues (152) failed at the stage of isolating a free 3-amino intermediate; homologous compounds (153) were synthesized but no biological data were reported.[98] The *N*-tetrazolyl monobactam analogues (154) and (155) have been synthesized by a similar strategy, but whereas compounds of the type (154) showed moderate to potent antibacterial activity[99] compounds of the type (155) were only weakly active.[100] This has been attributed to the difference in distance between the anionic group and the β-lactam carbonyl in the two structures, and this cyclization procedure has also been shown to be effective with *N*-aryl β-hydroxyamides[101] but not with the *N*-alkyl compounds. Cyclization of 2-acyloxy-3-chloro-2-chloromethyl propanamides with caesium fluoride

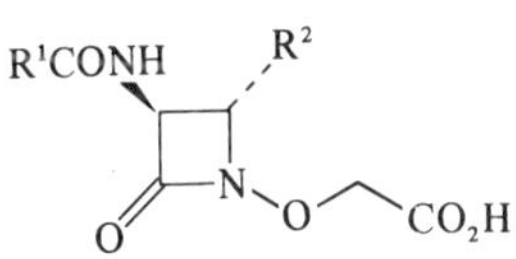

(150)

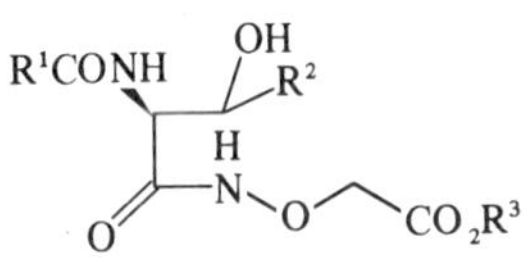

(151)

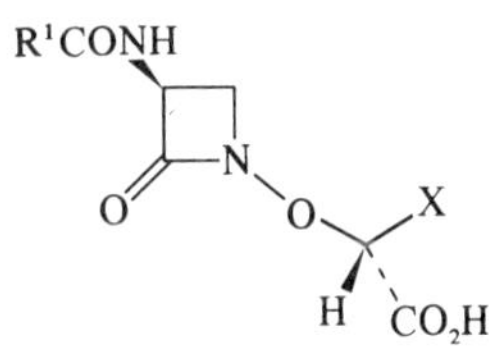

(152) X = Ph
(153) X = CH_2Ph

[95] M. Shiozaki, N. Ishida, T. Hiraoka, and H. Maruyama, *Heterocycles*, 1984, **22**, 1795.
[96] T. C. Smale, *Tetrahedron Lett.*, 1984, **25**, 2913.
[97] S. R. Woulfe and M. J. Miller, *Tetrahedron Lett.*, 1984, **25**, 3294.
[98] F. R. Atherton and R. W. Lambert, *Tetrahedron*, 1984, **40**, 1039.
[99] A. Andrus, B. Partridge, J. V. Heck, and B. G. Christensen, *Tetrahdron Lett.*, 1984, **25**, 911.
[100] A. Andrus, B. Partridge, J. V. Heck, B. G. Christensen, and J. P. Springer, *Heterocycles*, 1984, **22**, 1713.
[101] A. K. Bose, M. S. Manhas, D. P. Sahu, and V. R. Hege, *Can. J. Chem.*, 1984, **62**, 2498.

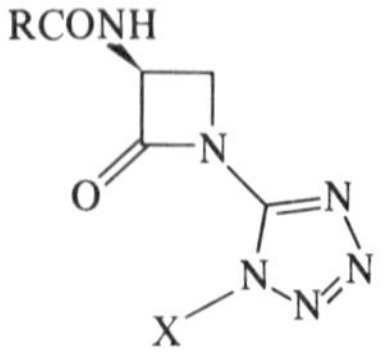

(154) X = H
(155) X = CH_2CO_2H

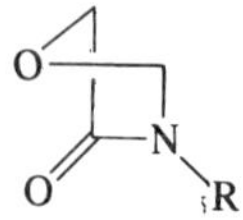

(156)

yielded 3-acyloxy-3-chloromethyl β-lactam derivatives, which upon reaction with potassium hydroxide formed spiroazetidinone epoxides (156).[102]

Reactions in Which Two Bonds are Formed. – The reactions discussed under this heading are formal [2+2] cycloadditions in which the bond-forming steps may be either concerted or merely taking place successively under the reaction conditions.

The ketene-plus-imine type of cycloaddition has formed the basis for a variety of reports, several with emphasis on the potential of these reactions for chiral induction. The chiral enolate (157) reacted with imine (158) to yield a mixture of free N–H β-lactams in good yield after work-up, in which (159) was the major product;[103] however, other diastereoisomers were present, and the difficulty in controlling selectivity was tentatively attributed to mixed geometry in the enolate. Similar studies with this enolate in its racemic form and 1,2-diphenyletheneimine gave a product (160) that had *trans* disubstitution across the 3- and 4-positions but with the hydroxyethyl side chain having the opposite stereochemistry (relative to C-4) to that found in thienamycin.[104] Utilizing a more conventional ketene–imine approach, a chiral auxiliary on a ketene precursor gave rise to up to 96% asymmetric induction, although the stereochemistry induced at C-3 and C-4 in this case (Scheme 4) was that normally associated with biologically inactive forms of antibacterials.[105] Reaction of racemic (*p*-tolylsulphinylacetyl)imidazolide with imines gave diastereoisomeric β-lactams in ratios of up to 87:13, although such reactions were only effective where a 4-aryl substituent was present in the product.[106] Imines that are *N*-substituted

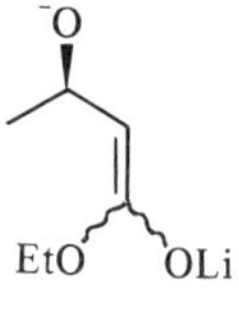

(157)

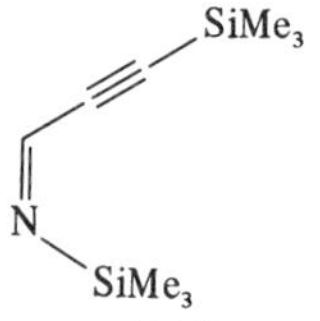

(158)

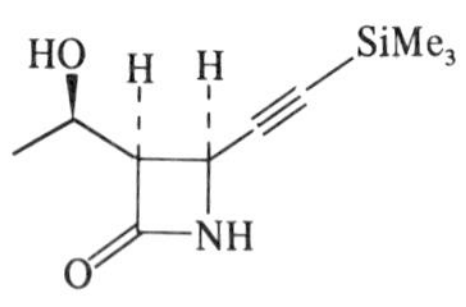

(159)

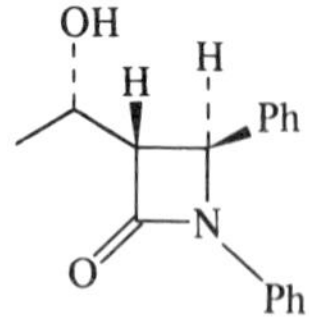

(160)

[102] S. Sebti and A. Foucaud, *Tetrahedron*, 1984, **40**, 3223.
[103] D.-C. Ha, D. J. Hart, and T.-K. Yang, *J. Am. Chem. Soc.*, 1984, **106**, 4819.
[104] G. I. Georg, *Tetrahedron Lett.*, 1984, **25**, 3779.
[105] N. Ikota and A. Hanaki, *Heterocycles*, 1984, **22**, 2227.
[106] G. Guanti, L. Banfi, E. Narisano, and S. Thea, *J. Chem. Soc., Chem. Commun.*, 1984, 861.

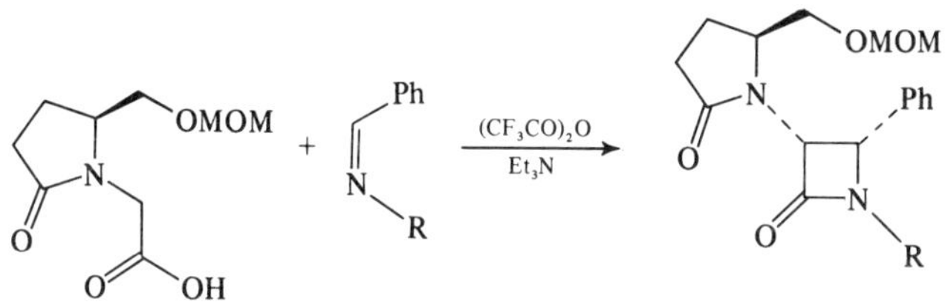

Scheme 4

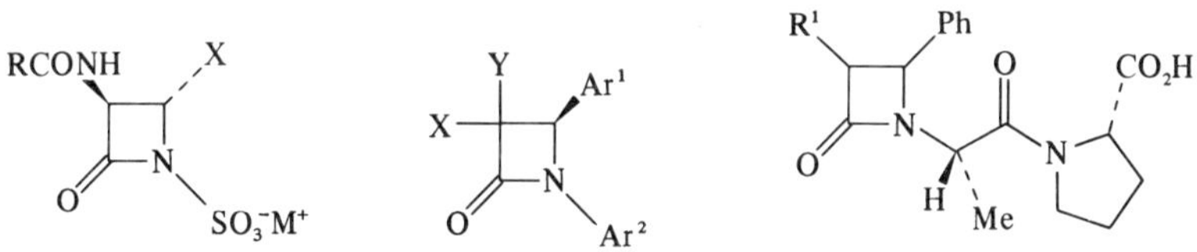

(161) X = CH_2F
(162) X = CO_2Me

(163) X = H, Y = SPh
(164) X, Y = =O

(165)

with a chiral auxiliary react with prochiral α-chloroimminium chlorides with significant induction and yield β-lactams upon work-up.[107] Hexynylcyanoketene reacted in the expected fashion with imines to yield β-lactam products in which the C-3 and C-4 relative stereochemistries were undefined.[108] 4-Fluoromethyl[109] and 4-methoxycarbonyl[110] sulphazecin derivatives (161) and (162) have also been synthesized by this approach and display significant biological activity. 3-Phenylthia-azetidinone derivatives (163) synthesized from the appropriate ketene and imine are chlorinated at the 3-position with sulphuryl chloride, and these products upon hydrolysis yield azetidin-2,3-diones (164).[111] Activation of acids to act as ketene equivalents has been achieved by formation of acyl saccharide,[112] *N*-methyl-*N*-phenylphosphoramidate,[113] *N,N*-dimethylchlorosulphitemethaminium,[114] and phenylchlorophosphate[115] compounds, all of which have been reported to react efficiently with imines; ketene equivalents formed by conventional routes have been reported in the synthesis of the potential (but inactive)

[107] E. Regalska and C. Belzecki, *J. Org. Chem.*, 1984, **49**, 1397.
[108] N. V. Nguyen, and H. W. Moore, *J. Chem. Soc., Chem. Commun.*, 1984, 1066.
[109] K. Yoshioka, T. Miyawaki, S. Kishimoto, T. Matsuo, and M. Ochiai, *J. Org. Chem.*, 1984, **49**, 1427.
[110] S. Kishimoto, M. Sendai, M. Tomimoto, S. Hashiguchi, T. Matsuo, and M. Ochiai, *Chem. Pharm. Bull.*, 1984, **32**, 2646.
[111] M. S. Manhas, S. S. Bari, B. M. Bhawal, and A. K. Bose, *Tetrahedron Lett.*, 1984, **25**, 4733.
[112] M. Miyake, N. Tokutake, and M. Kirisawa, *Synth. Commun.*, 1984, **14**, 353.
[113] D. R. Shridhar, B. Ram, V. L. Narayana, A. K. Awasthi, and G. J. Reddy, *Synthesis*, 1984, 846.
[114] A. Arrieta, J. M. Aizpurua, and C. Palomo, *Tetrahedron Lett.*, 1984, **25**, 3365.
[115] J. M. Aizpurua, I. Ganboa, F. P. Cossio, A. Gonzalez, A. Arrieta, and C. Palomo, *Tetrahedron Lett.*, 1984, **25**, 3905.

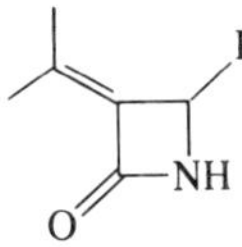

(166)

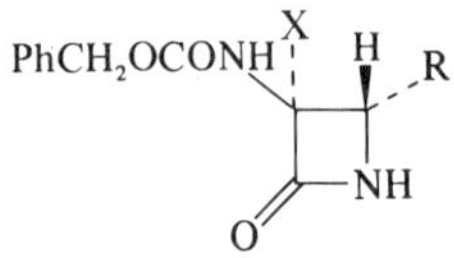

(167) X = H, R = Me
(168) X = OMe, R = H

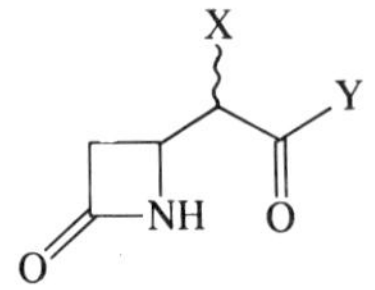

(169) X = H or Me
Y = Ph, SPh, or OR

protease inhibitor (165)[116] and of 4-alkoxyazetidin-2-ones where imidates replaced the imine components of the cycloaddition.[117]

The alternative 'concerted' strategy, that of addition of an isocyanate derivative, usually the chlorosulphonyl, to alkenes, has also been utilized in a number of cases. Addition to allenes to yield 3-methylene derivatives (166) has already been discussed with reference to its application in asparenomycin synthesis;[73] the adduct with vinyl acetate may be transformed into the chiral monobactam precursors (167) and (168).[118] A full paper has appeared describing the reactions of 4-acetoxyazetidin-2-one with a variety of silyl enol ethers using trimethylsilyl triflate catalysis; this yielded intermediates (169), which are of value in carbapenem synthesis.[119]

Reaction of azetidinone (170) with lithium di-isopropylamide and propargaldehyde yielded the corresponding aldol addition product, whereas the adduct with (171) underwent spontaneous Petersen elimination to yield the *E*- and *Z*-ene-ynes in 90% yield.[120] The synthesis of azetidin-2-ones from azetidine-2-carbonyl chlorides (172) and *m*-chloroperbenzoic acid has been shown to be sensitive to the nature of the 4-substituents, which, when interacting unfavourably with the acyl group, prevent the assumption of the correct geometry for formation of the iminium salt intermediates, leading to α-chloro-γ-lactam products.[121] The utility of allyl and 2,2-diethoxyethyl groups for the protection of the β-lactam nitrogen has been described; both types of protected derivative

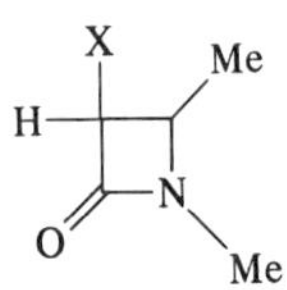

(170) X = H
(171) X = $SiMe_3$

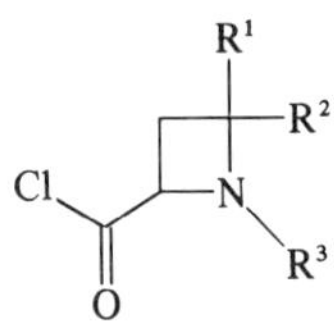

(172)

[116] C. J. Wharton, R. Wrigglesworth, and M. Rowe, *J. Chem. Soc., Perkin Trans. 1*, 1984, 29.
[117] M. Cardellini, F. Claudi, and F. Micheletti Moracci, *Synthesis*, 1984, 1070.
[118] A. Nishida, M. Shibasaki, and S. Ikegami, *Tetrahedron Lett.*, 1984, **25**, 765.
[119] R. P. Attrill, A. G. M. Barrett, P. Quayle, J. van der Westhuizen, and M. J. Betts, *J. Org. Chem.*, 1984, **49**, 1679.
[120] M. Thielmann and E. Winterfeldt, *Heterocycles*, 1984, **22**, 1161.
[121] H. H. Wasserman, W. T. Han, J. M. Schaus, and J. W. Faller, *Tetrahedron Lett.*, 1984, **25**, 3111.

may be converted into azetidin-2-on-1-yl acetaldehydes, which can be cleaved to the free N–H compounds.[122]

10 Major Structural Variants

This section will include descriptions of molecules possessing some characteristics of the better known groups of β-lactam antibacterials, whilst being sufficiently different not to merit inclusion under these headings. Penam-type structures reported include the 1-aza system (173)[123] and the isopenam (174) and its *S*,*S*-dioxide (175),[124] but none of these showed any biological activity.

Two almost identical approaches to the synthesis of the biologically active 1-oxacephem series have appeared, both utilizing the same penicillin derivative as a key intermediate and both effecting ring closure with the carbenoid from (176);[125,126] the use of clavulanic acid in the synthesis of similar systems has been described.[55] A substituted 1-azacepham system has been synthesized *via* ketene addition to 2-methylthio-3,4,5,6-tetrahydropyrimidine,[127] and approaches to the 1-phosphacephem system, starting from 4-acetoxyazetidinone derivatives, have been described; compounds synthesized include the simple ring system (177)[128] and the 7-unsubstituted (178),[129] 7-hydroxyethyl (179), and 7-acylamino (180) derivatives.[130] All of these compounds are apparently inactive, although it is significant that none possesses the C-7 structural characteristics necessary for activity in the cephem and oxacephem series. The 3-oxocarbacepham (181) has been synthesized *via* singlet-oxygen cleavage of an enamine[131] in a similar fashion to that described for synthesis of a 2-oxocarbapenam.[70] A synthesis of iso-oxacephem derivative (182) has been described, and the compound is reported

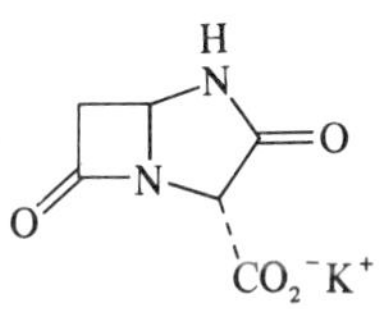

(173)

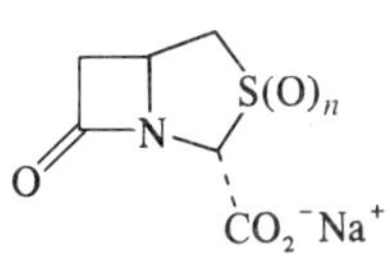

(174) $n = 0$
(175) $n = 2$

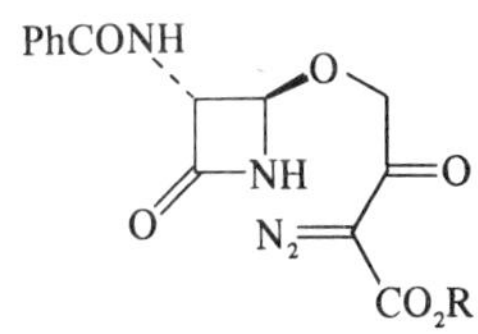

(176)

[122] T. Fukuyama, A. A. Laird, and C. A. Schmidt, *Tetrahedron Lett.*, 1984, **25**, 4709.
[123] G. Johnson, P. M. Rees, and B. C. Ross, *J. Chem. Soc., Chem. Commun.*, 1984, 970.
[124] P. H. Crackett, C. M. Pant, and R. J. Stoodley, *J. Chem. Soc., Perkin Trans. 1*, 1984, 2785.
[125] D. Habich and W. Hartwig, *Tetrahedron*, 1984, **40**, 3667.
[126] S. Yamamoto, H. Itani, H. Takahashi, T. Tsuji, and W. Nagata, *Tetrahedron Lett.*, 1984, **25**, 4545.
[127] S. D. Sharma and U. Mehra, *Tetrahedron Lett.*, 1984, **25**, 1849.
[128] M. M. Campbell, N. I. Carruthers, S. J. Mickel, and P. M. Winton, *J. Chem. Soc., Chem. Commun.*, 1984, 200.
[129] H. Sato and T. Tsuji, *Tetrahedron Lett.*, 1984, **25**, 1733.
[130] H. Sato and T. Tsuji, *Tetrahedron Lett.*, 1984, **25**, 1737.
[131] H. H. Wasserman and W. T. Han, *Tetrahedron Lett.*, 1984, **25**, 3743.

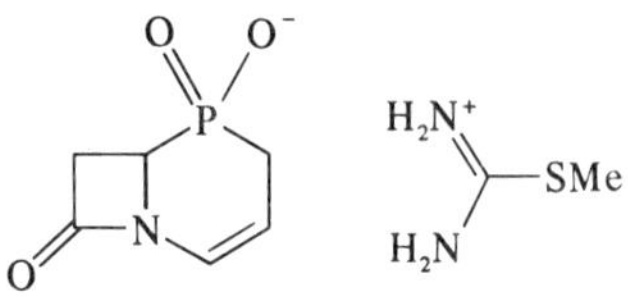

(177)

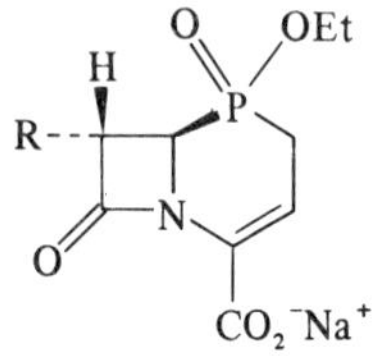

(178) R = H

(179) R = 

(180) R = $PhOCH_2CONH$

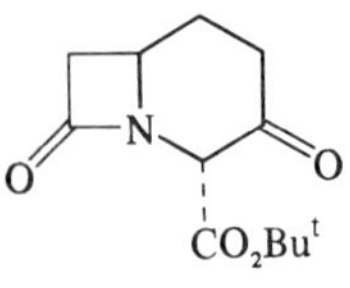

(181)

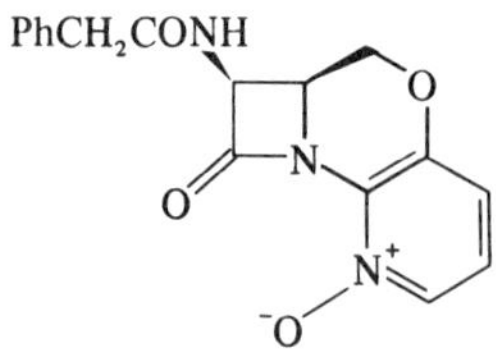

(182)

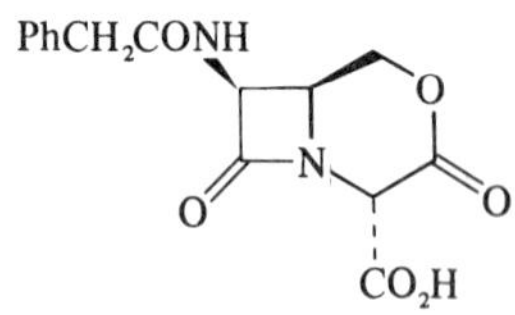

(183)

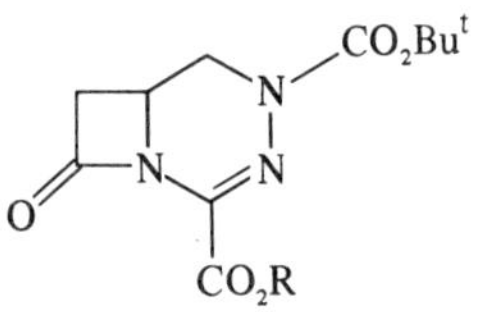

(184)

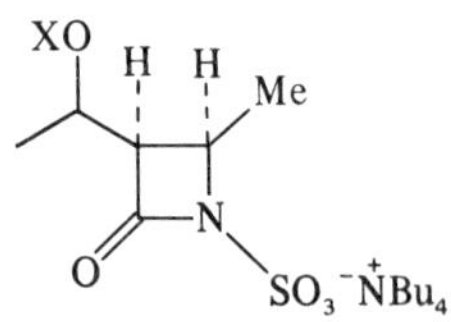

(185) X = H or $SO_3^-\overset{+}{N}Bu_4$

to possess 'interesting' antimicrobial activity;[132] an improved synthesis of compound (183) utilized carbenoid insertion in the closure of the six-membered ring.[133] Final closure to the 2,3-diaza compounds (184) proceeded *via* formation of the 1,2-bond from a hydrazine precursor with concurrent oxidation of the 3,4-bond.[134] Monobactam analogues (185), synthesized from 4-methyl-2-pyridone, displayed only weak *β*-lactamase inhibition.[135] Attempts to cyclize diazetidinone derivatives (186) by a Rh^{II}-catalysed carbenoid insertion reaction resulted in interception of the carbenoid by the more nucleophilic N-1 and rearrangement of the spiro-*N*-ylide;[136] *N,N'*-unsubstituted 1,2-diazetidin-3-one could be di-acylated under basic conditions to either N,N' or N-1,O derivatives, but N-1 alkyl

[132] G. H. Hakimelahi, *Helv. Chim. Acta*, 1984, **67**, 902.

[133] G. H. Hakimelahi and A. Khalafi-Nezhad, *Helv. Chim. Acta*, 1984, **67**, 18.

[134] C. M. Pant, R. J. Stoodley, A. Whiting, and D. J. Williams, *J. Chem. Soc., Chem. Commun.*, 1984, 1289.

[135] C. J. Ashcroft, J. Brennan, C. E. Newall, and S. M. Roberts, *Tetrahedron Lett.*, 1984, **25**, 877.

[136] E. C. Taylor and H. M. L. Davies, *J. Org. Chem.*, 1984, **49**, 113.

derivatives were readily acylated on N-2.[137] Diazetidinone 1,3-dipoles (187) reacted with nucleophiles such as borohydride and Grignard reagents to yield the corresponding 1-alkyldiazetidinones; 2,4-unsubstituted derivatives could be converted into 2,4-dianions that reacted at C-4 with electrophiles.[138] Ring contraction of 4-diazopyrazoline-3,5-diones (188), similar to that known for 3-diazopyrrolidin-2,4-diones, gave rise to diazetidinones.[139]

Analogues of β-lactam systems in which the β-lactam ring has been replaced by a γ-lactam have been described, and whereas compounds (189) and (190) were biologically inactive[140] the more strained system (191) was reported to possess activity.[141]

11 Mechanistic Studies Relating to Biological Activity

Labelling experiments involving the mechanism of β-lactamase inhibition by penicillanic acid sulphone (192) have added support to the hypothesis that the

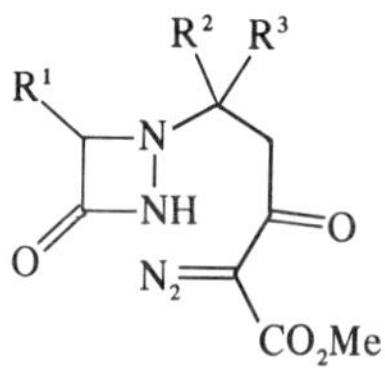

(186)

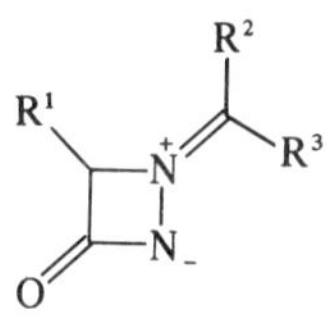

(187)

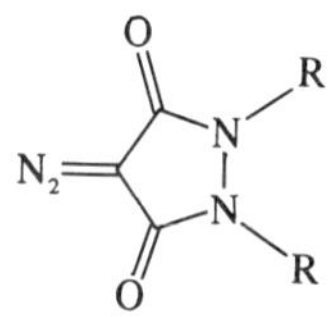

(188)

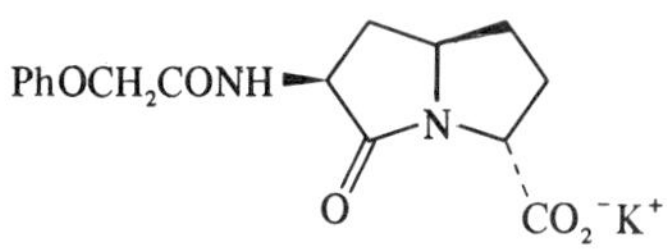

(189)

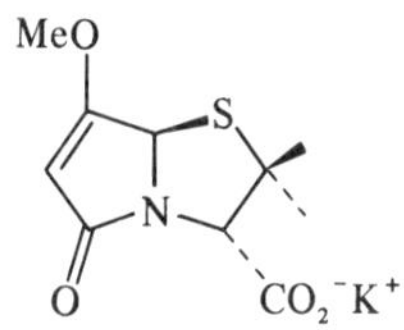

(190)

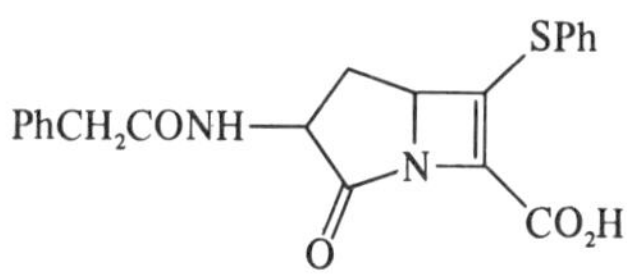

(191)

[137] E. C. Taylor, H. M. L. Davies, W. T. Lavell, and N. D. Jones, *J. Org. Chem.*, 1984, **49**, 2204.

[138] E. C. Taylor and H. M. L. Davies, *J. Org. Chem.*, 1984, **49**, 4415.

[139] G. Lawton, C. J. Moody, and C. J. Pearson, *J. Chem. Soc., Chem. Commun.*, 1984, 754.

[140] J. E. Baldwin, M. F. Chan, G. Gallacher, M. Otsuka, P. Monk, and K. Prout, *Tetrahedron*, 1984, **40**, 4513.

[141] U.S. Patent 4,428,960, 1984 (*Chem. Abstr.*, 1984, **100**, 191 655).

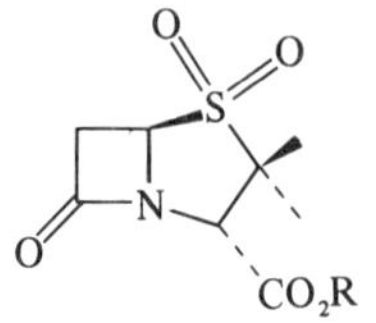

(192) R = H
(194) $R = CH_2Ph$

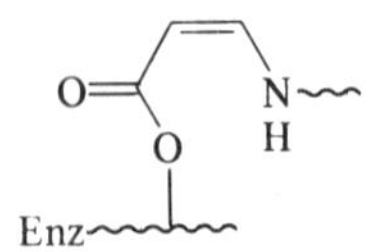

(193)

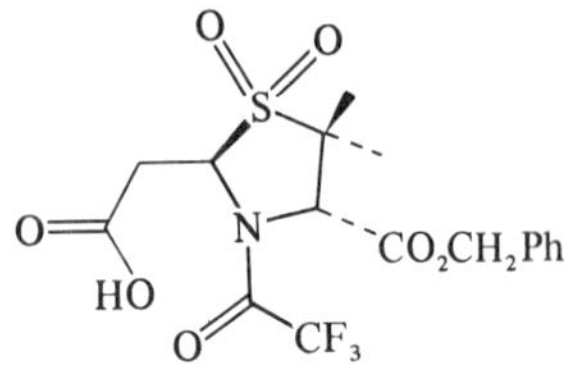

(195)

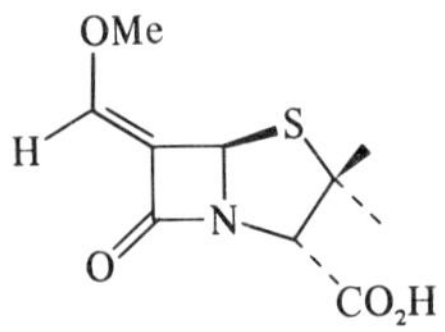

(196)

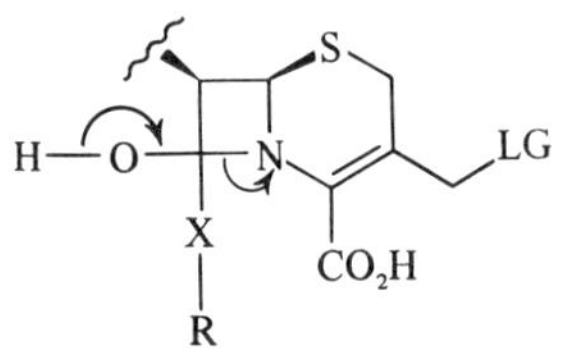

(197) X = O or NH

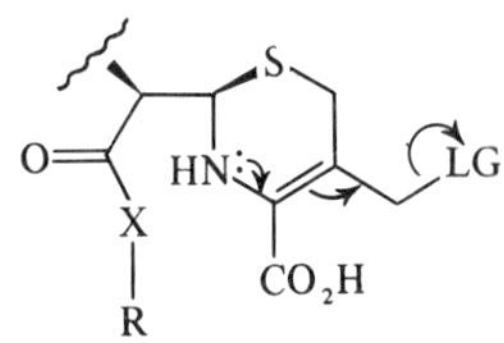

(198) X = O or NH

LG = leaving group

fragment generated by a suicide mechanism is one such as (193) derived from N, C-5, C-6, and C-7 of the substrate; this may be trapped by the enzyme as a β-aminoacrylate derivative, and model studies were also consistent with this proposal.[142] Additional model studies with the benzyl ester (194) have shown that in deuteriochloroform and trifluoroacetic acid the initial degradation product was (195), which indicated that C-7–N and C-5–S bond cleavages were not, in this case, and therefore need not be concerted; it was further argued that, since the relevant bonds are not antiperiplanar, such a concerted mechanism would be impossible, and it was suggested that similar steric constraints would apply to clavulanic acid.[143] 6-Methoxymethylene compound (196) has been shown to be an inhibitor of β-lactamases, and a mechanism thought to be effective for the related 6-acetonylidene compound was proposed.[144]

Studies on the enzymic hydrolysis[145] and aminolysis[146] of cephems bearing good leaving groups at C-3′ were in accord that cleavage of the β-lactam C-8–N-5 bond (197) and loss of the C-3′ substituent (198) are two distinct steps that are not necessarily concerted.

[142] D. G. Brenner and J. R. Knowles, *Biochemistry*, 1984, **23**, 5833.
[143] P. H. Crackett and R. J. Stoodley, *Tetrahedron Lett.*, 1984, **25**, 1295.
[144] D. G. Brenner and J. R. Knowles, *Biochemistry*, 1984, **23**, 5839.
[145] W. S. Faraci and R. F. Pratt, *J. Am. Chem. Soc.*, 1984, **106**, 1489.
[146] M. I. Page and P. Proctor, *J. Am. Chem. Soc.*, 1984, **106**, 3820.

6

Metal Complexes of Amino Acids and Peptides

BY R. W. HAY AND K. B. NOLAN

1 Introduction

This chapter describes work published during the years 1983 and 1984. Metalloproteins are not discussed, unlike in previous reviews. The area of metal complexes of amino acids and peptides will subsequently be reviewed on an annual basis.

Review articles of interest include a critical evaluation of formation constants of metal complexes of histidine, phenylalanine, tyrosine, L-dopa, and tryptophan,[1] long-range electron transfer in peptides and proteins,[2] haemerythrin,[3] *X*-ray structures of pyridoxal–amino acid Schiff bases as models for pyridoxal-catalysed reactions,[4] cytochromes P-450,[5] the biological chemistry of iron,[6] iron metabolism,[7] superoxide dismutase,[8] biochemistry of selenium,[9] biochemical and inorganic perspectives of copper co-ordination chemistry,[10] biological roles of copper,[11] iron–sulphur proteins,[12] copper deficiency and toxicity,[13] platinum, gold, and other metal chemotherapeutic agents,[14] and the co-ordination chemistry of metalloenzymes.[15]

[1] I.U.P.A.C. Commission on Equilibrium Data (U.K.), *Pure Appl. Chem.*, 1984, **56**, 247.
[2] 'Progress in Inorg. Chem.', Vol. 32, ed. S. J. Lippard, John Wiley, Chichester, 1984.
[3] I. M. Klotz and D. M. Kurtz, *Acc. Chem. Res.*, 1984, **17**, 16.
[4] M. D. Poojary, S. P. S. Rao, and H. Manohar, *Proc. Indian Acad. Sci., Chem. Sci.*, 1983, **92**, 485.
[5] F. P. Guengerich and T. L. MacDonald, *Acc. Chem. Res.*, 1984, **17**, 9.
[6] 'The Biological Chemistry of Iron', ed. H. B. Dunford, D. Dolphin, K. N. Raymond, and L. Sieker, NATO Adv. Stud. Inst. Ser., Reidel, The Netherlands, 1982.
[7] I. Bernát, 'Iron Metabolism', Plenum Press, New York, 1983.
[8] 'Superoxide Dismutase', Vols. I and II, ed. L. W. Oberley, CRC Press Inc., Boca Raton, Florida, 1982.
[9] R. J. Shamberger, 'Biochemistry of Selenium', Plenum Press, New York, 1983.
[10] 'Copper Coordination Chemistry; Biochemical and Inorganic Perspectives', ed. K. D. Karlin and J. Zubieta, Adenine Press, Guilderland, 1983.
[11] 'Biological Roles of Copper', Ciba Foundation Symposium No. 79 (new series), Excerpta Medica, Amsterdam, and Elsevier/North-Holland, New York, 1980.
[12] 'Iron–Sulphur Proteins', ed. T. G. Spiro, John Wiley and Sons, New York, 1982.
[13] C. A. Owen, 'Copper Deficiency and Toxicity. Acquired and Inherited in Plants, Animals, and Man', Noyes Publications, Park Ridge, New Jersey, 1981.
[14] 'Platinum, Gold and Other Metal Chemotherapeutic Agents', ed. S. J. Lippard, Am. Chem. Soc. Symp. Ser. No. 209, American Chemical Society, Washington, 1983.
[15] I. Bertini, R. S. Drago, and C. Luchinat, 'Coordination Chemistry of Metalloenzymes. The Role of Metals in Reactions Involving Water, Dioxygen and Related Species', D. Reidel, Dordrecht, Holland, 1983.

Recently published volumes of Sigel's 'Metal Ions in Biological Systems' deal with zinc and its role in biology and nutrition,[16] methods involving metal ions and complexes in clinical chemistry,[17] and properties of copper.[18]

Volume 5 (dealing with zinc enzymes) of Spiro's 'Metal Ions in Biology' series has been published.[19] This volume contains an interesting review article by Dixon and Sargeson that discusses the reactions of amino acids and peptides in the co-ordination sphere of cobalt(III), and it summarizes much of the work of the Australian group in this area. Previous volumes in this series deal with copper proteins (Volume 3) and iron–sulphur proteins (Volume 4).

Other reviews of interest include metalloproteins with phenolate co-ordination,[20] bioinorganic applications of magnetic circular-dichroism spectroscopy,[21] and the optical activity of Cr^{III}, Co^{III}, and Rh^{III} complexes of aminopolycarboxylate ligands.[22] An introductory text on bioinorganic chemistry has been published.[23]

2 Amino Acids

Synthetic and Spectroscopic Studies. – The cobalt(III)-promoted syntheses of the amino acids (*RS*)-2-cyclopropylglycine and (*R*)- and (*S*)-proline have been described,[24] using intramolecular imine formation between a co-ordinated aminate ion and a 2-keto acid, followed by borohydride reduction. Other work by the same group describes the cobalt(III)-promoted synthesis of β-carboxyaspartic acid[25] and the synthesis of the *C*-formylglycinate ion chelated to cobalt(III).[26]

Glycine co-ordinated to copper(II) reacts with a stoicheiometric quantity of aldehydes in basic solution to give β-hydroxyamino acids;[27] thus the use of acetaldehyde and benzaldehyde gives a 66% yield of *threo*-β-phenylserine. The use of the $[Co(NH_3)_5]^{3+}$ moiety as a useful C-terminal protecting group for sequential peptide synthesis was discussed in the previous review. Further papers in this area have dealt with the determination of the optical purity of amino acids and peptides bound to penta-amminecobalt(III)[28] and with the separation

16 'Metal Ions in Biological Systems', Vol. 15 (Zinc and Its Role in Biology and Nutrition), ed. H. Sigel, Marcel Dekker Inc., New York, 1983.
17 'Metal Ions in Biological Systems', Vol. 16 (Methods Involving Metal Ions and Complexes in Clinical Chemistry), ed. H. Sigel, Marcel Dekker Inc., New York, 1983.
18 'Metal Ions in Biological Systems', Vol. 12 (Properties of Copper), ed. H. Sigel, Marcel Dekker Inc., New York, 1981.
19 'Zinc Enzymes', Met. Ions Biol. Ser., Vol. 5, ed. T. G. Spiro, John Wiley and Sons Inc., New, York, 1983.
20 L. Que, *Coord. Chem. Rev.*, 1983, **50**, 73.
21 D. M. Dooley and J. H. Dawson, *Coord. Chem. Rev.*, 1984, **60**, 1.
22 D. J. Radanović, *Coord. Chem. Rev.*, 1984, **54**, 159.
23 R. W. Hay, 'Bio-Inorganic Chemistry', Ellis Horwood, Chichester, 1984.
24 P. J. Lawson, M. G. McCarthy, and A. M. Sargeson, *J. Am. Chem. Soc.*, 1982, **104**, 6710.
25 N. E. Dixon and A. M. Sargeson, *J. Am. Chem. Soc.*, 1982, **104**, 6716.
26 W. G. Jackson, G. M. McLaughlin, A. M. Sargeson, and A. D. Watson, *J. Am. Chem. Soc.*, 1983, **105**, 2426.
27 P. Sharrock, *Polyhedron*, 1983, **2**, 111.
28 S. S. Isied, J. Lyon, and A. Vassilian, *Int. J. Pept. Protein Res.*, 1982, **19**, 354.

of $(NH_3)_5Co^{III}$-amino acids and peptides by reversed-phase high-pressure liquid chromatography.[29] A recent investigation[30] has shown that, in the absence of base, the reaction of formaldehyde with bis(glycinato)zinc(II) monohydrate results in the formation of the complex bis-[*N*-(1,3-dioxa-5-azacyclohexyl)-acetato]zinc(II) dihydrate (1). This work has now been extended to the reactions of formaldehyde with the bis(glycinato) complexes of Ni^{II}, Co^{II}, and Cu^{II} at pH 4.5.[31] The Ni^{II} and Co^{II} complexes give the analogues of (1), but the Cu^{II} derivative lacks the water ligands. The copper(II) complex formed by the reaction of bis(glycinato)copper(II) with formaldehyde in the absence of base, when treated with $NaBH_4$, gives the sodium salt of *N*-(1,3-dioxa-5-azacyclohexyl)-acetic acid (2).[32] The reaction does not cause appreciable cleavage of the hetero-cyclic ring of the acid. Treatment of the copper complex with H_2S gives glycine and formaldehyde.

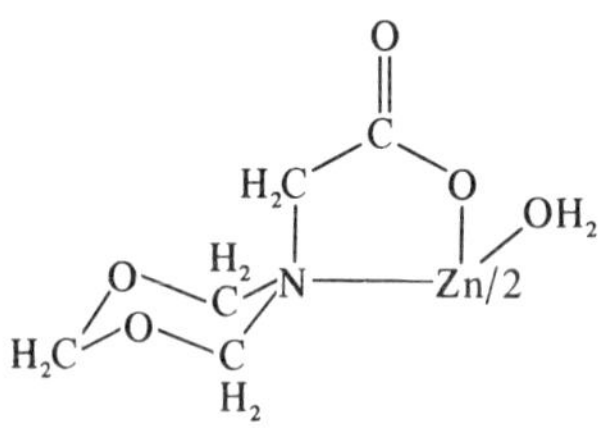

(1)

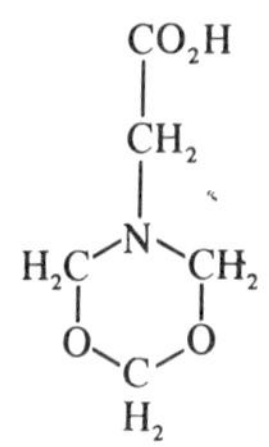

(2)

The asymmetric synthesis of alanine has been described using a dissymmetric cobalt(III) complex and α-amino-α-methylmalonate as the precursor of alanine.[33]

Amavadine is a vanadium natural product from the mushroom *Amanita muscaria*. A possible structure for amavadine is shown in (3) in which two

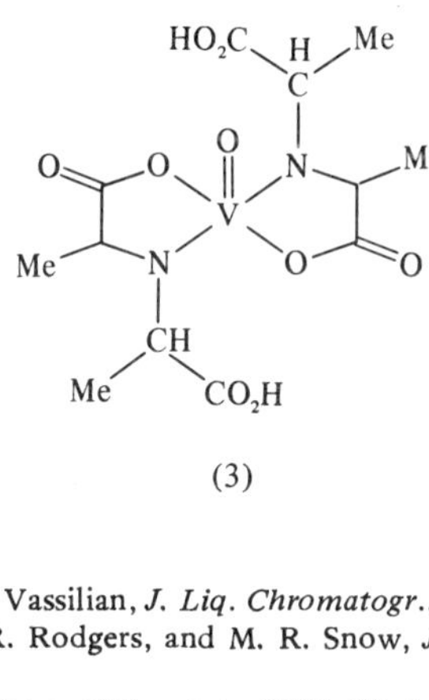

(3)

[29] S. S. Isied, J. Lyon, and A. Vassilian, *J. Liq. Chromatogr.*, 1982, **5**, 537.
[30] S.-B. Teo, S.-G. Teoh, J. R. Rodgers, and M. R. Snow, *J. Chem. Soc., Chem. Commun.*, 1982, 141.
[31] S.-B. Teo and S.-G. Teoh, *Inorg. Chim. Acta*, 1984, **91**, L17.
[32] S.-B. Teo and S.-G. Teoh, *Inorg. Chim. Acta*, 1983, **68**, 107.
[33] M.-J. Jun, N. M. Yoon, and C. F. Liu, *J. Chem. Soc., Dalton Trans.*, 1983, 999.

N-hydroxy-α,α-iminodipropionic acid ligands are co-ordinated to VO^{2+}. The synthesis of the model complexes bis(iminodiacetato)oxovanadium(IV) and bis-(α,α-iminodipropionato)oxovanadium(IV) has now been described.[34] Preliminary electrochemical data suggest reversible $V^{IV} \rightleftharpoons V^{III}$ couples.

Platinum(II) and palladium(II) complexes of amino acids have attracted considerable attention. Since the discovery of the antitumour activity of *cis*-$[Pt(NH_3)_2Cl_2]$ ('cisplatin') by Rosenberg there have been many attempts to synthesize more active and less toxic compounds for clinical use. Compounds of the type *cis*-$[PtCl_2(Xaa)(tba)]$ (Xaa = amino acid, tba = t-butylamine) have been prepared[35] with the aim of obtaining liposoluble analogues of cisplatin. The derivatives of Gly, D-Ala, L-Thr, and L-Ser were found to be moderately active against murine P388 and L1210 leukaemia models.

The synthesis of mixed-ligand complexes of Pt^{II} with glycine or alanine as primary ligands and the nucleosides adenosine, guanosine, inosine, cytidine, and uridine as secondary ligands has also been reported.[36] The reaction of palladium(II) chloride and L-proline (Pro) in aqueous solution gives the dimeric complex[37] (4), $[Pd(Pro)Cl]_2$.

The complex reacted further with the purine nucleosides inosine or guanosine (Nucl) to give $[Pd(Pro)(Nucl\text{-}H^+)]$. The insolubility of these complexes suggested a polymeric structure in which the nucleoside bridges two adjacent palladium atoms through N-7 and the exocyclic O-6 atoms.

A German patent[38] reports that complexes of type (5), where R^2 is greater than C_2 alkyl or a hydroxy-substituted group, display low toxicity and a strong antitumour effect. The complexes can be prepared by reaction of (6) with alcohols (R^2OH) and HCl or by treatment of K_2PtCl_4 with amino acid ester hydrochlorides.

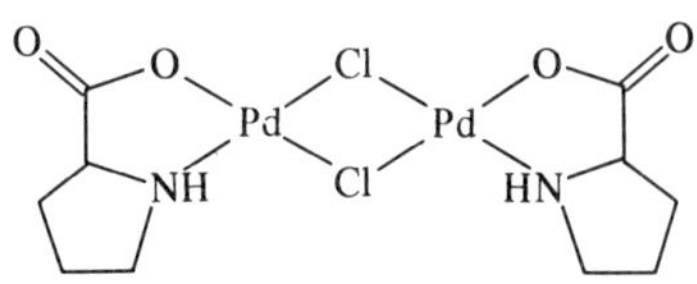

(4)

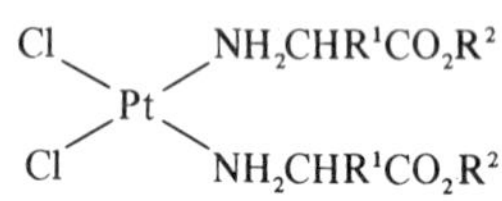

(5)

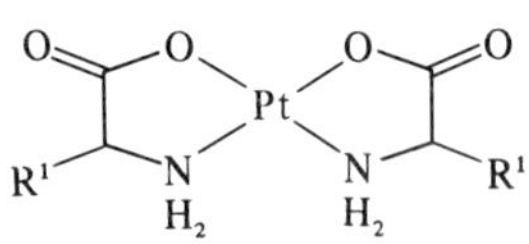

(6)

[34] M. Asri Nawi and T. L. Riechel, *Inorg. Chim. Acta*, 1984, **93**, 131.
[35] E. Bersanetti, A. Pasini, G. Pezzoni, G. Pratesi, G. Sava, R. Supino, and F. Zunino, *Inorg. Chim. Acta*, 1984, **93**, 167.
[36] B. T. Khan, G. N. Goud, and S. V. Kumari, *Inorg. Chim. Acta*, 1983, **80**, 145.
[37] G. Pneumatikakis, *Polyhedron*, 1984, **3**, 9.
[38] *Chem. Abstr.*, 1983, **99**, 33 123, 33 119.

Platinum(II) complexes with glycine as an oxygen-bonded unidentate ligand have been characterized.[39] Reaction of glycine (Gly) with *cis*-$[Pt(NH_3)_2(H_2O)_2]^{2+}$ in water gives the *O*-bonded *cis*-$[Pt(NH_3)_2(O\text{-Gly})(H_2O)]^{2+}$, which only slowly converts to the chelate complex.

Two types of Pd^{II} complexes have been prepared using the hydrazides of aspartic and glutamic acids.[40] One type has the general formula $[Pd_2(LH)_2Cl_2]X_2$ with two bridging chloride ions ($X = Cl^-$ or OH^- depending upon the pH), with thc ligands bonded *via* the aminocarboxylic NH_2 group and the oxygen of $CONHNH_2$, the carboxylic group being deprotonated and unco-ordinated. In the other type the ligands are bonded *via* NH_2 and the deprotonated hydrazide nitrogen. The reaction of both *C*-phenylglycine and its methyl ester with Pd^{II} acetate in acetic acid gives the acetato-bridged dimer $[PdL(OAc)_2]_2$ and the monomeric complex $[PdL_2(OAc)_2]$.[41] On the basis of i.r. and ^1H n.m.r. measurements the ligands are bonded in a monodentate fashion *via* the amine nitrogen. In addition, the *N,N*-dimethyl-*C*-phenylglycine ethyl ester reacts with Pd^{II} acetate in acetic acid to give the cyclopalladated species $[Pd\{Me_2NCH(CO_2Et)C_6H_4\}X]_2$ ($X = OAc$ or Cl).

The reactions of the binuclear complex $[Pd(Cys\text{-}OMe)Cl]_2$ (7) with nucleosides (Nucl) and adenosine-5′-monophosphate (AMP-Na_2) have been studied using aqueous and DMSO solutions.[42] The monomeric complex (8) was isolated from DMSO. The reaction with AMP-Na_2 in aqueous solution gave Na[Pd(Cys-OMe)(AMP)], in which the AMP is bidentate *via* N-7 of the purine and the phosphate group.

The complexes *cis*- and *trans*-$[PdCl_2(NH_2COCl)_2]$ and *trans*-$[Pd_2Cl_4(NH_2CH_2COCl)_2]$ can be obtained by the reaction of $[Pd(Gly\text{-}O)_2]$ with $SOCl_2$.[43] *cis*- and *trans*-$[PdCl_2(NH_2CO_2H)_2]$ were also characterized.

Unidentate amino acid platinum(II) complexes of the type *trans*-$[Pt(Xaa)_2(L)_2]Cl_2$ (Xaa = L-Ala, L-Val, L-Tyr, or L-hydroxyproline, L = thiocarbamide) have been synthesized and electronic and c.d. spectra studied.[44] ^1H n.m.r. and c.d. measurements have been used to study the chiral sulphur centres in Pd^{II} complexes of *S*-benzyl-L-cysteine and glycyl-*S*-benzyl-L-cysteine.[45]

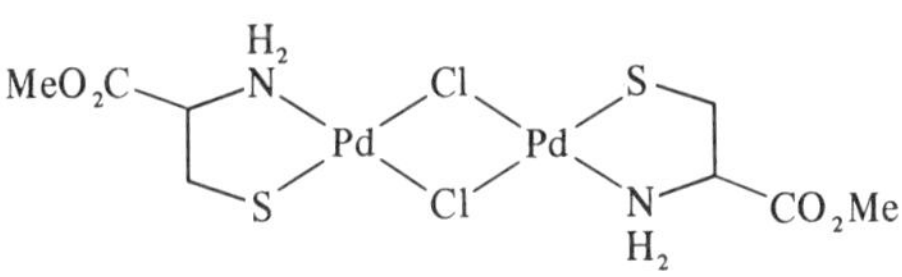

(7)

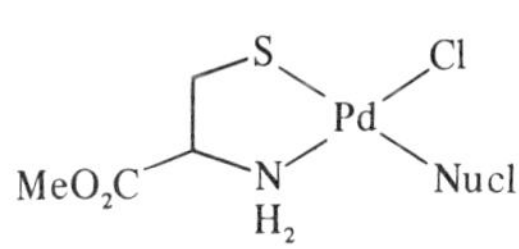

(8)

[39] T. G. Appleton and J. R. Hall, *J. Chem. Soc., Chem. Commun.*, 1983, 911.
[40] P. R. Bontchev, M. Boneva, M. Arnaudov, and V. I. Nefedov, *Inorg. Chim. Acta*, 1984, **81**, 75.
[41] A. D. Ryabov, V. A. Polyakov, and A. K. Yatsimirsky, *Inorg. Chim. Acta*, 1984, **91**, 59.
[42] G. Pneumatikakis, *Inorg. Chim. Acta*, 1983, **80**, 89.
[43] M. Castillo, A. Romero, and E. Ramirez, *Inorg. Chem.*, 1984, **23**, 2668.
[44] O. P. Slyudkin and B. J. F. Nordén, *Inorg. Chem.*, 1983, **22**, 2637.
[45] H. Kozlowski, B. Decock-le-Revérénd, J.-L. Delaruelle, C. Loucheux, and B. Ancian, *Inorg. Chim. Acta*, 1983, **78**, 31.

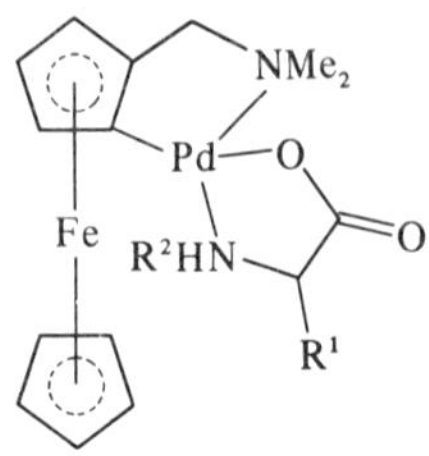

(9)

Intramolecular ligand-ligand interactions in ternary Pd^{II}- and Cu^{II}-amino acid complexes have been studied and various complexes characterized.[46] Ferrocenylpalladium derivatives of some amino acids and dipeptides (9) have been prepared.[47]

Chromium complexes are beginning to attract considerable attention, and in part this is due to current interest in the glucose-tolerance factor (GTF). Chromium was recognized as an essential trace element in 1955.[48] Rats fed a chromium-deficient diet developed an impaired tolerance for intravenous glucose, which could be reversed by an insulin-potentiating factor present in brewer's yeast, meat, and various foods. The topic of chromium(III) and the GTF has been reviewed.[49] Although no chemically defined complexes have been characterized, it has been suggested that the active species is a nitrogen-co-ordinated $[Cr^{III}$-(nicotinic acid)$_2(H_2O)_4]$ complex that is protected from olation by amino acids (Gly, Cys, Glu) occupying the equatorial sites. A number of octahedral chromium(III) complexes with amino acid ligands have been prepared with the general structure $Cr(Xaa)_2(H_2O)_2$, where the amino acids glycine, glutamic acid, and glutamine act as bidentate ligands.[50] The analogous compound with cysteine is stable at low pH, but at high pH a terdentate cysteine complex $[Cr(Cys)_2]^-$ is formed. Only $[Cr(glutamine)_2(H_2O)_2]^+$, Cr-nicotinic acid-glycine, and the mixture of complexes $[Cr(glycine)_n(H_2O)_{6-n}]^{3+}$ display significant biological activity. A thesis dealing with the evaluation of amino acids and nicotinic acid as biologically important ligands on Cr^{III} has been published.[51] A variety of neutral complexes of the type [Cr(dipeptide)(amino acid)$_2$] (involving only glycine and alanine) and [Cr(dipeptide)(amino acid)] (involving aspartic acid) have been characterized in the solid state and solution.[52] Chromium(III) complexes of the type CrL_3 (LH = picolinic acid, pipecolic acid, *o*-aminobenzoic acid, or sarcosine) and $[CrL_2(OH)]_2$ (LH = picolinic acid) have been prepared by the reaction of $CrCl_3\cdot 6H_2O$ with the respective aminocarboxy-

[46] O. Yamauchi, *J. Mol. Catal.*, 1984, **23**, 255.
[47] V. I. Sokolov, K. S. Nechaeva, and O. A. Reutov, *Zh. Org. Khim.*, 1983, **19**, 1103.
[48] W. Mertz and K. Schwarz, *Arch. Biochem. Biophys.*, 1955, **58**, 504.
[49] J. Barrett, P. O'Brien, and J. Pedrosa de Jesus, *Polyhedron*, 1985, **4**, 1.
[50] J. A. Cooper, L. F. Blackwell, and P. D. Buckley, *Inorg. Chim. Acta*, 1984, **92**, 23.
[51] L. A. Gerdom, *Diss. Abstr. Int. B*, 1984, **44**, 3772.
[52] K. Govindaraju, N. C. Kumar, C. N. Krishnan, and D. Ramaswamy, *Leather Sci. (Madras)*, 1983, **30**, 158 (*Chem. Abstr.*, 1983, **99**, 124402).

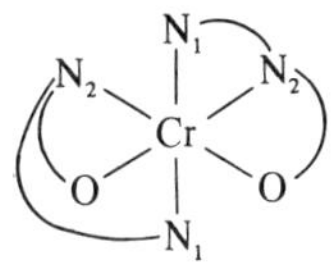

(10) *trans*-imidazole

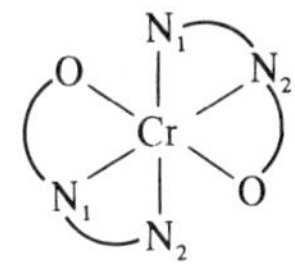

(11) *trans*-carboxylate

lic acids.[53] The electronic and luminescence spectra for the CrL_3 species indicate that they are all *fac* isomers.

The *trans*-imidazole and *trans*-carboxylate stereoisomers of $[Cr(L\text{-His})_2]NO_3$, (10) and (11), respectively, have been prepared and a novel binuclear complex $[Cr(L\text{-His})(OH)]_2$ has been characterized by magnetic susceptibility and e.s.r. studies.[54] Ferromagnetic behaviour at low temperature lends support to the proposed hydroxo-bridged structure for the dimer. The magnetic properties of the dimer $\Delta(-)_{546}$-$[Cr(L\text{-Pro})OH]_2 \cdot 4H_2O$ have also been investigated.[55]

Glycine in oxygen-free aqueous acid solutions of Cr^{II} halides co-ordinates in its zwitterionic form, $^{-}O_2CCH_2NH_3^{+}$, to give carboxylate-bridged Cr_2^{4+} complexes.[56] Stable purple salts of the type $Cr_2(O_2CCH_2NH_3)_4X_4 \cdot nH_2O$ have been isolated and their crystal structures determined (X = Cl or Br). Electronic-excitation spectroscopy and an angular-overlap-model analysis of *fac*-$[Cr(Gly)_3]$ have been carried out.[57]

A number of papers have appeared dealing with copper(II) complexes of *N*-tosylglycine. Two compounds of the type $[Cu(TsglyH)_2X_2 \cdot H_2O]$ [TsglyH = *N*-tosylglycinate monoanion, X = pyridine (py) or 4-methylpyridine (4-Mepy)] have been prepared and characterized by means of magnetic, e.p.r., electronic, and i.r. spectra.[58] For one of them, *catena*-(μ-aqua)bis-(*N*-tosylglycinato)bis-(4-methylpyridine)copper(II), the crystal structure was determined. The complex is a linear-chain water-bridged copper(II) polymer. Similar complexes $[Cu(TsglyH)_2X_2]$ (X = imidazole, *N*-methylimidazole, or pyridine) have been characterized and the crystal structure of the *N*-methylimidazole complex has been determined.[59] The interaction between *N*-tosylglycine and copper(II) at different pH values gives rise to four compounds: $Cu(TsglyH)_2 \cdot 4H_2O$, $[Cu(Tsgly)(H_2O)_3]$ (Tsgly = *N*-tosylglycinate dianion), and $K_2[Cu(Tsgly)_2]$, which exists in two forms, one blue and one violet.[60] The crystal structures of $[Cu(Tsgly)(H_2O)_3]$ and the blue form of $K_2[Cu(Tsgly)_2]$ were determined. At $pH < 5$, *N*-tosylglycine bonds *via* the oxygen of the carboxylate group [*e.g.* in $Cu(TsglyH)_2 \cdot 4H_2O$]; at $pH > 5$ amide deprotonation occurs. The interaction of

[53] G. Yuen, H. Heaster, and P. E. Hoggard, *Inorg. Chim. Acta*, 1983, **73**, 231.
[54] E. E. Eduok, J. W. Owens, and C. J. O'Connor, *Polyhedron*, 1984, **3**, 17.
[55] S. Kallesøe and E. Pedersen, *Acta. Chem. Scand., Ser. A*, 1982, **36**, 859.
[56] M. Ardon, A. Bino, S. Cohen, and T. R. Felthouse, *Inorg. Chem.*, 1984, **23**, 3450.
[57] W. M. Wallace and P. E. Hoggard, *Inorg. Chem.*, 1983, **22**, 491.
[58] L. P. Battaglia, A. B. Corradi, and L. Menabue, *Inorg. Chem.*, 1983, **22**, 3251.
[59] L. P. Battaglia, A. B. Corradi, G. Marcotrigiano, L. Menabue, and G. C. Pellacani, *Inorg. Chem.*, 1983, **22**, 1902.
[60] L. Antolini, L. P. Battaglia, G. B. Gavioli, A. B. Corradi, G. Grandi, G. Marcotrigiano, L. Menabue, and C. G. Pellacani, *J. Am. Chem. Soc.*, 1983, **105**, 4327.

copper(II) with *N*-tosylglycine in ethanol solution gives $[\{Cu(TsglyH)_2\}_n]$ (green) and $Na_2[Cu(Tsgly)_2]$ (blue).[61]

N-Acyl or *N*-benzoyl amino acids invariably interact with metal ions only *via* the carboxylate group. In basic solution the metal hydroxide precipitates before amide deprotonation occurs. In ethanolic solution it is possible to isolate dimeric compounds of the copper(II) acetate type with strong antiferromagnetic exchange. Recent work[62] has shown that *N*-benzyloxycarbonyl-substituted amino acids (HL) react with copper(II) in ethanol to give simple green complexes of formula $[\{CuL_2(H_2O)_n\}_2]$ (L = Z-glycinate, Z-valinate, or Z-leucinate, $n = 1$; L = Z-alaninate, $n = 0$). The complexes are dimers with strong antiferromagnetic interaction between the copper(II) ions ($-2J$ is 295–350 cm^{-1}). The magnetic and spectroscopic behaviour patterns of amine adducts of copper(II)–*N*-acetyl-β-alaninate have also been investigated.[63] These ternary complexes have the general formula $Cu(N\text{-acyl-}\beta\text{-Ala})_2B_2\cdot nH_2O$ (B = ethylenediamine, 4-methylpyridine, piperazine, bipy, *etc.*). A further paper in this area deals with imidazole-containing ternary copper(II) complexes of *N*-benzyloxycarbonyl-amino acids and the crystal structure of bis-(*N*-benzyloxycarbonyl-alaninato)bis-(*N*-methylimidazole)copper(II) ethanol solvate.[64]

Sulphur-containing amino acids have attracted considerable attention as ligands. The new mixed-ligand complex Cu(L-Trp)(DL-Met)·$1.5H_2O$ has been characterized in the solid state.[65] Currently few examples of such mixed complexes are known, although the crystal structure of [Cu(His)(Thr)] was determined in 1969.[66] Copper(II) complexes with polymer-bound *S*-methyl-L-cysteine have been studied by spectroscopic techniques and viscosity measurements.[67] The polymer chain brings several carboxylate groups close together, leading to the formation of $Cu(OOCR)_n$ species that are sufficiently stable to enable amide deprotonation to occur.

Amino acid amides of dithio-oxalic acid have been prepared [for example compound (12) derived from methionine] and their copper(II) complexes studied.[68] Thus (12) (= LH_2) gave the binuclear complex $Cu_2(L)_2(MeOH)\cdot 0.5H_2O$. The methyl ester of (12) gave $Cu_2\{C_2S_2[NCH(CH_2CH_2SMe)CO_2Me]_2\}X_2$ (X = Cl, Br, or ClO_4). The crystal structure of the bromide salt establishes that each copper has a distorted trigonal-bipyramidal stereochemistry; the bridging network $CuC_2S_2N_2Cu$ is almost planar with a Cu–Cu separation of

61 L. Antolini, L. Menabue, G. C. Pellacani, G. B. Gavioli, G. Grandi, L. P. Battaglia, A. B. Corradi, and G. Marcotrigiano, *J. Chem. Soc., Dalton Trans.*, 1984, 1687.
62 L. Antolini, L. Menabue, G. C. Pellacani, M. Saladini, L. P. Battaglia, A. B. Corradi, and G. Marcotrigiano, *J. Chem. Soc., Dalton Trans.*, 1984, 2325.
63 L. Menabue, P. Prampolini, M. Saladini, and P. Morini, *Inorg. Chim. Acta*, 1983, **68**, 157.
64 L. Antolini, L. Menabue, M. Saladini, M. Sola, L. P. Battaglia, and A. B. Corradi, *Inorg. Chim. Acta*, 1984, **93**, 61.
65 P. Gili and P. M. Zarza, *Inorg. Chim. Acta*, 1984, **89**, L23.
66 H. C. Freeman, J. M. Guss, M. J. Healy, R. P. Martin, and C. E. Nockholds, *J. Chem. Soc., Chem. Commun.*, 1969, 225.
67 M. Delporte and B. Decock-Le-Révérend, *Inorg. Chim. Acta*, 1984, **93**, L15.
68 R. Veit, J. J. Girerd, O. Kahn, F. Robert, Y. Jeannin, and N. El Murr, *Inorg. Chem.*, 1984, **23**, 4448.

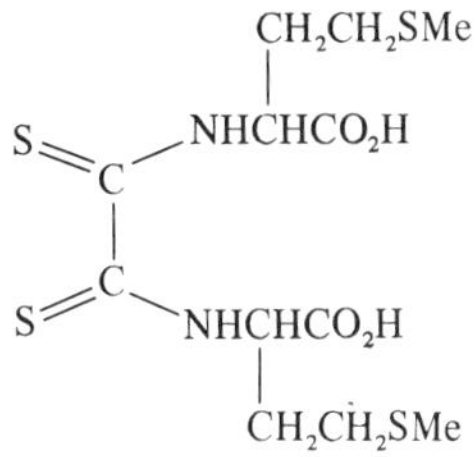

(12)

5.66 Å. Oxidation to $Cu^{II}Cu^{III}$ is totally reversible, but the oxidized species is chemically unstable.

Complexes of $[Cu^{I}(Me_2phen)_n]^+$ (Me_2phen = 2,9-dimethyl-1,10-phenanthroline, $n = 1$ or 2) with Cys, penicillamine, Met, *N*-acetyl-Cys, and *N*-acetylpenicillamine have been prepared.[69] ^{1}H n.m.r. and i.r. data indicate that the amino group is protonated in the complexes containing the sulphydryl amino acids. The Cu^{I}–S interaction is believed to be exclusively sulphur→copper(I) σ-donation. E.s.r. studies on copper(II) complexes of 12 α-amino acids have been published.[70]

Mixed-ligand clusters of the type $[Cu^{I}_8Cu^{II}_6L^1_nL^2_{12-n}Cl]$ can be readily distinguished by zone electrophoresis if the ligands L^1 and L^2 differ in formal charge.[71] Approximately random species distributions were observed when L^1 was a related ligand lacking the negative charge of a carboxylate group and L^2 was penicillamine. The cluster containing only penicillamine appears to form addition products with carbodi-imides and aziridines involving reactions with the peripheral carboxylate groups of the cluster.

A 1:1 complex of mercury(II) chloride with D-penicillamine has been characterized and its crystal structure determined.[72] A triply bridging chloride ion links three equivalent $[HgSCMe_2CH(NH_3)COO]^+$ units. Methylmercury-DL-selenocysteinate monohydrate, a key model for methylmercury(II)-selenoprotein interaction *in vivo*, has been prepared by the reaction of seleno-DL-cysteine with methylmercury(II) hydroxide.[73] *X*-Ray data establish that the selenoamino acid is co-ordinated to mercury *via* a deprotonated selenohydryl group. The binding of Hg^{II} to DL-selenomethionine has also been investigated by ^{1}H n.m.r. spectroscopy,[74] as has the binding of Au^{III} to the ligand.[75] Gold(III) oxidizes DL-selenomethionine to DL-methionine-selenoxide (13).

Widely different types of inorganic arsenic, antimony, and bismuth derivatives have been used against protozoal infections. The pharmacological effects appear to

[69] W.-L. Kwik, K.-P. Ang, and P.-C. Lau, *J. Chem. Soc., Dalton Trans.*, 1983, 2269.
[70] T. Szabo-Planka and L. I. Horvath, *J. Coord. Chem.*, 1984, **13**, 163.
[71] M. E. Cooke, M. E. McDaniel, S. R. James, S. L. Jones, N. Trobak, B. C. Craytor, D. R. Buschman, and J. R. Wright, *J. Inorg. Biochem.*, 1983, **18**, 318.
[72] L. Book and T. C. W. Mak, *Inorg. Chim. Acta*, 1984, **92**, 265.
[73] A. J. Carty, S. F. Malone, N. J. Taylor, and A. J. Canty, *J. Inorg. Biochem.*, 1983, **18**, 291.
[74] A. A. Isab, *Inorg. Chim. Acta*, 1984, **91**, L35.
[75] A. A. Isab, *Inorg. Chim. Acta*, 1983, **80**, L3.

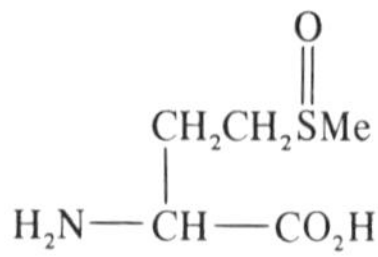

(13)

result mainly from the formation of insoluble complexes with sulphydryl groups of enzymes. The synthesis of the L-cysteine complexes $As(CysH)_3$, $Sb(CysH)_3 \cdot H_2O$, and $Bi(CysH)_3 \cdot H_2O$ (CysH = a monoanionic form of L-cysteine) has recently been described,[76] and possible structures have been considered. The formation of zinc(II) cysteinate complexes has been monitored by ^{15}N n.m.r.[77]

The (diethylenetriamine) (L-penicillaminato)cobalt(III) complex [Co(L-Pen)-(dien)]$^+$ has been chromatographically separated[78] into three isomers, *trans*(*N,N*), *trans*(*N,O*), and *trans*(*N,S*). Each of these gave the corresponding isomer of the (diethylenetriamine) (*S*-methyl-L-penicillaminato)cobalt(III) complex on reaction with dimethyl sulphate. Three [Co(terdentate-*N,O,S*)$_2$] complexes have been synthesized, where terdentate-*N,O,S* denotes D-penicillaminate (D-Pen), *S*-methyl-D-penicillaminate (D-Smp), and *S*-methyl-L-cysteinate (L-Smc).[79] The D-Smp and L-Smc complexes were separated into the three possible isomers and the D-Pen derivative was separated into two isomers.

Cobalt(III) complexes with *S*-(2-aminoethyl)-L-homocysteinate and a bidentate ligand (Gly, oxalate, ethylenediamine) have been characterized.[80] The preparation and stereochemistry of cobalt(III) complexes with *S*-alkyl-L-cysteinate, *N,N'*-trimethylenebis-(*S*-methyl-L-cysteinate), and *S,S'*-ethylene-(or -trimethylene)bis-(L-cysteinate)[81] and of (L-histidinato)(L-methioninato)-cobalt(III) bromide[82] have also been described.

N.m.r. has been used to study the structures of the dimeric peroxybis-(histidinato)cobalt(III) complexes formed by the oxygenation of solutions containing L-His and Co^{II}.[83] ^{59}Co n.m.r. only confirms the presence of peroxy complexes, but ^1H and ^{13}C n.m.r. provides more detailed structural information. At low pH (4.0–5.8) a single isomer predominates, identified as a complex containing one tridentate and one bidentate histidine per Co ion. Above pH 5.8 the two carboxyl ligands are successively replaced by water molecules that ionize with pK_as of about 6.2 and 7.2.

The cobalt(III) complexes [Co(bipy)(Xaa)]I_2 and [Co(phen)(Xaa)]I_2 (Xaa = anions of L-Ala, Gly, L-Leu, L-Phe, and L-Pro) have been prepared and

[76] G. Alonzo, N. Bertazzi, and M. Consiglio, *Inorg. Chim. Acta*, 1984, **85**, L35.

[77] B. P. Bammel and R. F. Evilia, *Inorg. Chim. Acta*, 1984, **81**, L5.

[78] K. Wakayama, K. Okamoto, H. Einaga, and J. Hidaka, *Bull. Chem. Soc. Jpn.*, 1983, **56**, 1995.

[79] K. Okamoto, K. Wakayama, H. Einaga, S. Yamada, and J. Hidaka, *Bull. Chem. Soc. Jpn.*, 1983, **56**, 165.

[80] M. Suzuki, O. Arisato, K. Okamoto, H. Einaga, and J. Hidaka, *Bull. Chem. Soc. Jpn.*, 1984, **57**, 2751.

[81] T. Konno, K. Okamoto, and J. Hidaka, *Bull. Chem. Soc. Jpn.*, 1983, **56**, 2631.

[82] K. Okamoto, H. Maki, and J. Hidaka, *Bull. Chem. Soc. Jpn.*, 1984, **57**, 595.

[83] D. R. Eaton and J. D. Weegar, *Inorg. Chim. Acta*, 1984, **93**, 73.

their c.d. spectra analysed.[84] All six possible geometrical isomers of $[Co(edma)_2]^+$ (edma = ethylenediaminemonoacetate) have been isolated and the five racemates resolved.[85]

Several compounds containing chiral amino acids co-ordinated to the Mo_2^{4+} unit have been prepared recently. Reaction of DL-Phe, DL-tyrosine, and DL-*C*-phenylglycine (DL-*C*-Phgly) with Mo_2^{4+} in aqueous solution gives $[Mo_2(D\text{-Phe})_2(L\text{-Phe})_2]I_4 \cdot 6H_2O$, $[Mo_2(D\text{-Tyr})(L\text{-Tyr})_2]I_4 \cdot 6H_2O$, and $[Mo_2(D\text{-}C\text{-Phgly})_2(L\text{-}C\text{-Phgly})_2](p\text{-toluenesulphonate})_4 \cdot 4H_2O$.[86] Crystallography establishes that the four chiral ligands are arranged around the Mo–Mo unit in the cyclic order of DDLL.

The complex $[MoO_2\{(S)\text{-Pen-OMe}\}_2]$ (Pen-OMe = methyl penicillaminate) has been synthesized and its crystal structure determined.[87] Absorption and c.d. spectra of some trinuclear molybdenum(IV) and dinuclear molybdenum(III) complexes of optically active amino acids have been described.[88] The binuclear complex $[Mo_2O_4(GlyO)_2(H_2O)_2]$ (GlyO = $NH_2CH_2CO_2^-$) has been prepared and characterized.[89] The complex contains dioxo-bridged molybdenum(V), which in aqueous solution, below pH 5.5, undergoes proton-assisted oxidation to an oxo-bridged Mo^V–Mo^{VI} species.

Although numerous cobalt(III) complexes with amino acid ligands are known, only a few Rh^{III} complexes have been characterized. The complex $Na[Rh(L\text{-Asp})_2]$ has now been prepared[90] and separated into three isomers on Sephadex A-25. ^{15}N n.m.r. has been used[91] to study the three isomers of $[Co(L\text{-Asp})_2]^-$.

Other studies dealing with cobalt(III) complexes include the absolute configurations and c.d. spectra of *fac*-$[Co(NH_3)_3(Gly)X]^{n+}$ complexes (X = CN^-, NO_2^-, CNS^-, or N_3^-),[92] optical resolution and c.d. spectra of $[Co(Adao)(Gly)]^+$ isomers (Adao = 8-amino-3,6-diazaoctanate),[93] and the kinetically controlled stereoselective synthesis of $[Co(phen)_2(S\text{-aminoacidate})]^{n+}$-type complexes.[94]

Complexes of the types $[CoLX_2]$ and $[CoLXY]$ [L is the quadridentate ligand ethylenediamine-*N*,*N*′-di-(*S*)-α-isovalerate derived from *S*-valine, X = H_2O, Y = NO_2^-] have been prepared.[95] In principle, three geometrical isomers are possible, *cis*-α, *cis*-β, and *trans*, but the *cis*-α isomer predominates. Cobalt(III) complexes of the analogous ligand derived from *S*-proline have also been studied.[96]

84 M. B. Monroe, D. R. Boone, and R. N. Kust, *Polyhedron*, 1984, **3**, 49.
85 T. Yasui, H. Kawaguchi, and T. Ama, *Chem. Lett.*, 1983, 1277.
86 F. Apfelbaum-Tibika and A. Bino, *Inorg. Chem.*, 1984, **23**, 2902.
87 I. Buchanan, C. D. Garner, and W. Clegg, *J. Chem. Soc., Dalton Trans.*, 1984, 1333.
88 Y. Sasaki, T. Tani, and T. S. Morita, *Chem. Uses Molybdenum*, Proc. Int. Conf., 4th, 1982 (*Chem. Abstr.*, 1983, **99**, 202 636).
89 M. Chaudhury, *J. Chem. Soc., Dalton Trans.*, 1983, 857.
90 M. Watabe, K. Furihata, and Y. Odaka, *Bull. Chem. Soc. Jpn.*, 1984, **57**, 2669.
91 M. Watabe, M. Takahashi, and A. Yamasaki, *Inorg. Chem.*, 1983, **22**, 2650.
92 S. Fujinami, A. Hattori, and M. Shibata, *Bull. Chem. Soc. Jpn.*, 1983, **56**, 2420.
93 K. Watanabe, *Bull. Chem. Soc. Jpn.*, 1983, **56**, 2839.
94 J. A. Chambers, R. D. Gillard, P. A. Williams, and R. S. Vagg, *Inorg. Chim. Acta*, 1983, **70**, 167.
95 M. Strasak and J. Majer, *Inorg. Chim. Acta*, 1983, **70**, 231.
96 M. Strasak and F. Bachraty, *J. Coord. Chem.*, 1984, **13**, 105.

Other investigations have dealt with co-ordination shifts in the ^{15}N n.m.r. spectra of glycinato ligands in $[Co(ox)_x(Gly)_y(en)_z]$ complexes[97] and pK_a and isomer determinations of cobalt(III) imidazole and histidine complexes by ^1H n.m.r. and *X*-ray crystallography.[98] In the latter paper the pK_as for *cis*-$[Co(en)_2(H_2O)(ImH)]^{3+}$, *cis*-$[Co(en)_2(H_2O)(N\text{-MeIm})]^{3+}$, and $[Co(en)(H_2O)(HisH)]^{2+}$ were found to be 5.85, 5.95, and 6.20, respectively, for ionization of the water ligand to give the hydroxo complexes. Ionization of the imidazole hydrogen has a p*K* of 10.5–10.8; crystallographic data for [Co(HisH)(en)Cl]Cl indicate that histidine is tridentate.

The *meso* and racemic diastereoisomers of the iron(III) complex of *N,N'*-ethylenebis-(*o*-hydroxyphenyl)glycinate have been separated and characterized.[99] These complexes were evaluated as models for the iron-binding site in human serum transferrin. Both isomers undergo hydrolysis at high pH, and the hydrolysed species is a much closer spectral match to di-iron(III) transferrin than the unhydrolysed form.

The antibiotic D-cycloserine (14) has a potential to chelate metal ions. A reappraisal of the structures of the copper(II) complexes has been given, and the new complex $[Cu\text{-cyclo(-D-Ser-)}]ClO_4{\cdot}4H_2O$ has been isolated.[100] Complexes of deprotonated cycloserine with Cr^{III}, Mn^{II}, Fe^{II}, Fe^{III}, Co^{II}, Ni^{II}, Zn^{II}, Zr^{IV}, Pd^{II}, Ag^{I}, Cd^{II}, Os^{III}, Pt^{II}, and Hg^{II} have been isolated and tentative structures suggested.[101] The antibiotic appears to act as a uninegative bidentate ligand forming five-membered chelate rings in which the O^- and NH_2 groups act as donors.

A variety of studies with copper(II) complexes have been carried out, including e.n.d.o.r. investigations of powdered copper amino acid complexes,[102] an e.s.r. study on copper(II) complexes of α-amino acids in frozen solutions,[103] and an e.s.r. study of copper(II) ions in powder and crystals of aquatris-(L-glutamato)-cadmium(II) monohydrate.[104]

Structures of various complexes in solutions of copper(II)-L-methylhistidine and copper(II)-L-N^{α},N^{α}-dimethylhistidine have been deduced by investigating

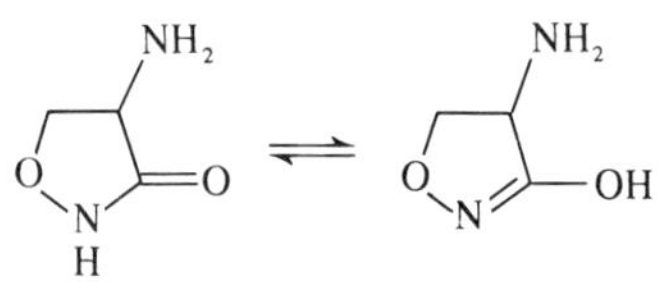

(14)

[97] N. Juranic and R. L. Lichter, *J. Am. Chem. Soc.*, 1983, **105**, 406.
[98] N. R. Brodsky, N. M. Nguyen, N. S. Rowan, C. B. Storm, R. J. Butcher, and E. Sinn, *Inorg. Chem.*, 1984, **23**, 891.
[99] M. G. Patch, K. P. Simolo, and C. J. Carrano, *Inorg. Chem.*, 1983, **22**, 2630.
[100] P. O'Brien, *Inorg. Chim. Acta*, 1983, **78**, L37.
[101] F. Forghieri, C. Preti, G. Tosi, and P. Zannini, *Aust. J. Chem.*, 1983, **36**, 1125.
[102] K. A. Kraft, *Diss. Abstr. Int. B*, 1983, **44**, 1842.
[103] T. Szabo-Planka and L. I. Horvath, *Acta Chim. Hung.*, 1983, **114**, 15.
[104] R. P. Bonomo, J. R. Pilbrow, and G. R. Sinclair, *J. Chem. Soc., Dalton Trans.*, 1983, 489.

the pH dependence of the c.d. spectra.[105] Copper(II) complexes with L-proline, pipecolinic acid, picolinic acid, and a variety of aminocarboxylic acids have been prepared and their e.s.r. spectra studied.[106] The complex with proline displays a frozen-solution e.s.r. spectrum with axial resonance parameters.

Binuclear copper(II) complexes with amides derived from 3-amino-1-propanol and various amino amides (15) display strong antiferromagnetic coupling. The structure of (15; R = Me) has been determined.[107] Complexes of the type [Cu(bipy)(*N*-acetyl-DL-leucinate)$_2$] and [Cu(bipy)(*N*-tosyl-β-alaninate)$_2$] have been prepared and their structures discussed.[108]

Interesting 12-, 15-, and 18-membered diaza-crown-*N,N'*-dialkanoic acids (16) have been prepared. The complexes of 15- and 18-membered ligands contain Cu^{II} inside the ring.[109] Zwitterionic or anionic amino acids occur in the recently characterized[110] complexes $[MCl_3(GlyH)_3]$ (M = Ti^{III}, V^{III}, or Fe^{III}), $[CrCl_3(GlyH)_2H_2O]$, $[MCl_2(GlyH)_2(H_2O)_2]$ (M = Co^{II} or Cu^{II}), and $[NiCl_2(GlyH)_3H_2O]$.

Sixteen rhena β-keto imine derivatives (17) of twelve amino acids have been prepared by the route shown in Scheme 1.[111] Induced Cotton effects observed in solutions of magnesium protoporphyrin and magnesium mesoporphyrin containing the amino acids D- and L-Pro, L-Ser, L-Thr, and L-Try are attributed to six-co-ordinate magnesium porphyrin(amino acid)$_2$ species.[112] Similar effects are also observed with L-histidine.[113] Nickel(II) complexes of ligands of type (18) have been characterized;[114] *X*-ray work confirms the planar stereochemistry on nickel.

A series of *S*-valinatobis-(1,10-phenanthroline)zinc(II) complexes have been prepared and their conductivities and o.r.d. spectra studied; evidence for ion pairing was obtained in methanol solvent.[115] Complexes of nickel(II), copper(II),

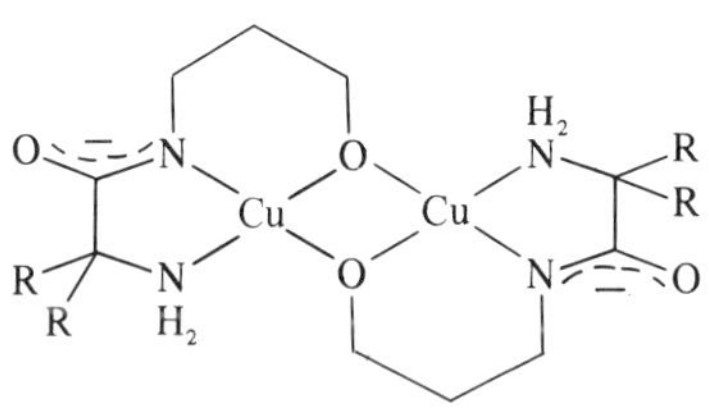

(15)

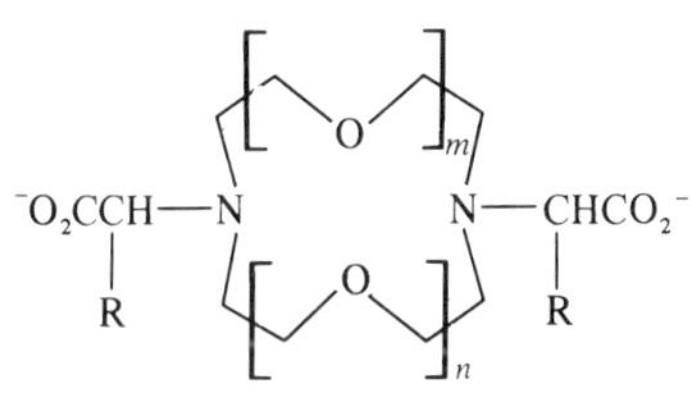

(16)

[105] L. Casella and M. Gullotti, *Inorg. Chem.*, 1983, **22**, 242.
[106] N. F. Albanese and H. M. Haendler, *Polyhedron*, 1983, **2**, 1131.
[107] M. Mikuriya, T. Harada, H. Okawa, and S. Kida, *Inorg. Chim. Acta*, 1983, **75**, 1.
[108] L. Antolini, L. Menabue, M. Saladini, L. P. Battaglia, and A. B. Corradi, *Inorg. Chim. Acta*, 1984, **90**, 97.
[109] R. A. Kolinski and J. Mrozinski, *Polyhedron*, 1983, **2**, 1217.
[110] M. Castillo and E. Ramirez, *Transition Met. Chem.*, 1984, **9**, 268.
[111] D. Afzal and C. M. Lukehart, *Inorg. Chem.*, 1983, **22**, 3954.
[112] O. C. Choon and G. A. Rodley, *Inorg. Chim. Acta*, 1983, **80**, 177.
[113] G. A. Rodley and O. C. Choon, *Inorg. Chim. Acta*, 1983, **78**, 171.
[114] Y. Fukuda, H. Miyamae, K. Yamagata, and K. Sone, *Chem. Lett.*, 1984, 1309.
[115] A. Decinti and G. Larrazabel, *Polyhedron*, 1983, **2**, 1075.

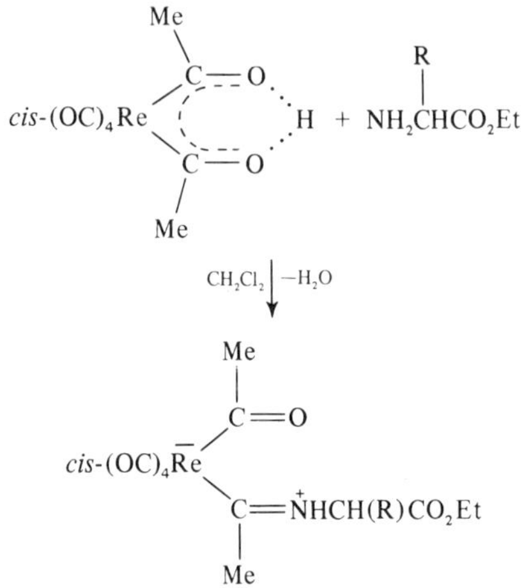

(17)

Scheme 1

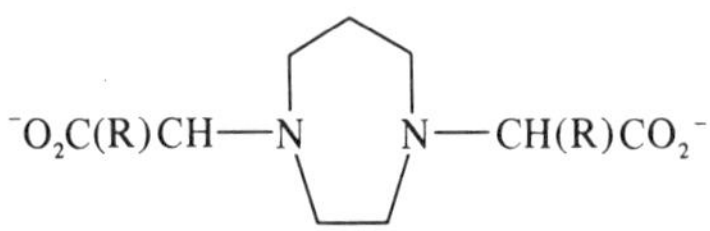

(18)

zinc(II), and cadmium(II) with 4-hydroxy-L-proline have been prepared and characterized by a variety of physical techniques.[116] The complexes are of three distinct types: $ML_2 \cdot nH_2O$ (M = Ni, Zn, or Cu, L = 4-hydroxy-L-prolinato anion), $CuClL \cdot H_2O$, and $CdCl_2HL$, in which the cadmium is bonded through the oxygen atom of the carboxylato group. The i.r. spectra of some metal-valine chelates have been studied[117] from 3600 to 250 cm^{-1}; the trend in metal-nitrogen bond strengths is shown to be $Co^{II} < Ni^{II} < Cu^{II} > Zn^{II}$.

A variety of investigations have been carried out with rare-earth complexes of amino acids, and this area is attracting considerable attention. The Nd^{III}, Ho^{III}, and Eu^{III} complexes with glycine, alanine, and glutamic acid have been characterized and their absorption and luminescence spectra recorded.[118] Electronic spectra of Nd^{III} and Pr^{III} complexes of glycine, alanine, and valine have also been studied in the range 360–920 nm in water.[119] The energies and intensities of the various transitions calculated using the Judd/Ofelt relations are in good agreement with those obtained experimentally. The structure of aqueous L-proline at

[116] Y. Inomata, T. Takeuchi, and T. Moriwaki, *Inorg. Chim. Acta*, 1983, **68**, 187.
[117] C. W. Moszczenski and R. J. Hooper, *Inorg. Chim. Acta*, 1983, **70**, 71.
[118] J. Legendziewicz, E. Huskowska, G. Argay, and A. Waskowska, *Inorg. Chim. Acta*, 1984, **95**, 57.
[119] M. P. Bhutra and A. K. Gupta, *Indian J. Pure Appl. Phys.*, 1983, **21**, 374.

pH 3 has been investigated using the lanthanide-induced-shift technique.[120] The relaxation-rate data indicate that isostructural complexes are formed between proline and all ten lanthanide cations. The i.r. and luminescence spectra of $M(hfaa)_3 \cdot 2L$ complexes (M = La, Eu, Tb, Dy, or Lu, hfaa = the anion of hexafluoroacetylacetone, L = glycine, α-alanine, β-alanine, valine, norvaline, aspartic acid, histidine, or proline) establish that the amino acids are neutral ligands co-ordinated *via* the carboxylic O atom.[121] A spectral study of ternary complexes of rare-earth ions with a variety of amino acids has also been published.[122] ^1H n.m.r. data indicate[123] that Tl^I in the pH range 7–14 is co-ordinated to β-alanine *via* the O atom, by aspartic acid *via* the O and N atoms, and by methionine through the N and S atoms. Glycine and α-alanine bind Tl^I *via* the N and O atoms, and Tl^{III} forms chelates with amino acids. Triorganotin and triorganolead derivatives of *N*-acetyl amino acids have been prepared[124] from R_3MOH (R = Me or Ph) or $(Bu_3Sn)_2O$ and the appropriate *N*-acetyl amino acid. The compounds are 5-co-ordinate and polymeric in the solid state.

By using Δ-tris-(1,10-phen)nickel(II) montmorillonite as a column material it was possible to resolve at least partially ten cobalt(III) complexes $[Co(acac)_2L]$ (Hacac = acetylacetone, HL = amino acid) into two configurational isomers.[125] Reversed-phase h.p.l.c. has been successfully applied to the separation of cobalt(III) complexes.[126] The phenomenon is not restricted to chiral complexes but is due to the interaction between $\geqslant 2+$ complex cations and the pairing anions.

The nickel(II) and cobalt(II) complexes of DL-Phe, Gly, β-Ala, and other amino acids have been shown to be only weakly active against herpes virus.[127] An undergraduate experiment involving the preparation and characterization of cobalt(III) complexes containing *N*-bonded monodentate, *O*-bonded monodentate, and *N,O*-chelated glycine ligands has been described.[128]

Equilibrium Studies. – Solution equilibrium studies between metal ions and amino acid ligands continue to attract considerable attention. Simple binary systems studied include iron(III) with histidine-hydroxamic acid,[129] nickel(II), cobalt(II), zinc(II), and cadmium(II) with *N*-(2-hydroxy-4-nitro)benzylglycine,[130] uranyl(VI) and thorium(IV) with glycine,[131] formation of protonated

[120] M. Singh, J. J. Reynolds, and A. D. Sherry, *J. Am. Chem. Soc.*, 1983, **105**, 4172.
[121] V. E. Karasev, N. I. Steblevskaya, and R. N. Schelokov, *Koord. Khim.*, 1983, **9**, 199.
[122] M. P. Bhutra and A. K. Gupta, *Indian J. Pure Appl. Phys.*, 1983, **21**, 674.
[123] Yu. B. Yakovlev and V. G. Ushakova, *Zh. Neorg. Khim.*, 1983, **28**, 2142.
[124] G. Roge, F. Huber, H. Preut, A. Silvestri, and R. Barbieri, *J. Chem. Soc., Dalton Trans.*, 1983, 595.
[125] A. Yamagishi, *J. Chem. Soc., Dalton Trans.*, 1983, 679.
[126] D. A. Buckingham, C. R. Clark, and R. F. Tasker, *Inorg. Chem.*, 1983, **22**, 2772.
[127] A. E. Milgrom, A. S. Chegolya, A. V. Fillippova, G. V. Vladyko, and N. I. Karako, *Khim.-Farm. Zh.*, 1984, **18**, 316.
[128] R. G. Harrison and K. B. Nolan, *J. Chem. Educ.*, 1982, **59**, 1054.
[129] D. A. Brown and B. S. Sekhon, *Inorg. Chim. Acta*, 1984, **91**, 103.
[130] M. Chandra, *Transition Met. Chem.*, 1983, **8**, 276.
[131] A. Bismondo, L. Rizzo, G. Tomat, D. Curto, P. di Bernardo, and A. Cassol, *Inorg. Chim. Acta*, 1983, **74**, 21.

polynuclear complexes between Cd^{II} and D-penicillamine in aqueous solution,[132] lead(II) with D-penicillamine,[133] iron(III) with dopamine,[134] cobalt(II) with L-cysteine and its derivatives,[135] and copper(II) with aspartic acid.[136]

The thermodynamic quantities relating to the formation of Mn^{II}, Co^{II}, Ni^{II}, Cu^{II}, and Zn^{II} complexes of tyrosine, *o*-tyrosine, and *m*-tyrosine have been determined pH-metrically and calorimetrically at 25 °C.[137] Thermometric titrimetry has been employed to determine K, $\Delta H^{\ominus}$, and $\Delta S^{\ominus}$ for the complexation of alkaline-earth cations by linear poly(aminocarboxylic) acids.[138]

Formation constants have been obtained for bis complexes of *O*-phospho-DL-serine and *O*-phospho-L-serine with Co^{II}, Zn^{II}, Cu^{II}, and Ni^{II}.[139] The LL and DL complexes have virtually the same stabilities, but in the case of zinc(II) stereoselectivity favouring the LL complex approaches the theoretical maximum. Complexes of L-dopa (L-3,4-dihydroxyphenylalanine) have been investigated in detail.[140] The tendency for L-dopa complexes containing both amino acid-like and pyrocatechol-like binding to rearrange to species containing only (*O*,*O*) bonds varies with the metal ion: $Cu^{II} \sim Zn^{II} > Co^{II} > Mn^{II} \sim Ni^{II}$. The formation reactions of complexes of *N*-tosylglycine with copper(II) have been studied by polarography.[141] In the pH range 8–10 $Cu(TsglyH)_2$, $[Cu(Tsgly)_2]^{2-}$, and $[Cu(Tsgly)(OH)]^-$ are present.

Other studies in the area of binary complexes include copper(II) with glycinehydroxamic acid,[142] uranyl(VI) and thorium(IV) with aminopolycarboxylic acids,[143] silver(I) with aminocarboxylate ligands,[144] and uranyl(VI) and thorium-(IV) with a range of amino acids.[145] In the latter study a comparison of the constants for alanine with those for serine, cysteine, and methionine suggests only limited bonding between the metal ions and S in cysteine and methionine, but somewhat stronger bonding with the O in serine.

Formation constants for the interaction of $MeHg^+$ with 2-mercaptoethanol, mercaptoacetic acid, L-Cys, DL-penicillamine, *N*-acetyl-DL-penicillamine, glutathione, and a variety of other mercapto compounds have been determined by potentiometric titration.[146]

Ternary-complex formation is a field of current interest, and many investigations have been published. A quantitative study of the copper(II)–histidine ternary complexes with Leu, Glu, Met, Try, and Ala has been carried out to

[132] A. Avdeef and D. L. Kearney, *J. Am. Chem. Soc.*, 1982, **104**, 7212.
[133] M. J. Willes and D. R. Williams, *Inorg. Chim. Acta*, 1983, **80**, L35.
[134] G. Crisponi, A. Lai, M. Monduzzi, and G. Saba, *Inorg. Chim. Acta*, 1983, **80**, 85.
[135] B. Harman and I. Sóvágó, *Inorg. Chim. Acta*, 1983, **80**, 75.
[136] Y.-C. Liang and A. Olin, *Acta Chem. Scand., Ser. A*, 1984, **38**, 247.
[137] T. Kiss and A. Gergely, *J. Chem. Soc., Dalton Trans.*, 1984, 1951.
[138] G. Ewin and J. O. Hill, *J. Chem. Soc., Dalton Trans.*, 1983, 865.
[139] M. S. Mohan, D. Bancroft, and E. H. Abbott, *Inorg. Chem.*, 1983, **22**, 714.
[140] T. Kiss and A. Gergely, *Inorg. Chim. Acta*, 1983, **78**, 247.
[141] L. Antolini, L. P. Battaglia, G. B. Gavioli, A. B. Corradi, G. Grandi, G. Marcotrigiano, L. Menabue, and G. C. Pellacini, *J. Am. Chem. Soc.*, 1983, **105**, 4333.
[142] E. B. Paniago and S. Carvalho, *Inorg. Chim. Acta*, 1984, **92**, 253.
[143] M. Nourmand, I. Bayat, and S. Yousefi, *Polyhedron*, 1982, **1**, 827.
[144] J. S. Redinha and J. M. C. Costa, *Rev. Port. Quim.*, 1981, **23**, 175.
[145] M. Nourmand and N. Meissami, *J. Chem. Soc., Dalton Trans.*, 1983, 1529.
[146] A. P. Arnold and A. J. Canty, *Can. J. Chem.*, 1983, **61**, 1428.

determine trace-metal requirements in nutrition.[147] Other studies in this general area include mixed-ligand complex formation of copper(II)-glycyl-DL-serine with the amino acids Gly, Ala, Val, Thr, Ser, Tyr, Asp, and Glu,[148] simple and mixed complexes of Cu^{II} and Zn^{II} with adenosine 5′-triphosphate and L-tryptophan or L-alanine,[149] and mixed-ligand adenosine 5′-triphosphate-metal(II)-L-leucinate and related ternary complexes ($M^{2+} = Mn^{II}$, Cu^{II}, Zn^{II}, Cd^{II}, or Pb^{II}).[150]

A study of copper(II) ternary complexes involving histidine and tertiary amines suggests that at low pH histidine is bonded *via* the N and O^- donors, but at higher pH it is 'histamine like' *via* the N and N donors.[151] Ternary copper(II) complexes with ethylenediamines and amino acids are favoured if the amino acid contains an aromatic side chain.[152] Radioactive indium complexes possessing short half-lives have been employed as radiopharmaceuticals as they easily undergo urinary elimination. In the In^{III}–L-histidine system polarographic measurements establish that 1:1 and 1:2 complexes occur at pH 4 whereas 1:2 complexes predominate at pH $\geqslant$ 5.9.[153] The log values of overall formation constants for $[In(L\text{-}His)]^{2+}$, $[In(L\text{-}His)_2]^+$, $[In(L\text{-}His)(L\text{-}Glu)]^+$, and $[In(L\text{-}His)(L\text{-}Pro)]^+$ are 10.05, 17.86, 16.37, and 18.14, respectively.

Binary, ternary, and quaternary complexes involved in the systems pyridoxamine-glycine-imidazole with Co^{II}, Ni^{II}, Cu^{II}, Zn^{II}, and Cd^{II} have been investigated using the MINIQUAD-75 program.[154] Complexation of L-aspartic acid by a series of mixed-ligand Tb^{III} complexes has been studied by circularly polarized luminescence spectroscopy, and formation constants were obtained.[155] Other studies deal with mixed complexes of Ni^{II} with taurine, DL-methionine and DL-ethionine in aqueous solution,[156] histidine-containing mixed-amino acid and -dipeptide copper(II) complexes,[157] and mixed complexes of cadmium(II) with *N*-(2-hydroxyethyl)ethylenediamine and a range of amino acids.[158]

Plutonium is an increasing environmental hazard: some 5000 kg of the element has been released into the environment since the mid-1940s. The nature of Pu^{IV} binding to low-molecular-weight ligands in human blood plasma has been studied, and chelation therapy using polyaminopolycarboxylic acids has been investigated.[159] Specific metal deficiencies have been reported to affect patients receiving total parenteral nutrition (TPN). Formation constants for copper(II)-histidine-amino acid complexes (amino acid = Thr, Lys, Gly, Phe,

[147] G. Berthon, M. Piktas, and M.-J. Blais, *Inorg. Chim. Acta*, 1984, **93**, 117.
[148] D. N. Shelke, *Inorg. Chim. Acta*, 1983, **80**, 255.
[149] G. Arena, R. Cali, V. Cucinotta, S. Musumeci, E. Rizzarelli, and S. Sammartano, *J. Chem. Soc., Dalton Trans.*, 1983, 1271.
[150] H. Sigel, B. E. Fischer, and E. Farkas, *Inorg. Chem.*, 1983, **22**, 925.
[151] V. K. Patel and P. K. Bhattacharya, *Inorg. Chim. Acta*, 1984, **92**, 199.
[152] A. Odani and O. Yamauchi, *Inorg. Chim. Acta*, 1984, **93**, 13.
[153] S. L. Jain and R. C. Kapoor, *Inorg. Chim. Acta*, 1983, **78**, 93.
[154] M. S. El-Ezaby, M. Rashad, and N. M. Moussa, *Polyhedron*, 1983, **2**, 245.
[155] H. G. Brittain, *Inorg. Chim. Acta*, 1983, **70**, 91.
[156] J. Maslowska and L. Chruściński, *Polyhedron*, 1984, **3**, 1329.
[157] G. Thomas and P. S. Zacharias, *Transition Met. Chem.*, 1984, **9**, 377.
[158] C. P. S. Chandel and C. M. Gupta, *Bull. Chem. Soc.*, 1984, **57**, 2303.
[159] J. R. Duffield, P. M. May, and D. R. Williams, *J. Inorg. Biochem.*, 1984, **20**, 199.

Val, or Cys) have been determined in order to deal with the problem of TPN-induced copper deficiency and its remedy.[160]

Formation constants of ternary complexes M(A)L (M = Cu^{II} or Ni^{II}, A = bipy or phen, L = dopa, Tyr, or Phe) have been obtained.[161] The visible spectrum of Cu–A–dopa is similar to that of Cu–A–Phe or Cu–A–Tyr over the entire pH range, confirming *N,O* co-ordination. Mixed-ligand complexes of the type $[M(ATP)(L)]^{3-}$ (M = Cu^{II} or Zn^{II}, ATP = adenosine 5′-triphosphate, L = L-histidinate) and $[M(ATP)(L)]^{2-}$ (M = Cu^{II} or Zn^{II}, L = histamine) have been studied by potentiometric and calorimetric titration,[162] an extension of previous work in this area.[149]

The structures of various species in the copper(II)–L-histidine (1:2) system have been studied by investigating the pH dependence of the electronic and c.d. spectra.[163] The contributions to the spectra of the glycine- and histamine-like binding modes of L-histidine were determined by recording the spectra of the ternary systems Cu^{II}–histamine–L-histidine (1:1:1) and Cu^{II}–amino acid–L-histidine (1:1:1), respectively. Apical binding to copper(II) by the donor atom in the histidine side chain can contribute significantly to the stabilization of each of the two basic histidine-binding modes.

The statistics of ternary-complex formation with special reference to biological fluids have been discussed.[164,165] Thus, for the equilibrium $MA_2 + MB_2 \rightleftharpoons 2MAB$ the treatment verifies that the distribution constant $K_d = 4$, assuming the absence of any intraligand interactions or other factors favouring the formation of any of the species. Analysis of published data for Cu^{II} and Ni^{II} complexes with amino acids and *N*-donor ligands shows that ligand hardness and ionization potential affect complex stability.[166] Formation of interligand bonds in mixed complexes can increase or decrease mixed-complex stability depending upon ligand type and the nature of the interligand association. However, interligand association is considered to be a second-order effect.

Other studies dealing with mixed-ligand complexes include nickel(II)–, copper(II)–, and zinc(II)–diaminomonocarboxylates with glycine, bipy, and ATP,[167] copper(II)–nitrilotriacetic acid with amino acids,[168] nickel(II)–histidine (or –nitrilotriacetic acid) with aspartic acid, glycine and α-alanine,[169] copper(II)–valine with glycine, alanine, serine, and threonine,[170] cadmium(II)–1,10-phenan-

[160] G. Berthon, M.-J. Blais, M. Piktas, and K. Houngbossa, *J. Inorg. Biochem.*, 1984, **20**, 113.

[161] V. K. Patel and P. K. Bhattacharya, *J. Inorg. Biochem.*, 1984, **21**, 169.

[162] G. Arena, R. Cali, V. Cucinotta, S. Musumeci, E. Rizzarelli, and S. Sammartano, *J. Chem. Soc., Dalton Trans.*, 1984, 1651.

[163] L. Casella and M. Gullotii, *J. Inorg. Biochem.*, 1983, **18**, 19.

[164] S. H. Laurie and C. James, *Inorg. Chim. Acta*, 1983, **78**, 225.

[165] S. H. Laurie, *Inorg. Chim. Acta*, 1983, **80**, L27.

[166] Ya. D. Fridman, *Koord. Khim.*, 1984, **10**, 1034.

[167] E. Farkas, A. Gonczy, and A. Gergely, *Magy. Kem. Foly*, 1984, **90**, 211.

[168] J. M. C. Pingarron, I. M. Del Bairio, and L. M. D. Polo, *An. Quim., Ser. B*, 1984, **80**, 141.

[169] J. D. Joshi, *Indian J. Chem., Sect. A*, 1984, **23**, 611.

[170] S. J. Mandloi, P. K. Chitale, K. S. Verma, and H. L. Nigam, *Indian J. Chem., Sect. A*, 1984, **23**, 224.

throline with amino acids,[171] zinc(II)/cadmium(II)-iminodiacetic acid with amino acids,[172] neodymium(III)-L-histidine or -L-cysteine with various diols,[173] copper(II)-malonate with amino acids,[174] zinc(II)-oxalate or -ethylenediamine with amino acids,[175] and copper(II)-orotic acid with amino acids.[176]

Thermodynamic parameters for the formation of Cu^{II} and Zn^{II} mixed complexes with ATP and L-phenylalanine or L-tyrosine have been determined by potentiometric and calorimetric measurements, and the results have been evaluated.[177]

Copper(II) and nickel(II) complexes of a variety of sulphur-containing α,ω-amino acids $[NH_2(CH_2)_nS(CH_2)_{m-1}CO_2^-]$ have also been studied potentiometrically and calorimetrically.[178] In $[Cu(n,m\text{-NSO})]^+$ and $[Ni(n,m\text{-NSO})]^+$ the aminocarboxylates act as tridentate ligands. In the bis complexes $[Ni(n,m\text{-NSO})_2]$ and $[Cu(n,m\text{-NSO})_2]$ the nickel is six co-ordinate and the copper is five co-ordinate (in the latter complex the second ligand is bound only *via* the N and S donors).

Potentiometry and differential pulse polarography have been used to determine formation constants of Cd^{II} complexes with Ala, Ser, Val, and Glu in 0.70 mol dm^{-3} $NaClO_4$ to simulate the ionic strength of sea water.[179] The complexes of the first three ligands are completely labile in polarographic terms, but the dissociation of the 1:1 complex with glutamic acid is slower. Formation constants of dioxygen complexes of Ru^{III} with a series of aminopolycarboxylic acids have been determined.[180] Complexes of edta, hedta, dtpa, and ttha were most thoroughly investigated, and the rates of formation of the complexes at 1 atm O_2 were determined. Paramagnetic superoxo complexes $K[Ru(L)O_2^-]$ (L = edta or hedta) were characterized.

π-Acid effects on the polarographic electrode behaviour of ternary $[Cu_x(phen)_yL_z]$ complexes (L represents a series of ligands including amino acids) have been considered in two papers.[181,182] Such effects become more pronounced if the secondary ligand also contains a π-system. Complexation between copper(II) and the following pairs of α-amino acids, proline + serine, proline + threonine, proline + tyrosine, proline + valine, serine + valine, and tyrosine + valine, has been studied potentiometrically.[183] The relative stabilities of the

171 Z. Zhu, J. Zhu, L. Qin, and Y. An, *Shanghai Keji Daxue Xuebao*, 1982, 60.
172 J. D. Joshi, *J. Indian Chem. Soc.*, 1984, **61**, 257.
173 A. Kothari, R. K. Jain, and S. N. Misra, *Proc. Indian Natl. Sci. Acad., Part A*, 1983, **49**, 398.
174 V. V. Ramanujam and U. Krishnan, *Acta Cienc. Indica, Ser. Chem.*, 1982, **8**, 243.
175 A. R. Aggarwal, K. B. Pandeya, and R. P. Singh, *Electrochem. Soc. India*, 1983, **32**, 273.
176 N. N. Vlasova and N. K. Davidenko, *Koord. Khim.*, 1983, **9**, 1470.
177 G. Arena, R. Cali, V. Cucinotta, S. Musumeci, E. Rizzarelli, and S. Sammartano, *Thermochim. Acta*, 1984, **74**, 77.
178 C. T. Huys, J. Tombeux, and A. M. Goeminne, *Thermochim. Acta*, 1983, **63**, 191.
179 M. L. S. G. Simoes and M. M. Correia dos Santos, *J. Electroanal. Chem. Interfacial Electrochem.*, 1984, **163**, 315.
180 M. M. Taqui Khan, *Pure Appl. Chem.*, 1983, **55**, 159.
181 V. Srivastava and H. L. Nigam, *Bioelectrochem. Bioenerg.*, 1982, **9**, 627.
182 V. Srivastava and H. L. Nigam, *Bioelectrochem. Bioenerg.*, 1982, **9**, 639.
183 N. Al-Ani and A. Olin, *Chem. Scr.*, 1984, **23**, 165.

ternary complexes are discussed with reference to the values of the equilibrium constants for the reactions $CuA_2 + CuB_2 \rightleftharpoons 2CuAB$ and $CuA + CuB \rightleftharpoons CuAB + Cu$ (A and B denote the amino acid ligands). In all systems containing proline the diastereoisomer with ligands of the same chirality forms the most stable complex. A further paper on the stereoselectivity issue from the same group[184] concludes that in solutions of the bis complexes with racemic forms of serine, threonine, and proline the deviations from the statistical concentration ratios, $[CuL_2]:[CuD_2]:[CuDL] = 1:1:2$, are less than 2%. Stereoselectivity was only detected for complexes with tyrosine and for deprotonated complexes with threonine at high pH.

Ternary-complex formation between copper(II)–noradrenaline and a range of amino acids has also been studied.[185] Complexes of the type $[Cu(nadr)LH]^{2-P}$ occur with L = Orn, Lys, Asp, or Glu, and complexes of the type $[Cu(nadr)L]^{1-P}$ occur with L = Gly, Ala, Arg, Asp, or Glu (for both types of complex $P = 0$ for Arg, $P = 1$ for Gly, Ala, Orn, and Lys, and $P = 2$ for Asp and Glu). A final paper deals with complex formation between Zn^{II}, Cd^{II}, and Hg^{II} with D-penicillamine.[186] Complexes of the types ML, MH_2L_2, MHL^{2-}, and ML_3^{4-} occur. The magnitude of the respective constants cannot, by itself, account for the ineffectiveness of penicillamine treatment for mercury and cadmium poisoning.

Diffraction Studies. – The reaction of bis(glycinato)nickel(II) dihydrate with formaldehyde and ammonia at pH 8.5 gives orange crystals, for which *X*-ray analysis establishes the structure (19).[187]

When the quadridentate ligand *S*-(2-aminoethyl)-L-homocysteinate (L-aehc) and glycinate co-ordinate to cobalt(III), four possible geometric isomers can occur. The crystal structure of *trans*(O)-*mer*(N)-$[Co(L\text{-aehc})(Gly)]^+$ has been determined by *X*-ray analysis.[188, 189] This isomer has the Δ-configuration, with the sulphur donor having an *R*-configuration and the six-membered N–S chelate

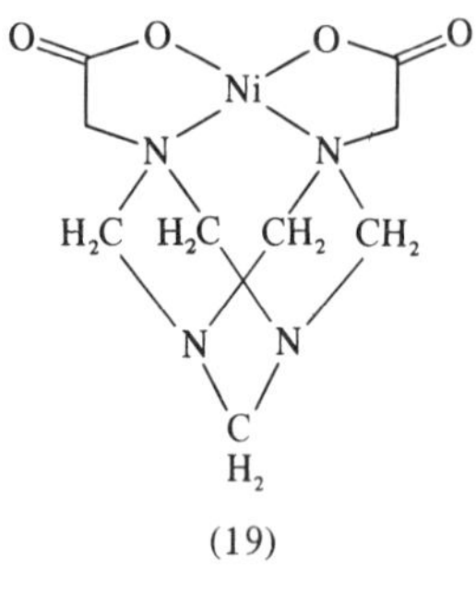

(19)

[184] N. Al-Ani and A. Olin, *Chem. Scr.*, 1984, **23**, 161.
[185] P. G. Daniele, P. Amico, and G. Ostacoli, *Ann. Chim. (Rome)*, 1984, **74**, 105.
[186] R. Strand, W. Lund, and J. Aaseth, *J. Inorg. Biochem.*, 1983, **19**, 304.
[187] S.-B. Teo, S.-G. Teoh, and M. R. Snow, *Inorg. Chim. Acta*, 1984, **85**, L1.
[188] K. Okamoto, M. Suzuki, H. Einaga, and J. Hidaka, *Bull. Chem. Soc. Jpn.*, 1983, **56**, 3513.
[189] K. Okamoto, M. Suzuki, and J. Hidaka, *Chem. Lett.*, 1983, 401.

ring adopting a chair conformation. A variety of structures relating to complexes of glycine have been determined. In the neodymium complex $Nd_2(Gly)_6 \cdot (ClO_4)_6 \cdot 9H_2O$, the co-ordination polyhedron of neodymium consists of seven oxygen atoms from glycine and two from water molecules.[190] The structures of mono(glycinato)nickel(II) and tris(glycinato)nickelate(II) in *aqueous* solution have been determined by *X*-ray diffraction.[191] The mono complex is combined with four water molecules at a distance of 2.08 Å, and the Ni–O and Ni–N distances of the glycinato ligand are essentially the same (2.09 Å), leading to a regular octahedral structure. The $[Ni(Gly)_3]^-$ ion has a slightly distorted octahedral structure with Ni–O and Ni–N bond lengths of 2.03 Å and 2.14 Å, respectively.

Structures of mono(glycinato)copper(II) and tris(glycinato)cuprate(II) in aqueous solution have also been determined.[192] The $[Cu(Gly)(OH_2)_4]^+$ ion has an axially elongated octahedral structure; $[Cu(Gly)_3]^-$ is a regular octahedron with both Cu–O and Cu–N bond distances of 2.02 Å.

The crystal structure[193] of [*N*-benzenesulphonyl-DL-alaninato(2—)] copper(II) monohydrate establishes that the alanine moiety forms a five-membered chelate ring in the equatorial square plane *via* the carboxylate oxygen and the deprotonated nitrogen, the third and fourth corners of the square being occupied by a water molecule and the 'free' carboxylate oxygen of a screw-related complex. The complex bis-(*N*-benzyloxycarbonylglycinato)(2,2′-bipyridine)(propan-2-ol)copper(II) has two crystallographically independent, but chemically equivalent molecules, linked in dimeric units by hydrogen-bonding interactions.[194] Each copper atom has a slightly distorted square-pyramidal stereochemistry with the bidentate 2,2′-bipyridine and two *N*-benzyloxycarbonylglycinate ions in the equatorial plane and the propan-2-ol molecule in an apical position. In aqua-(L-aspartato)(2,2′-bipyridine)copper(II) trihydrate a similar distorted square-pyramidal geometry is also observed.[195] The two organic ligands are bidentate in the equatorial plane, with a water molecule occupying the apical position.

The complex Δ,Λ-$[Ru(bipy)_2(L\text{-}Ala)]ClO_4 \cdot 0.5H_2O$ contains both Δ,L and Λ,L diastereoisomeric cations in the one crystalline form.[196] The bipyridine and L-alanine molecules are bidentate, but in the Λ,L isomer the amino acid ring is puckered, whereas in the Δ,L form this ring is close to planar.

The structure of dichloro-(L-proline)cadmium(II) hydrate[197] is similar to that of dichloro-(4-hydroxy-L-proline)cadmium(II), consisting of a one-dimensional polymer bridged by chlorine atoms and carboxyl oxygen atoms.

[190] J. Legendziewicz, E. Huskowska, A. Waskowska, and G. Argay, *Inorg. Chim. Acta*, 1984, **92**, 151.
[191] K. Ozutsumi and H. Ohtaki, *Bull. Chem. Soc. Jpn.*, 1983, **56**, 3635.
[192] K. Ozutsumi and H. Ohtaki, *Bull. Chem. Soc. Jpn.*, 1984, **57**, 2605.
[193] S. Chaudhuri, *J. Chem. Soc., Dalton Trans.*, 1984, 779.
[194] L. Antolini, L. Menabue, G. C. Pellacani, M. Saladini, M. Sola, L. P. Battaglia, and A. B. Corradi, *J. Chem. Soc., Dalton Trans.*, 1984, 2319.
[195] L. Antolini, G. Marcotrigiano, L. Menabue, and G. C. Pellacani, *Inorg. Chem.*, 1983, **22**, 141.
[196] F. S. Stephens, R. S. Vagg, and P. A. Williams, *Inorg. Chim. Acta*, 1983, **72**, 253.
[197] Y. Yukawa, Y. Inomata, and T. Takeuchi, *Bull. Chem. Soc. Jpn.*, 1983, **56**, 2125.

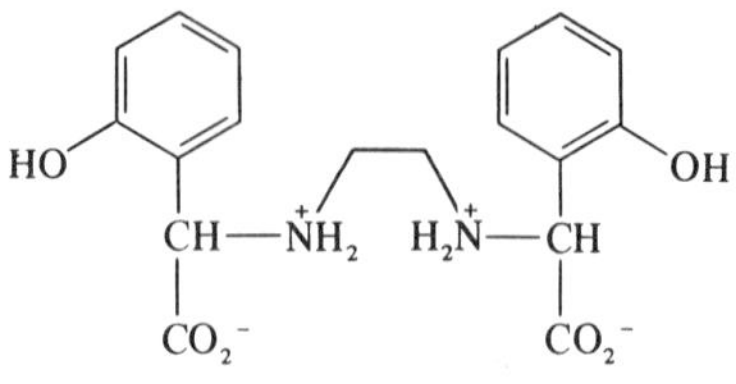

(20)

The structures of Co^{III}, Ga^{III}, and Cu^{II} complexes of the hexadentate ligand ethylenebis-[(*o*-hydroxyphenyl)glycine] [EHPG = (20)] have been determined.[198] These complexes may be regarded as models for the metal-binding site of the human iron-transport protein transferrin. In $Na_2[Cu(EHPG)]\cdot 5.5H_2O$ the metal is bound by the two nitrogen atoms, two phenolate oxygen atoms, and two carboxylate oxygens in a pseudo-octahedral geometry with the two carboxylate ligand *trans* to each other.

Stereochemistry and Stereoselectivity. – Optical resolution of DL-alanine *via* formation of the ternary complex [Cu(D-Ala)(L-Ile)] has been described.[199] The complex crystallizes selectivity at 0 °C from an aqueous solution containing DL-Ala, L-Ile, and copper(II) acetate monohydrate in the molar ratio 1:1:1. The claim by Astanina *et al.*[200] that there are differences in complex formation between L- and D-amino acids and Fe^{III} has been refuted by Williams.[201] Such processes cannot be viewed as potential reactions leading to the generation of optical activity in prebiotic systems.

The four catatopic complexes of general form Δ,Λ-[Ru(di-imine)$_2$(aa)]-$ClO_4\cdot nH_2O$ (di-imine = bipy or phen, aa = *S*-Thr or *S*-*allo*-Thr) have been isolated, and each complex has been resolved into its two diastereoisomeric forms.[202] Each isomer is photolabile, equilibrating to a definite Λ/Δ ratio on light irradiation. A crystalline form of β_1-[Co($\pm$picpn)(*S*-Ala)]ClO_4 [picpn = (21)] has been shown by *X*-ray analysis to contain four diastereoisomeric cations in a single crystal.[203] Several isomers of $[Co(R\text{-picpn})(Ala)]^{2+}$ (Ala = *R*- or

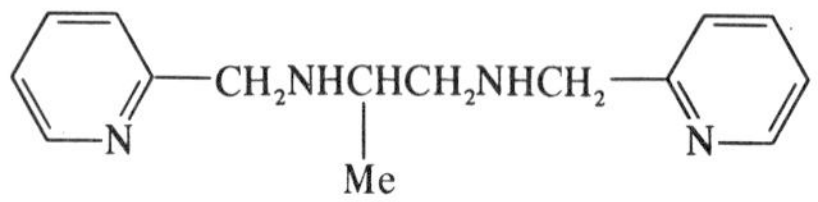

(21)

[198] P. E. Riley, V. L. Pecoraro, C. J. Carrano, and K. N. Raymond, *Inorg. Chem.*, 1983, **22**, 3096.

[199] T. Shiraiwa, H. Fukuoka, M. Yoshida, and H. Kurokawa, *Bull. Chem. Soc. Jpn.*, 1984, **57**, 1675.

[200] A. N. Astanina, A. P. Rudenko, M. A. Ismailova, E. Y. Offengenden, and H. M. Yakubov, *Inorg. Chim. Acta*, 1983, **79** (B7), 284.

[201] P. A. Williams, *Inorg. Chim. Acta*, 1984, **93**, L13.

[202] T. J. Goodwin, P. A. Williams, F. S. Stephens, and R. S. Vagg, *Inorg. Chim. Acta*, 1984, **88**, 165.

[203] M. W. Mulqi, P. A. Williams, F. S. Stephens, and R. S. Vagg, *Inorg. Chim. Acta*, 1984, **88**, 183.

S-alanine) have been prepared and studied by ^{1}H n.m.r. and c.d. spectroscopy.[204] Of the 40 possible isomers only five are observed when Λ-*cis*-α-[Co(*R*-picpn)-Cl_2]$^+$ is used as the starting complex.

Substitution reactions of Δ-*cis*-α-[Rh(*SS*-EDDP)Cl_2]$^-$ (*SS*-EDDP = ethylenediamine-*N*,*N'*-di-*S*-α-propionate) with *R*-alanine, *S*-alanine, *R*,*S*-alanine, and glycine give the corresponding amino acid rhodium(III) complexes of *SS*-EDDP with retention of configuration.[205] Stereoselective deuterium exchange at the co-ordinated NH_2 group of L-alanine in the Δ,L and Λ,L diastereoisomers of [Ru(bpy)$_2$(L-Ala)]$^+$ has been observed.[206] The relationship between torsion angles in Co^{III}–(α-amino acidato) five-membered chelate rings and pseudo-rotational co-ordinates has been established on the basis of experimental data.[207] Conformational studies on copper(II) complexes of threoninates and isoleucinates by the 'consistent force-field' method have been reported.[208, 209] Conformations of six-membered cobalt(III)–amino acid chelate rings have also been discussed on the basis of crystallographic data.[210]

The stereoselectivity and the diastereoselectivity in the formation of [CoL-(en)]$^+$ and [RhL(en)]$^+$ [H_2L = ethylenebis(amino acids) of the type (22)] have been considered in terms of the bulk of the substituents on N and intramolecular interaction between ethylenediamine and the ethylenebis(amino acids).[211] Other investigations in this general area deal with n.m.r. studies of conformational equilibria in cobalt(II) and nickel(II) complexes of amino acids,[212] the effects of the absolute configuration of cobalt(III) complexes (with amino acid ligands) on their chromatographic behaviour (paper, silica gel G, and aluminium oxide G),[213] and stereoselectivity in [CoL_2(aa)]$^{2+}$ complexes (L = en, phen, or bipy, aa = *S*-serine, *S*-threonine, *S*-glutamic acid, or *S*-valine).[214]

Kinetics and Reactivity. – Polarimetric data show that the base-catalysed reaction of bis-(L-serinato)copper(II) with excess formaldehyde proceeds *via*

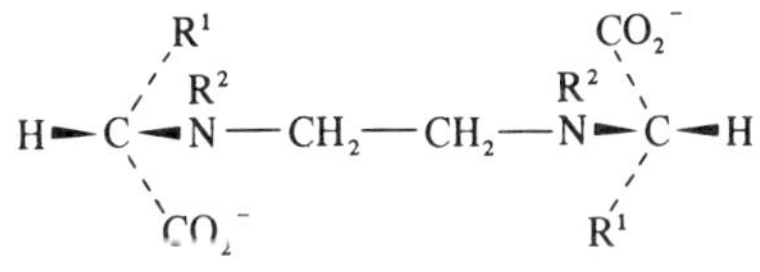

(22)

204 J. A. Chambers, T. J. Goodwin, M. W. Mulqi, P. A. Williams, and R. S. Vagg, *Inorg. Chim. Acta*, 1984, **88**, 193.
205 M. E. F. Sheridan, M.-J. Jun, and C. F. Liu, *Inorg. Chim. Acta*, 1983, **69**, 183.
206 R. S. Vagg and P. A. Williams, *Inorg. Chem.*, 1983, **22**, 355.
207 F. Pavelčik, *J. Coord. Chem.*, 1984, **13**, 299.
208 N. Raos and V. Simeon, *J. Inorg. Biochem.*, 1983, **18**, 133.
209 N. Raos, S. R. Niketic, and V. Simeon, *J. Inorg. Biochem.*, 1982, **16**, 1.
210 R. Herak and S. Kalman, *Glas. Hem. Drus., Beograd*, 1983, **48**, 343.
211 M. Strasak, *Inorg. Chim. Acta*, 1984, **83**, L57.
212 V. P. Tikhonov and N. A. Kostromina, *Teor. Eksp. Khim.*, 1983, **19**, 244.
213 M. J. Malinar, G. Vuckovic, P. N. Radivojsa, T. J. Janjic, and M. P. Celap, *J. Chromatogr.*, 1982, **249**, 65.
214 A. Pasini, *Gazz. Chim. Ital.*, 1983, **113**, 793.

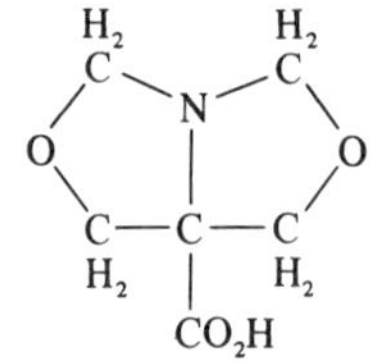

(23)

the initial dissociation of the proton on the nitrogen atom of the amino acid chelate to give the copper(II) complex of (23).[215] A bis(oxazolidine)copper(II) complex appears as an intermediate, but this species is not detected polarimetrically at 50 °C and above. The catalytic effects of transition-metal ions on the reactivity of amino acids and their derivatives are well established; such reactions include base-catalysed aldol-type condensations, isotope exchange, racemization, the formation and hydrolysis of esters and peptides, and Schiff-base formation. A full discussion of the effects of nickel(II), cobalt(III), and other metal ions on the racemization of free and bound L-alanine has been published.[216] Contrary to previous results, it is reported that Ni^{II} when complexed to L-alanine and Co^{III} when complexed to Gly-Ala or Ala-Gly *retard* the rate of racemization of the L-alanine.

Stereoselective hydrolysis of enantiomeric *N*-acyl-L(or D)-phenylalanine *p*-nitrophenyl esters $H(CH_2)_{n-1}CONHCH(CH_2Ph)CO_2C_6H_4NO_2$-*p* ($n = 2$, 10, or 16) has been observed in the presence of pentammine-L-histidine-ruthenium-(III) and ionic surfactants.[217] The highest selectivity of 4.2 was obtained for $n = 10$. Rate constants for the hydrolysis of *p*-nitrophenyl picolinate at 25 °C in the pH range 6.5–8.5 have been determined in the absence and presence of divalent metal ions (Ni^{II}, Zn^{II}, Co^{II}, Ca^{II}, and Mg^{II}) and substituted imidazoles or pyridines as ligands having hydroxyl groups in their side chains.[218] In some cases, in the presence of both a metal ion (Ni^{II} or Zn^{II}) and a ligand, high rate accelerations were observed due to the formation of catalytically active ternary complexes.

It has been previously shown[219] that amino acid esters react with $[Pd(en)(OH_2)_2]^{2+}$ to give complexes of type (24). The chelate ester complexes are

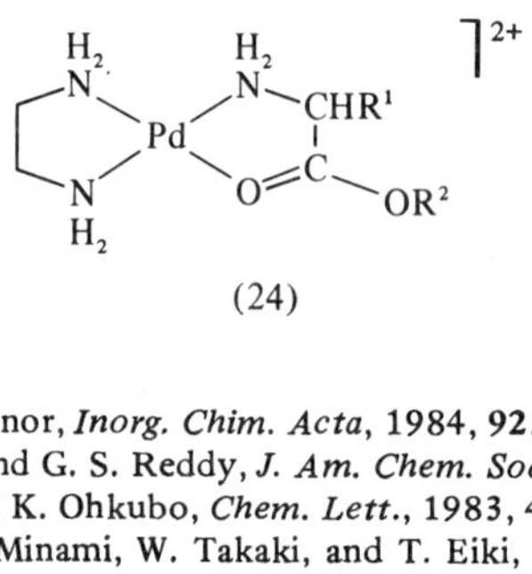

(24)

[215] S.-B. Teo and M. J. O'Connor, *Inorg. Chim. Acta*, 1984, **92**, 57.
[216] G. G. Smith, A. Khatib, and G. S. Reddy, *J. Am. Chem. Soc.*, 1983, **105**, 293.
[217] S. Sakaki, Y. Nakano, and K. Ohkubo, *Chem. Lett.*, 1983, 413.
[218] K. Ogino, K. Shindo, T. Minami, W. Takaki, and T. Eiki, *Bull. Chem. Soc. Jpn.*, 1983, **56**, 1101.
[219] R. W. Hay and P. Banerjee, *J. Chem. Soc., Dalton Trans.*, 1981, 362.

susceptible to nucleophilic attack by water and hydroxide ion. Substantial rate accelerations (factors of 4×10^4 for Gly-OEt to 1.4×10^7 for ethyl picolinate) were observed in comparisons with base hydrolysis rates of the free esters. A confirmatory study of the formation and hydrolysis of a diaqua-(1,2-diamino-ethane)palladium(II) complex of ethyl glycinate has been published.[220] Similar kinetic studies have been reported using 2,2′-bipyridyl as the inert ligand on palladium.[221]

When *cis*-$[Ru(NH_3)_4(OH)(OH_2)]^{2+}$ reacts with glycylglycine, glycinamide, or *N*′-ethylglycinamide, the *N,N*′-bound amide is formed (25). The uncatalysed transformation of these chelates to the *N,O* forms is extremely slow, but the latter can be produced by reducing ruthenium to the 2+ state and reoxidizing it.[222] In weakly acidic solutions (pH 3) penta-ammineruthenium(III) complexes of glycinamide, *N*-ethylglycinamide, glycylglycine, glycylglycinamide, and ethyl glycylglycinate undergo reaction leading to *N,O*-bound tetra-ammineruthen-ium(III) chelates, ammonia being released into solution.[223] At higher acidities (up to 0.1 mol dm^{-3} H^+) the chelation is accompanied by an aquation reaction producing $[Ru(NH_3)_5OH_2]^{3+}$ and the free ligand.

The kinetics of oxidation of the cysteinatobis(ethylenediamine)cobalt(III) ion $[Co(Cys\text{-}OS)(en)_2]^+$ (26) by peroxodisulphate have been studied in detail, and rate constants and activation parameters have been obtained.[224]

The stoicheiometry, kinetics, and mechanism of the Cr^{VI} oxidation of L-cysteine at neutral pH have also been studied.[225] A 3 : 1 L-cysteine: Cr^{VI} redox stoicheiometry was established, with L-cysteine and $[Cr(\text{L-cysteinato-}N,O,S)_2]^-$ as the sole products.

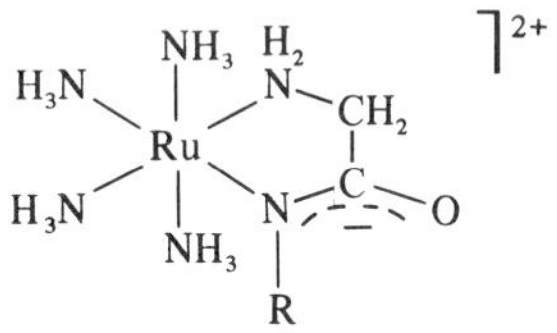

(25)

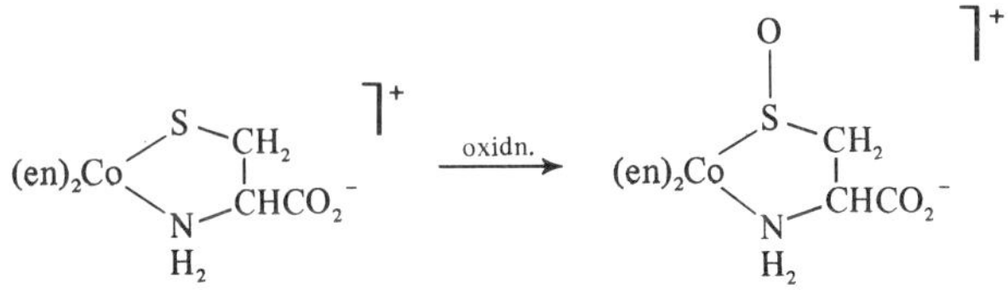

(26)

[220] M.-C. Lim, *J. Chem. Soc., Dalton Trans.*, 1983, 1675.
[221] R. W. Hay and A. K. Basak, *J. Chem. Soc., Dalton Trans.*, 1982, 1819.
[222] Y. Ilan and H. Taube, *Inorg. Chem.*, 1983, **22**, 1655.
[223] Y. Ilan and H. Taube, *Inorg. Chem.*, 1983, **22**, 3144.
[224] O. Vollárová and J. Benko, *J. Chem. Soc., Dalton Trans.*, 1983, 2359.
[225] D. W. J. Kwong and D. E. Pennington, *Inorg. Chem.*, 1984, **23**, 2528.

Stopped-flow kinetic studies of the oxidations of excess cysteine, cysteine methyl ester, and penicillamine by the copper(II)–2,9-dimethyl-1,10-phenanthroline (dmp) complex indicate a first-order dependence on the concentration of Cu^{II}, in contrast to the second-order behaviour typical of weaker Cu^{II} oxidants.[226] The kinetics of decay in absorbance at 610 nm in the reaction of cysteine with ceruloplasmin copper are biphasic under anaerobic conditions.[227] Admission of oxygen to the bleached ceruloplasmin restores the blue colour to about 75% of the original value. Evidence is also presented that ceruloplasmin catalyses the oxidation of cysteine with a one-electron reduction of oxygen and the formation of superoxide ion, which is then converted to H_2O_2 by ceruloplasmin. A 1:1 *S*-bonded adduct rapidly forms on mixing of cysteine with [tris-(2-pyridylmethyl)amine]copper(II) [$Cu(tmpa)^{2+}$] at pH 4.0–11.2, resulting in an intense absorption maximum at 396 nm.[228] The formation constant of the RS^-–Cu^{II} complexation reaction is $4.7 \times 10^5\ dm^3\ mol^{-1}$. The remarkable stability of the *S*-bonded cysteine–$Cu(tmpa)^{2+}$ adduct in aqueous solution can be accounted for in terms of the weak oxidizing nature of the Cu^{II} centre and hindrance of mercaptide radical coupling due to steric crowding about the co-ordinated sulphur atoms.

When excess cysteine is added to a V^{II} solution at pH 6.0–9.5, dihydrogen is evolved according to the following equations:[229]

$$V^{II} + Cys \rightarrow V^{II}\text{–}Cys$$

$$V^{II}\text{–}Cys + H_2O \rightarrow V^{III}\text{–}Cys + OH^- + \tfrac{1}{2}H_2$$

Currently few homogeneous metal-ion systems are known that are capable of reducing water thermally under mild conditions in basic solution.

The thermal decomposition in the solid phase of bis-(L-tryptanato)copper(II) has been studied[230] in an attempt to develop a viable synthesis of tryptamine. In solution the reaction proceeds according to the equation in Scheme 2, but this was not found to occur in the solid phase, suggesting that the solvent must actively intervene in the thermal-decomposition mechanism. The dehydration kinetics of $Cu(Gly)_2\cdot H_2O$, $Cu(\alpha\text{-Ala})_2\cdot H_2O$, $Cu(\beta\text{-Ala})_2\cdot 6H_2O$, and $Cu(Pro)_2\cdot$

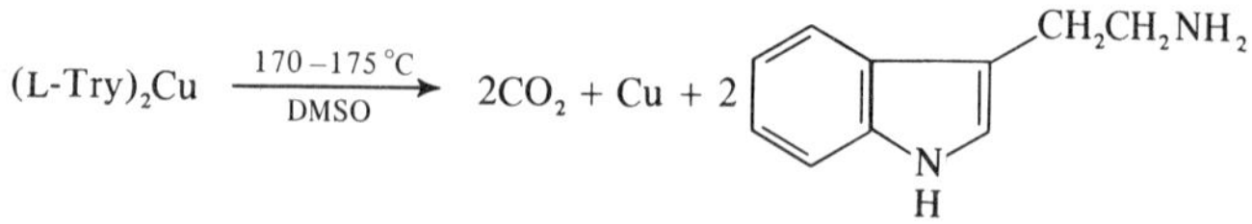

Scheme 2

[226] G. D. Stevens and R. A. Holwerda, *Inorg. Chem.*, 1984, **23**, 2777.
[227] M. V. Chidambaram, A. Zgirski, and E. Frieden, *J. Inorg. Biochem.*, 1984, **21**, 227.
[228] H. K. Baek and R. A. Holwerda, *Inorg. Chem.*, 1983, **22**, 3452.
[229] G. Kalatzis, J. Constantatos, E. Vrachnou-Astra, and D. Katakis, *J. Am. Chem. Soc.*, 1983, **105**, 2897.
[230] P. Gili and K. de la Fuente, *Inorg. Chim. Acta*, 1983, **78**, L5.

$2H_2O$ have also been investigated in detail, and activation parameters have been obtained.[231]

The rates of formation of pentacyanoferrate(II) complexes of amino acids have been studied.[232] The rates of formation are 240–370 $dm^3\ mol^{-1}\ s^{-1}$ for zwitterionic amino acids, 18–30 $dm^3\ mol^{-1}\ s^{-1}$ for mononegatively charged and 9 $dm^3\ mol^{-1}\ s^{-1}$ for dinegatively charged α-aminocarboxylate ligands at 25 °C. The reactions of $[Fe(CN)_5NO]^{2-}$ with glycine, α-alanine, β-alanine, γ-aminobutyric acid, ornithine, and lysine lead to the evolution of N_2 in weakly basic solution.[233] An interesting example of amino acidate dechelation upon amide deprotonation in bis-[*N*-2-acetamidoiminodiacetato]copper(II), $[Cu(ADA)_2]^{2-}$ (27), has been noted.[234] Ionized amide groups may thus labilize amino acidate binding to metal ions.

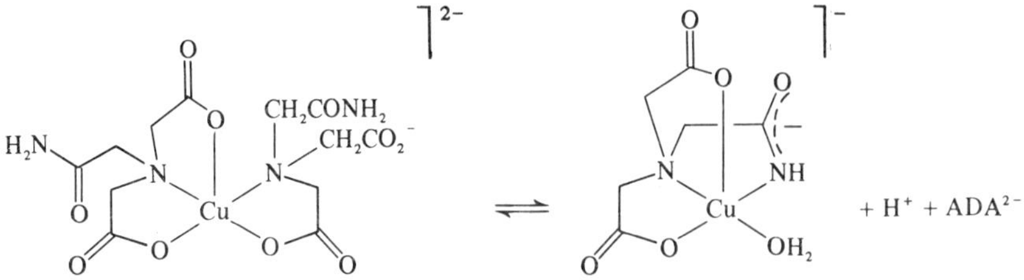

(27)

The equilibrium relations and the dynamics of equilibria in aqueous solutions of copper(II)-*N*-methylglycine have been studied by n.m.r. techniques.[235] The species CuL, CuL_2, CuL_2OH, and CuL_3 occur. The paramagnetic relaxation times determined for the different protons in the co-ordination sphere suggest that the OH^- ligand is axially co-ordinated to copper(II). N.m.r. relaxation studies have also been carried out on the formation, dissociation, and exchange rate of Cr^{II}-glycine complexes.[236]

Kinetic data have been reported[237] for the reversible second-order substitution of Cl^- in $[Pt(amino\ acid)Cl_2]$ (amino acid = glycine, sarcosine, *N,N*-dimethylglycine, or proline) by DMSO to form *cis*-(*N,S*)- and *trans*-(*N,S*)-[Pt(amino acid)-(DMSO)Cl]. A pseudo-rotation mechanism involving five co-ordinate $[Pt(amino\ acid)(DMSO)Cl_2]^-$ appears to predominate. The acid hydrolysis of ethylenediaminetetra-acetate, ethylenediaminediacetate, and nitrilotriacetate complexes of dioxovanadium(V) has been investigated by stopped-flow techniques in aqueous DMSO.[238] A series of tris(imidazole)-containing phosphines (L) have

[231] V. V. Shelkovnikov, V. I. Eroshkin, A. B. Tronov, and V. B. Durasov, *Izv. Vyssh. Uchebn. Zaved. Khim. Tekhnol.*, 1982, **25**, 1198.
[232] H. E. Toma, A. A. Batista, and H. B. Gray, *J. Am. Chem. Soc.*, 1982, **104**, 7509.
[233] A. Katho, Zs. Bodi, L. Dozsa, and M. T. Beck, *Inorg. Chim. Acta*, 1984, **83**, 145.
[234] D. P. Parr, C. Rhodes, tert., and R. Nakon, *Inorg. Chim. Acta*, 1983, **80**, L11.
[235] F. Debreczeni, J. Polgar, and I. Nagypál, *Inorg. Chim. Acta*, 1983, **71**, 195.
[236] I. Nagypál, K. Micskei, and F. Debreczeni, *Inorg. Chim. Acta*, 1983, **77**, L161.
[237] L. E. Erickson, T. A. Ferrett, and L. F. Buhse, *Inorg. Chem.*, 1983, **22**, 1461.
[238] J. Lagrange, K. Aka, and P. Lagrange, *J. Chim. Phys. Phys.-Chim. Biol.*, 1983, **80**, 755.

been prepared and their Zn^{II} and Co^{II} complexes studied as biomimetic catalysts for the hydrolysis of *p*-nitrophenyl picolinate.[239] Evidence for the importance of LM^{2+}–OH complexes in the hydrolysis was obtained. Equilibrium constants have been determined for the various steps involved in the oxygenation of Co^{II}–1,10-phenanthroline–amino acid mixed-ligand complexes.[240] The formation constant for the oxygenation step increases as the alkyl chain length or degree of branching increases in the amino acid.

The S_N2 reaction between glycine and ammonia with Mg^{II} as a catalyst has been studied as a model reaction for Mg^{II}-catalysed peptide-bond formation using the *ab initio* Hartree–Fock molecular-orbital method.[241] Substantial decreases in free energies of activation were found for both two-step and concerted mechanisms in Mg^{II}-catalysed amide-bond formation when comparisons were made with the uncatalysed and amine-catalysed reactions. The catalytic effect of the Mg^{II} is to stabilize both the transition states and the tetrahedral intermediate. Stabilization is attributed to the neutralization of the developing negative charge on the electrophile and to the formation of a conformationally flexible non-planar five-membered chelate ring structure.

Site-exchange reactions of $[PtMe_3(Gly)(D_2O)]$ and isomers of $[PtMe_2Br(Gly)(D_2O)]$ with D_2O co-ordinated *trans* to methyl have been studied,[242] and rate constants have been determined by complete lineshape analysis of ^1H n.m.r. spectra in the methyl region. The reactions of these Pt^{IV} complexes are believed to involve exchange of the labile water molecule with solvent water, with an associated migration of a donor atom (nitrogen or oxygen).

The oxidative deamination of α-amino acids occurs with various oxidizing agents including metal ions;[243] however, until now, the oxidative dehydrogenation of α-amino acids co-ordinated to metal ions has not been reported. A recent paper now describes the formation of (2-iminocarboxylato)cobalt(III) complexes of the type $[Co\{NH{=}C(R)COO\}(tetramine)]^{2+}$ (28) by $KMnO_4$ oxidation of the appropriate amino acid complexes.[244] The preparation of several 2-iminocarboxylato complexes has previously been described[245] by the intramolecular condensation of (2-ketocarboxylato)penta-amminecobalt(III) complexes.

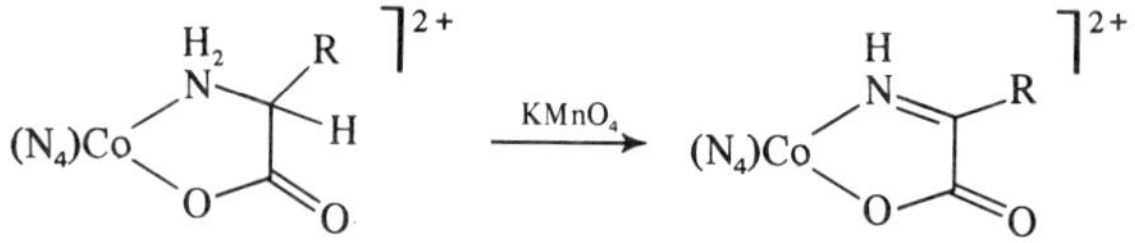

R = Ph, Me, or $CHMe_2$ (28)

[239] R. S. Brown, M. Zamkanei, and J. L. Cocho, *J. Am. Chem. Soc.*, 1984, **106**, 5222.
[240] D. M. Palade, V. V. Shapovalov, and V. S. Semykin, *Zh. Neorg. Khim.*, 1983, **28**, 2156.
[241] T. Oie, G. H. Loew, S. K. Burt, and R. D. MacElroy, *J. Am. Chem. Soc.*, 1984, **106**, 8007.
[242] T. G. Appleton, J. R. Hall, N. S. Ham, F. W. Hess, and M. A. Williams, *Aust. J. Chem.*, 1983, **36**, 673.
[243] See for example A. Kumar and P. Neta, *J. Am. Chem. Soc.*, 1980, **102**, 7284.
[244] M. Yamaguchi, M. Saburi, and S. Yoshikawa, *J. Am. Chem. Soc.*, 1984, **106**, 8293.
[245] See for example J. M. Harrowfield and A. M. Sargeson, *J. Am. Chem. Soc.*, 1974, **96**, 2634; J. M. Harrowfield and A. M. Sargeson, *J. Am. Chem. Soc.*, 1979, **101**, 1514.

Schiff Bases. – Zinc(II) complexes derived from the condensation of (1*R*)-3-hydroxymethylenebornan-2-one and a series of L-amino acids undergo tautomeric equilibria in solution between enolimine and ketoenamine species.[246] Optically active alanine, valine, and leucine have been obtained by a transamination reaction between pyridoxamine and the corresponding α-keto acid in the presence of a Cu^{II} complex with the tridentate ligand 2,6-bis-[(3*S*)-3-phenyl-2-azabutyl)]-pyridine.[247] In each case the amino acid with the (*R*)-configuration was formed preferentially and the maximum enantiomeric excesses were 54% (alanine), 48% (leucine), and 29% (valine). A similar enantioselective synthesis of phenylalanine from phenylpyruvic acid has also been described.[248]

The kinetics of isomerization between Δ_L-β_2- and Λ_L-β_2-diastereoisomers of [Co(α-Me-sal$_2$en)(L-aa)] [Me-sal$_2$en = *N,N'*-ethylenebis-(α-methylsalicylidene-aminato), L-aa = L-Phe, L-Met, L-Ile, or L-Pro] have been studied in methanol solvent;[249] possible mechanisms were considered.

The mechanism of enzymatic and non-enzymatic vitamin B_6-catalysed transamination of α-amino and α-keto acids proposed by Metzler, Ikawa, and Snell[250] involves the formation and interconversion of aldimine and ketimine Schiff bases (29) and (30). Evidence for the α,α'-carbanion intermediate (31) in an Al^{III}-catalysed reaction has now been obtained.[251] 1-(*N,N*-Dihexadecyl-carbamoylmethyl)-2-methyl-3-hydroxy-4-formyl-5-hydroxymethylpyridinium chloride (32) undergoes a transamination reaction with L-phenylalanine in single-walled bilayer vesicles formed from two different peptide lipids. Co-ordination of copper(II) to the Schiff-base intermediate results in a marked rate acceleration.[252]

Four complexes of Cu^{II} with Schiff-base ligands derived from salicylaldehyde and glycine, ε-acetyl-L-lysine, histamine, and L-histidine have been prepared and

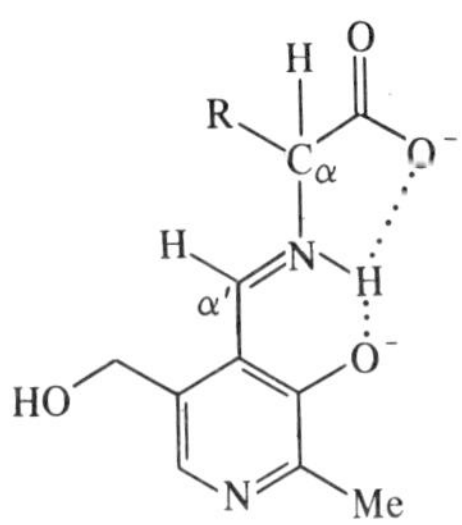

(29)

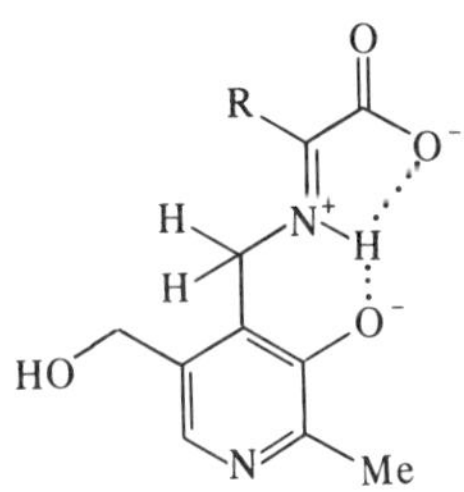

(30)

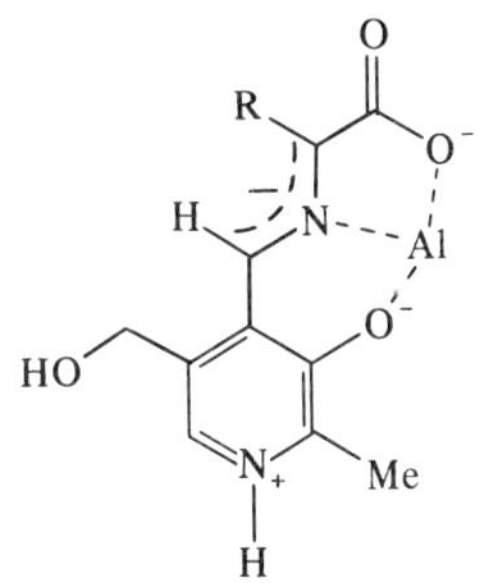

(31)

[246] L. Casella, M. Gullotti, and A. Rockenbauer, *J. Chem. Soc., Dalton Trans.*, 1984, 1033.
[247] R. Deschenaux and K. Bernauer, *Helv. Chim. Acta*, 1984, **67**, 373.
[248] K. Bernauer, R. Deschenaux, and T. Taura, *Helv. Chim. Acta*, 1983, **66**, 2049.
[249] Y. Fujii, T. Kobayashi, M. Matsufuru, and S. Takahashi, *Bull. Chem. Soc. Jpn.*, 1983, **56**, 3608.
[250] D. E. Metzler, M. Ikawa, and E. E. Snell, *J. Am. Chem. Soc.*, 1954, **76**, 648.
[251] A. E. Martell and P. Taylor, *Inorg. Chem.*, 1984, **23**, 2734.
[252] Y. Murakami, J. Kikuchi, T. Imori, and K. Akiyoshi, *J. Chem. Soc., Chem. Commun.*, 1984, 1434.

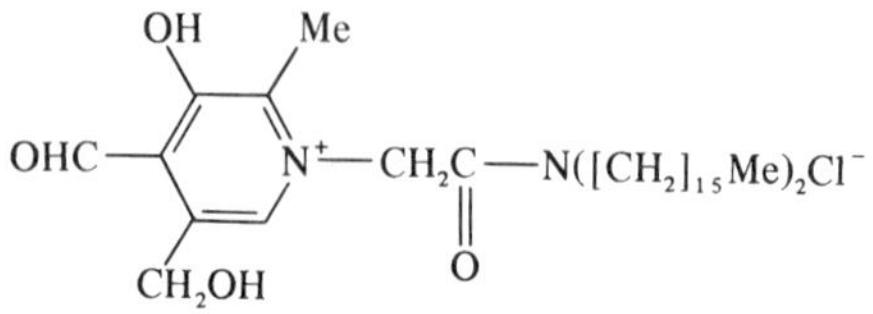

(32)

characterized and their spectroscopic properties investigated.[253] The nickel(II) complexes of a number of Schiff bases derived from substituted *o*-hydroxy-acetophenones and glycine have also been characterized.[254] Spectroscopic data suggest that the complexes are four co-ordinate with the donor atoms being the imino nitrogen and the carboxylate oxygen.

Asymmetric synthesis of threonine and partial resolution and retroracemization of α-amino acids *via* the copper(II) complexes of their Schiff bases with (*S*)-2-*N*-(*N'*-benzylprolyl)aminobenzaldehyde and (*S*)-2-*N*-(*N'*-benzylprolyl)-aminoacetophenone have been described, and the crystal structure of a copper complex of the glycine Schiff base with (*S*)-2-*N*-(*N'*-benzylprolyl)aminoacetophenone has been obtained.[255]

Formation constants of divalent-metal complexes with Schiff bases formed from salicylaldehydes and amino acids have been determined by pH titration at 30 °C and $I = 0.2$ mol dm^{-3} in water or aqueous ethanol.[256] Copper(II) shows little tendency to form 1:2 complexes in aqueous media. In aqueous ethanol, 1:1 and 1:2 complexes are formed with each transition metal. Other solution studies deal with the thermodynamic equilibria in the system zinc(II)–pyridoxal 5'-phosphate–2-amino-3-phosphonopropionic acid in aqueous solution,[257] mixed-ligand complexes of some divalent-metal ions (Co^{II}, Ni^{II}, Cu^{II}, and Zn^{II}) with pyridoxamine and histidine,[258] and quaternary complexes involving pyridoxamine, glycine, and ethylenediamine with Co^{II}, Ni^{II}, Cu^{II}, and Zn^{II}.[259]

As discussed in the introduction, a review of crystal structures of divalent-metal complexes with pyridoxal–amino acid Schiff bases has been published.[4]

Miscellaneous. – *E. coli* cells grown in the presence of L-serine and gallium(III) nitrate display ultrastructural changes when observed under the electron microscope.[260] No cell-surface changes were detected when mixtures of L-serine and $K_2[PdCl_4]$ were used as modifying agents; in this case all changes observed

[253] M. R. Wagner and F. A. Walker. *Inorg. Chem.*, 1983, **22**, 3021.
[254] T. M. Aminabhavi, N. S. Biradar, G. V. Karajagi, and W. E. Rudzinski, *Inorg. Chim. Acta*, 1984, **91**, 49.
[255] Yu. N. Belokon, I. E. Zel'tzer, V. I. Bakhmutov, M. B. Saporovskaya, M. G. Ryzhov, A. I. Yanovsky, Yu. T. Struchkov, and V. M. Belikov, *J. Am. Chem. Soc.*, 1983, **105**, 2010.
[256] E. S. Jayadevappa and S. C. Galgali, *J. Indian Chem. Soc.*, 1983, **60**, 1098.
[257] B. Szpoganicz and A. E. Martell, *Inorg. Chem.*, 1984, **23**, 4442.
[258] M. S. El-Ezaby and F. M. Al-Sogair, *Polyhedron*, 1982, **1**, 791.
[259] H. M. Marafie, M. S. El-Ezaby, and A. S. Shawali, *Polyhedron*, 1983, **2**, 775.
[260] N. T. McArdle, A. J. Charlson, C. D. Shorey, R. Arnold, and N. Barker, *Inorg. Chim. Acta*, 1984, **92**, 113.

were intracellular. The antitumour activities of potassium *cis*-$[PdCl_2(Gly)]^-$ and caesium *cis*-$[PdCl_2(Ser)]^-$ have been assessed, and both compounds were found to be active.[261]

In recent years joint replacement for patients with arthritis has become widespread; materials used have included metal alloys, plastics, and ceramics. As a result, the reaction of D-penicillamine with some metals and alloys has been investigated.[262] Possible mechanisms for the adsorption of amino acids on homoionic smectite clays have been considered, where Cu^{II} and Ni^{II} are the probable binding sites.[263]

The antiviral activity of copper(II) complexes of amino acids has been tested against avian influenza A virus, Newcastle disease virus, and Ayeskii disease virus.[264] Copper(II)-glycine and copper(II)-DL-serine displayed significant activity. A series of copper(II)-amino acid complexes has also been tested *in vitro* against bacteria, actinomycetes, and fungi.[265] The complexes display weak to moderate antimicrobial activity but are highly active against *Azobacter* species. Complexes containing racemic amino acids were more active than those with D- or L-amino acids.

The effects of amino acids on the rate of iron adsorption have been investigated, the most significant effects being observed with asparagine and glycine.[266]

3 Peptides

A large number of papers dealing with syntheses, structures, equilibrium studies, and reactions of metal-peptide complexes have been published during 1983/84. Reviews have appeared on substitution, redox, and proton-transfer reactions of metal-peptide complexes[267] and on the applications of complex formation to peptide synthesis.[268] The reactions of copper(II/III)-peptide complexes with O_2,[269] axial-substitution and electron-transfer reactions in nickel(III)-peptide complexes,[270] and studies of copper(II/III) complexes with peptide ligands containing α-aminoisobutyrate residues[271] are the subjects of dissertations.

Complexes of Silver(I), Thallium(I), and Alkaline and Alkaline-earth Metal Cations. – A number of silver(I) complexes containing cyclopeptide ligands have been synthesized and studied by both ^1H and ^{13}C n.m.r. spectroscopy. In

261 A. J. Charlson and W. A. Shorland, *Inorg. Chim. Acta*, 1984, **93**, L67.
262 D. H. Brown and W. E. Smith, *Inorg. Chim. Acta*, 1984, **93**, L29.
263 A. Gupta, G. H. Loew, and J. Lawless, *Inorg. Chem.*, 1983, **22**, 111.
264 N. I. Mitin, N. A. Lagutkin, L. F. Chapurina, M. M. Zubairov, T. K. Petracheva, and T. N. Arkhipova, *Khim.-Farm. Zh.*, 1983, **17**, 565.
265 A. A. Tumanov, L. F. Chapurina, and I. A. Filimonova, *Izv. Akad. Nauk Mold. SSSR, Ser. Biol. Khim. Nauk*, 1983, 44.
266 J. M. Christensen, M. Ghannam, and J. W. Ayres, *J. Pharm. Sci.*, 1984, **73**, 1245.
267 D. W. Margerum, *Pure Appl. Chem.*, 1983, **55**, 23.
268 Y. Kobayashi and S. Inoue, *Kagaku (Kyoto)*, 1983, **38**, 810 (in Japanese).
269 R. A. Read, *Diss. Abstr. Int. B*, 1983, **43**, 2550.
270 C. K. Murray, *Diss. Abstr. Int. B*, 1982, **43**, 1842.
271 A. W. Hamburg, *Diss. Abstr. Int. B*, 1983, **43**, 2537.

the complexes [Ag{cyclo(-Gly-L-His-)}$_2$]X (X = NO_3^- or ClO_4^-) and [Ag{cyclo(-L-His-L-His-)}]$_n$Y$_n$ (Y = ClO_4^- or PF_6^-) the metal ion is linearly co-ordinated by two imidazole nitrogen atoms, whereas in the dimeric complex [Ag{cyclo(-L-Met-L-His-)}]$_2$(ClO_4^-)$_2$ it is also linearly co-ordinated but in this case by an imidazole nitrogen and a thioether sulphur.[272] Results of ^{1}H n.m.r. spectroscopic studies in D_2O at 50 and 65 °C suggest that in the silver(I) complex of cyclo(-L-Met-L-Met-) the metal is co-ordinated to two sulphur atoms.[273] *X*-Ray studies show that the crystalline complex consists of a helical polymer (33) in which each silver is also linearly co-ordinated by two sulphur atoms. The structure of the complex [Ag{cyclo(-L-Met-Gly-)}]NO_3 (34) consists of an infinite linear array of silver ions that are doubly bridged by methionine sulphur

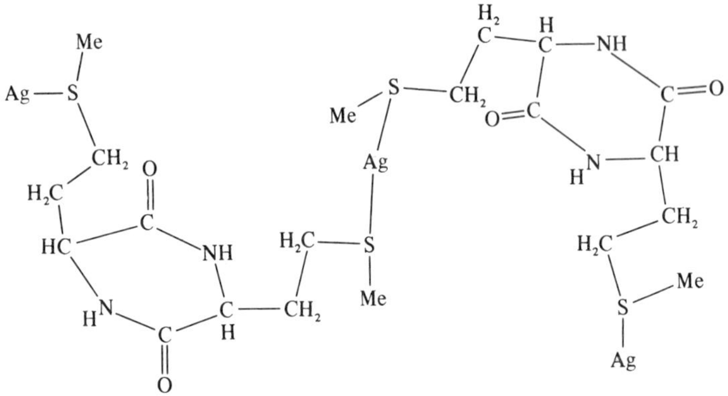

(33)

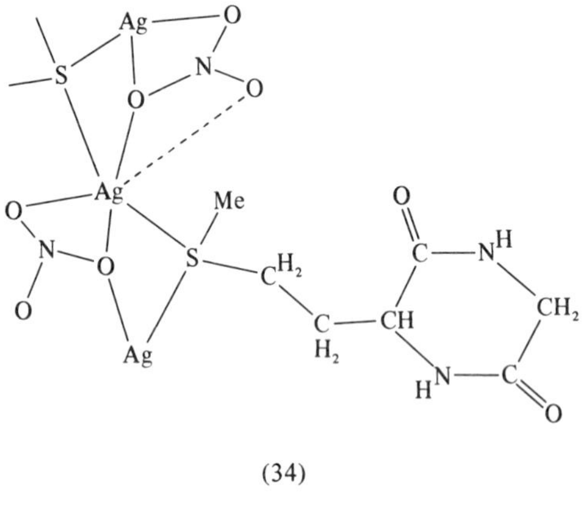

(34)

[272] Y. Kojima, T. Yamashita, Y. Ishino, T. Hirashima, and T. Miwa, *Bull. Chem. Soc. Jpn.*, 1983, **56**, 3841.

[273] Y. Kojima, T. Yamashita, Y. Ishino, T. Hirashima, and K. Hirotsu, *Chem. Lett.*, 1983, 453.

atoms and by nitrate anions.[274] Formation constants of nine silver(I) complexes with sulphur-containing dipeptide ligands have been measured at 25 °C, $I = 0.1$M KNO_3; results are reported for monomeric and dimeric complexes in various states of protonation.[275] Complexes (1:1) of peptides containing two sulphur atoms, *e.g.* L-Met-L-Met, are considerably more stable than those with one, *e.g.* L-Met-Gly, which is indicative of the preferred linear co-ordination of silver(I) to both sulphur atoms of the dipeptide.

The valinomycin analogue cyclo(-L-Ala-Gly-D-Phe-L-Pro-) has been synthesized, and its interaction with various mono- and di-valent cations in CD_3CN solutions has been investigated by n.m.r. spectroscopy.[276] While K^+, Rb^+, Cs^+, Tl^+, and Ba^{2+} all form predominantly 1:1 complexes, Na^+ and Li^+ give mainly 2:1 (peptide : cation) complexes. Complex formation between *s*-block metal cations and the cyclic octapeptides cyclo(-Phe-Pro-)$_4$, cyclo(-Leu-Pro-)$_4$, and cyclo-(-Boc-Lys-Pro-)$_4$ in ethanol and acetonitrile solution has been investigated.[277]

Complexes of Calcium(II), Cobalt(II), Nickel(II), Zinc(II), Cadmium(II), and Mercury(II). – *Ab initio* calculations on complexes of calcium(II) with Gly-Gly, Gly-Ala, and Ala-Ala show that the anionic peptides ($NH_2 \sim CO_2^-$) form their most stable complexes when co-ordinated to the metal through the three oxygen (peptide and two carboxylate) atoms.[278] For the neutral peptides ($NH_3^+ \sim CO_2^-$) this co-ordination mode and another involving the carboxylate oxygen atoms only are of comparable stability. Similar calculations have been carried out for complexes of zinc(II) with Gly-Gly and for complexes of zinc(II) and calcium(II) with Gly-Pro and Pro-Gly.[279]

Complex formation between cobalt(II), nickel(II), and zinc(II) and the dipeptides Gly-L-His, L-His-Gly, and β-Ala-L-His (carnosine) in 0.2 M KCl solution has been investigated by potentiometric and spectrophotometric methods (including ^{13}C n.m.r. in the case of zinc).[280] While all of these ions promote peptide-group ionization in Gly-L-His ($Ni^{II} > Zn^{II} > Co^{II}$), none of them do so in L-His-Gly, which appears to be complexed through the amino and imidazole N-3 groups. Formation of ternary complexes between these binary systems and the ligands Gly, Gly-Gly, L-His, and 2,2′-bipyridyl has been investigated, and formation constants are reported for a wide range of complexes.

Cyclo(-L-Met-L-His-) (L) forms the complex $[ZnL_4](SO_4)_2 \cdot 10H_2O$, in which the ligands are tetrahedrally co-ordinated to the metal through the imidazole N-1 atoms (Zn—N bond distances 197—204 pm, N—Zn—N bond angles 106.1—112.2°).[281] The complex $[ZnL]^{2+}$ in solution is of the same stability as $[NiL]^{2+}$ ($\log K = 2.4$, 25 °C, $I = 0.2$M) but is much less stable than $[CuL]^{2+}$ ($\log K = 3.6$) under the same conditions.[282] Complexes of this peptide containing various

[274] G. Valle and R. Ettorre, *J. Chem. Soc., Dalton Trans.*, 1983, 453.
[275] A. Q. Lyons and L. D. Pettit, *J. Chem. Soc., Dalton Trans.*, 1984, 2305.
[276] J. P. Degelaen, P. Pham, and E. R. Blout, *J. Am. Chem. Soc.*, 1984, **106**, 4882.
[277] S. Kimura and Y. Imanishi, *Biopolymers*, 1983, **22**, 2383.
[278] M. M. Probst and B. M. Rode, *Inorg. Chim. Acta*, 1983, **78**, 135.
[279] M. M. Probst and B. M. Rode, *Inorg. Chim. Acta*, 1984, **92**, 75.
[280] E. Farkas, I. Sovago, and A. Gergely, *J. Chem. Soc., Dalton Trans.*, 1983, 1545.
[281] Y. Kojima, N. Ishio, T. Yamashito, and K. Hirotsu, *Chem. Lett.*, 1983, 1365.
[282] M. Bressan, F. Marchiori, and G. Valle, *Int. J. Pept. Protein Res.*, 1984, **23**, 104.

ligand-to-metal ratios have been isolated for cobalt(II), nickel(II), and zinc(II). Potentiometric and ^1H n.m.r. methods have been used to determine formation constants and to propose structures for complexes of zinc(II) with Gly-L-His-L-Lys and the related peptides L-Ala-L-His, Gly-L-His, Gly-L-Lys, L-His-L-Lys, L-Leu-L-His, L-Val-L-Lys, and human serum albumin.[283] In aqueous solution nickel(II) forms a 1:1 square-planar complex with the 4N donor Gly-Gly-L-Tyr-NHMe, a tripeptide amide that mimics the NH_2 terminal binding site of dog serum albumin.[284]

Complexes of glutathione (35) ($= H_3L$) continue to attract interest. The formation and dissociation of its 1:1 and 2:1 complexes with zinc(II) in 0.3M $NaNO_3$ solution have been studied by the temperature-jump method in the pH range 4.58–4.98. Reaction of zinc(II) with HL^{2-}, in which only the amino group is protonated and which is the dominant form of the ligand in the above pH range, proceeds 2×10^4 times faster than its reaction with H_2L^- (the sulphur is also protonated) for both mono and bis complexation.[285] The decreased rates for H_2L^- are attributed to blockage of the metal-complexing sites by intramolecular hydrogen bonding. The recognition of the central role played by erythrocytes in binding $MeHg^+$ has prompted investigations into the interaction of this ion with glutathione, ergothioneine, and haemoglobin, the most abundant intracellular sulphydryl-containing molecules.[286] These have been carried out using ^1H n.m.r. spectroscopy, and formation constants of glutathione and ergothioneine complexes have been determined by competitive complexation with mercaptoacetic acid as a function of pH. Formation constants for complexes of $MeHg^+$ and ten sulphydryl-containing ligands, including glutathione, have been determined.[287]

The mechanism by which His stabilizes the cobalt(II)–carnosine (β-Ala-L-His) complex against oxidation in aqueous solution depends on whether it is coordinated as a monodentate or as a bidentate ligand.[288] In the former case it occupies an equatorial site that would otherwise be available to H_2O and thus prevents an interaction favouring oxidation of cobalt(II). As a bidentate ligand, His causes carnosine to behave likewise, thus precluding strain associated with

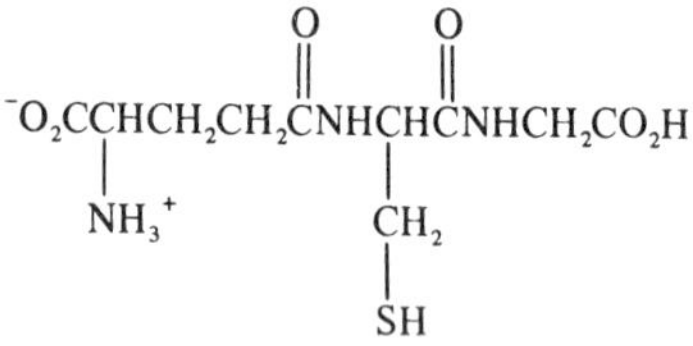

(35)

[283] M. J. A. Rainer and B. M. Rode, *Inorg. Chim. Acta*, 1984, **93**, 109.
[284] J. D. Glennon, D. W. Hughes, and B. Sarkar, *J. Inorg. Biochem.*, 1983, **19**, 281.
[285] L. A. Dominey and K. Kustin, *J. Inorg. Biochem.*, 1983, **18**, 153.
[286] R. S. Reid and D. L. Rabenstein, *J. Am. Chem. Soc.*, 1982, **104**, 6733.
[287] A. P. Arnold and A. J. Canty, *Can. J. Chem.*, 1983, **61**, 1428.
[288] C. E. Brown, D. W. Vidrene, R. Czernuszewicz, and K. Nakamoto, *J. Inorg. Biochem.*, 1982, **17**, 247.

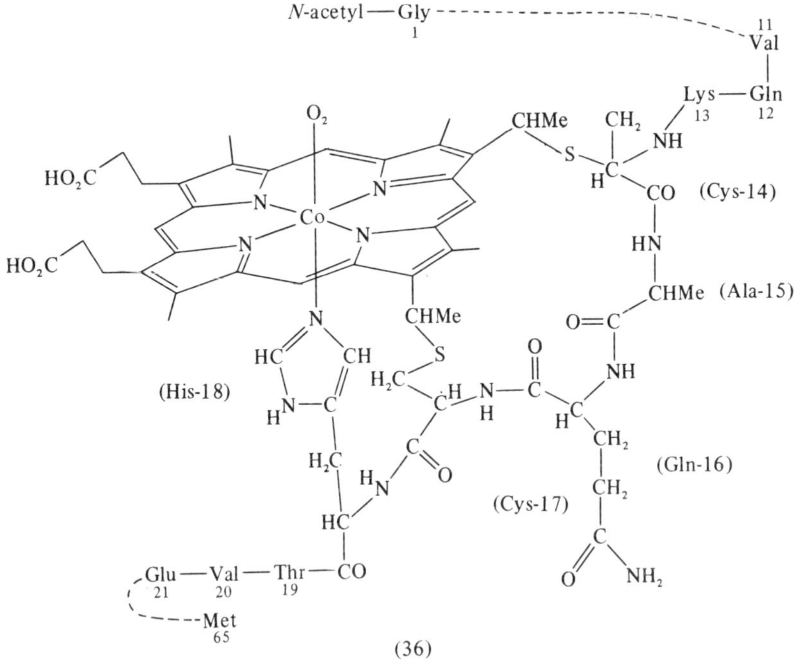

(36)

tridentate co-ordination of the latter. The cobalt(II)-cytochrome *c* 1–65 peptide complex (36) gives a stable monomeric oxygenated complex at room temperature even in aqueous solution.[289] The e.s.r. spectrum of (36) at 77 K in 15% acetic acid solution shows a strong, axially symmetric signal with $g = 2.276$. Exposure of this solution to air causes the signal to be replaced by another at $g = 2.003$, consistent with formation of a cobalt(III)-($O_2^{\bullet-}$) species.

Complexes of sulphur-containing peptide ligands have been investigated as possible structural models for metallothioneins. The electronic-absorption, m.c.d., and c.d. spectra of cobalt(II) complexes containing the ligands Boc-Cys-Ala-Ala-Cys-OMe and Boc-Ala-Cys-OMe point to complicated core structures having bridging and terminal thiolate ligands.[290] Comparisons are made with cobalt(II)-substituted methallothionein and with cobalt(II)-thiolate complexes such as $[Co(SPh)_4]^{2-}$ and $[(CoSPh)_4(\mu\text{-SPh})_6]^{2-}$. The structure of $(Et_4N)(Et_3NH)[Cd_4(SPh)_{10}]$ has been determined by *X*-ray-diffraction methods, and its relationship to the cadmium(II)-cysteinate aggregates in metallothionein is discussed.[291] Evidence from ^{113}Cd n.m.r. spectroscopy suggests that the seven cadmium(II) ions in mammalian metallothioneins are split into $Cd_3(Cys)_9$ and $Cd_4(Cys)_{11}$ aggregates that contain distorted CdS_4 units and $Cd_3(\mu\text{-S})_3$ rings, features that are both found in $[Cd_4(SPh)_{10}]^{2-}$.

[289] S. Kawanishi and S. Sano, *J. Chem. Soc., Chem. Commun.*, 1984, 1628.
[290] M. Nakata, N. Ueyama, A. Nakamura, T. Nozawa, and M. Hatano, *Inorg. Chem.*, 1983, **22**, 3028.
[291] K. S. Hagen and R. S. Holm, *Inorg. Chem.*, 1983, **22**, 3171.

Complexes of Copper(II). – As in the past, the number of publications on copper(II)-peptide complexes far exceeds that on any other metal ion. Formation constants for complexes of copper(II) with 13 aliphatic dipeptides containing combinations of Gly, Ala, Leu, and Pro residues have been determined by e.s.r. spectroscopy.[292] This method, although of lower accuracy than potentiometric methods, offers an advantage in being able to distinguish between complexes having similar species distribution and pH dependence that cannot be distinguished potentiometrically. The e.s.r. method has also been used to obtain stability constants for binary and ternary complexes in solutions containing copper(II), ATP, and the dipeptides Gly-Gly,[293] Gly-L-Pro,[293] Gly-L-Trp,[294] and L-Trp-Gly.[294] Proton- and copper(II)-association constants are reported for eight peptides and four peptide amides containing the α-aminoisobutyric acid (Aib) residue, and the effects of the methyl substituents are discussed.[295]

The interaction of copper(II) and the tripeptide Gly-L-His-L-Lys in 0.15 M NaCl solution at 37 °C, both in the absence of and in the presence of His, has been investigated, and the distribution of the resulting copper(II) complexes in human blood plasma and in cell culture media has been computed.[296] The structure of the complex Cu(H_{-1}-Gly-L-His-L-Lys) (37) has been determined by *X*-ray crystallography.[297]

In solutions containing copper(II) and Gly-L-His-Gly (A) the principal species in the pH range 4–9 is Cu(H_{-1}-A), although low concentrations of the 2:1 complexes CuA_2 and $[Cu(H_{-1}\text{-}A_2)]^-$ also exist.[298] In the case of Gly-Gly-L-His (B) only the highly stable species $[Cu(H_{-2}\text{-}B]^-$ is formed. For these and copper(II) complexes of L-pyroGlu-L-His-L-Pro-NH_2 (C), Gly-L-His, and L-pyroGlu-L-His-OMe (D), the acidity of the imidazole NH groups decreases in the sequence A > Gly-L-His > D > C > B. In order to investigate the influence of the L-Tyr residue in the third position of dog serum albumin, copper(II) complexes of Gly-Gly-L-Tyr-NHMe have been investigated and five 1:1 species with various degrees of protonation have been identified.[299] Evidence based on e.s.r. spectroscopy is consistent with the existence of dimeric species in alkaline solutions (pH > 9) containing equimolar amounts of copper(II) and $(Gly)_n$ ($n \geqslant 3$).[300]

The role of Pro and Sar (X) as 'break points' in copper(II)-peptide interactions is the subject of a number of papers. With Gly-X-Gly_2 and Gly_2-X-Gly (L) as ligands, the complexes Cu(H_{-1}-L) and Cu(H_{-2}-L), respectively, are destabilized and the peptide segments on either side of X co-ordinate independently to the metal ion.[301] Further evidence for the existence of Pro 'break

[292] W. S. Kittl and B. M. Rode, *J. Chem. Soc., Dalton Trans.*, 1983, 409.
[293] E. R. Werner and B. M. Rode, *Inorg. Chim. Acta*, 1984, **91**, 217.
[294] E. R. Werner and B. M. Rode, *Inorg. Chim. Acta*, 1984, **93**, 27.
[295] A. W. Hamburg, M. T. Nemeth, and D. W. Margerum, *Inorg. Chem.*, 1983, **22**, 3535.
[296] P. M. May, J. Whittaker, and D. R. Williams, *Inorg. Chim. Acta*, 1983, **80**, L5.
[297] C. M. Perkins, N. J. Rose, B. Weinstein, R. E. Stenkamp, L. H. Jensen, and L. Pickart, *Inorg. Chim. Acta*, 1984, **82**, 93.
[298] E. Farkas, I. Sovago, T. Kiss, and A. Gergely, *J. Chem. Soc., Dalton Trans.*, 1984, 611.
[299] D. Muller, B. D. Reverend, and B. Sarkar, *J. Inorg. Biochem.*, 1984, **21**, 215.
[300] H. Yokoi and A. Hanaki, *Chem. Lett.*, 1983, 1319.
[301] M. Bataille, G. Formicka-Kozlowska, H. Kozlowski, L. D. Pettit, and I. Steel, *J. Chem. Soc., Chem. Commun.*, 1984, 231.

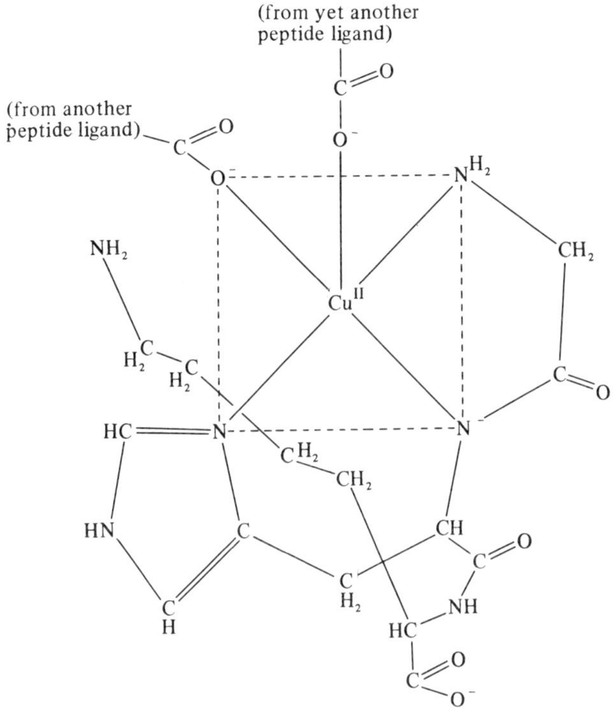

(37)

points' is seen (i) in species-distribution curves for copper(II) complexes with L-Phe-L-Pro-Gly_2, Gly-L-Pro-L-Phe-Gly, Gly-L-Pro-D-Phe-Gly, and Gly-L-Pro-Gly-L-Phe[302] and with the corresponding set of tetrapeptides in which Phe is replaced by Tyr[303] and (ii) in the solution structures of copper(II) complexes of Thr-Lys-Pro-Lys-Thr-Lys-Pro-Lys (canine tuftsinyltuftsin octapeptide).[304]

Stability constants and species-distribution curves are reported for copper(II) complexes of Arg-Lys-Asp-Val-Tyr, a pentapeptide fragment of the thymus hormone thymopoietin.[305] Stability constants are also reported for ternary species involving Cu(H_{-1}-Gly-Gly) and various nucleosides[306] and for ternary and quaternary species in solutions containing this complex, pyridoxamine, and imidazole.[307]

[302] M. Bezer, L. D. Pettit, I. Steel, M. Bataille, S. Djemil, and H. Kozlowski, *J. Inorg. Biochem.*, 1984, **20**, 13.
[303] H. Kozlowski, M. Bezer, L. D. Pettit, M. Bataille, and B. Hecquet, *J. Inorg. Biochem.*, 1983, **18**, 231.
[304] G. Formicka-Kozlowska, D. Konopinska, H. Kozlowski, and B. D. Reverend, *Inorg. Chim. Acta*, 1983, **78**, L47.
[305] I. Sovago, T. Kiss, and A. Gergely, *Inorg. Chim. Acta*, 1984, **93**, L53.
[306] S. V. Deshpande, R. K. Sharma, and T. S. Srivastava, *Inorg. Chim. Acta*, 1983, **78**, 13.
[307] H. M. Marafie, M. S. El-Ezaby, M. Rashad, and N. M. Moussa, *Polyhedron*, 1984, **3**, 787.

The superoxide dismutase activities of a number of copper(II)-peptide complexes have been compared. Whereas copper(II)-(His-AA) (AA = Phe, Ala, Val, or Tyr) complexes show high activity towards O_2^-, copper(II)-(AA-His) species are almost inactive.[308] Polarographic and e.s.r. data for the partial reduction of the imidazolate-bridged complex $[Cu_2(Gly\text{-}Gly)_2Im]^-$ by sodium dithionite under anaerobic conditions are consistent with the reaction mechanism outlined in Scheme 3.[309] The relevance of this system to the active site of bovine superoxide dismutase is discussed. The reduction of copper(II)-(Gly-AA) complexes (AA = Gly, Ser, Phe, Ile, Trp, Tyr, or Thr) in aqueous solution has been studied by cyclic voltammetry, and evidence for the formation of copper(I)-(Gly-AA) complexes has been obtained.[310] These short-lived species give copper(0) at the mercury electrode or are reoxidized to the original copper(II) complexes. Similar behaviour is observed for the mixed-ligand complexes copper-(II)-His-AA and copper(II)-His-(Gly-AA).[311]

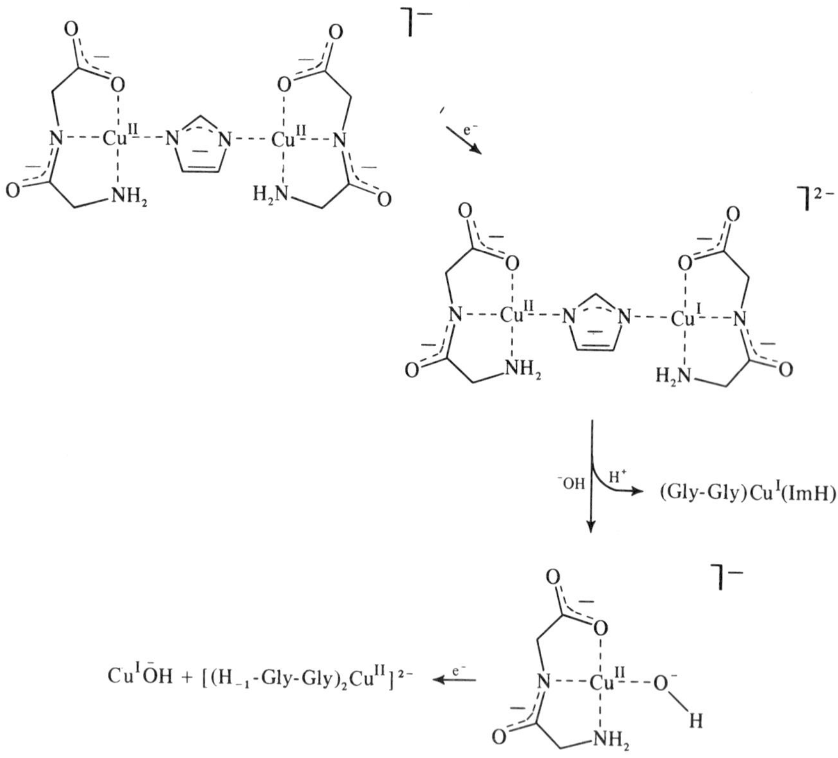

Scheme 3

[308] C. Amar, E. Vilkas, and J. Foos, *J. Inorg. Biochem.*, 1982, **17**, 313.
[309] M. Sato, M. Ikeda, and J. Nakaya, *Inorg. Chim. Acta*, 1984, **93**, L61.
[310] G. Thomas and P. S. Zacharias, *Polyhedron*, 1984, **3**, 861.
[311] G. Thomas and P. S. Zacharias, *Transition Met. Chem.*, 1984, **9**, 377.

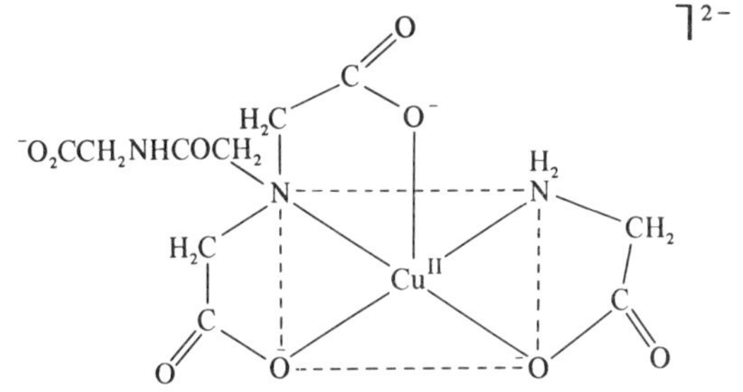

(38)

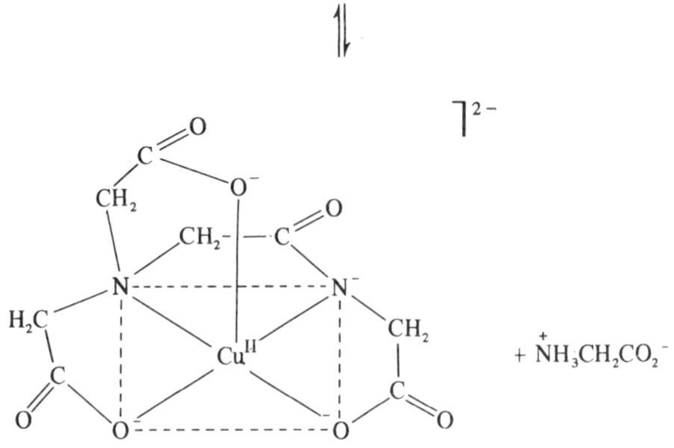

Scheme 4

The main species in weakly acidic solutions containing equimolar quantities of copper(II), glycine, and *N,N*-bis(carboxymethyl)glycylglycine is the 1:1:1 mixed ligand complex (38).[312] Addition of base to this solution causes peptide-group ionization ($pK_a = 11.68$ at 25 °C, $I = 0.1$ M KNO_3) with concomitant 'dechelation' and removal of the glycinate ligand (Scheme 4). The application of this reaction as a model for substrate removal from the active sites of metallo-enzymes is discussed. Ligand substitution in the complex [Cu(H_{-2}-Gly-en-Gly)] (39) by trien and edta occurs by nucleophilic and proton-transfer limited pathways.[313] Rate constants for these reactions are compared with those for similar reactions involving copper(II)–peptide complexes.

Eight square-pyramidal complexes of the type Cu(H_{-1}-Gly-AA)(HL)H_2O (AA = Gly, L-Tyr, or L-Phe, HL = 1- or 2-methylimidazole or benzimidazole) in which the peptides behave as tridentate ligands have been prepared and characterized.[314] The reaction of $K_3[FeCN_6]$ and Cu(H_{-1}-Gly-AA) (AA = Gly, L-Tyr, or L-Trp) in aqueous solution gives polymeric cyano-bridged complexes of

[312] R. Nakon and C. R. Krishnamoorthy, *J. Am. Chem. Soc.*, 1984, **106**, 5193.
[313] P. K. Mitchell and C. K. Pagenkopf, *Inorg. Chem.*, 1984, **23**, 1330.
[314] S. V. Deshpande and T. S. Srivastava, *Polyhedron*, 1983, **2**, 761.

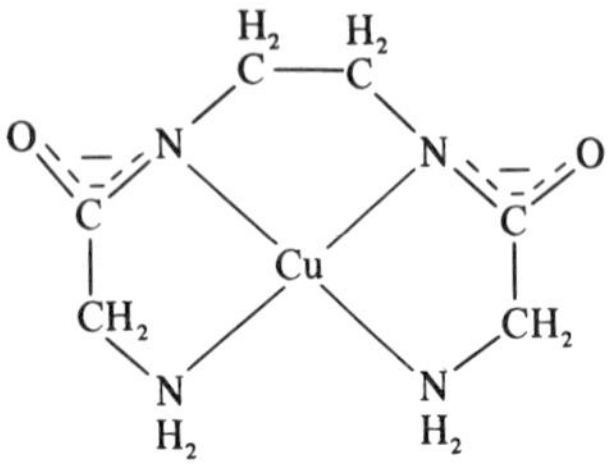

(39)

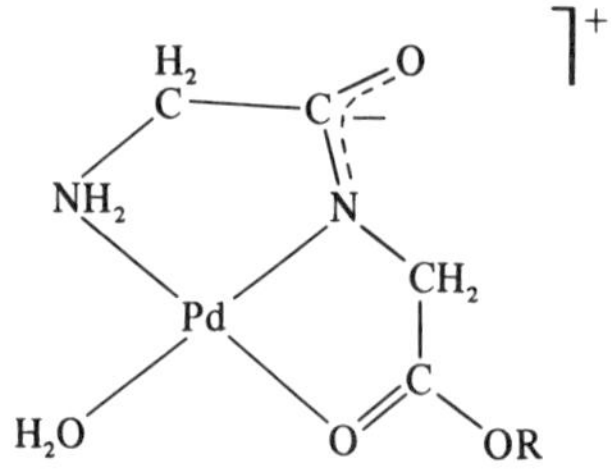

(40) R = Me or Pr^i

stoicheiometry $K_3[Cu(H_{-1}\text{-Gly-AA})FeCN_6]$.[315] Antiferromagnetic coupling between the metal ions through the bridging ligands accounts for the observed magnetic moment of 2.83 BM at 27 °C.

The recent literature also contains reports on solution spectra of copper(II) complexes with tosylated and benzoylated amino acids, peptide-amides,[316] -esters,[317] and -hydrazides,[317] n.m.r. studies on copper(II) co-ordination in $[Cu(H_{-1}\text{-Gly-Gly})(\text{Gly-Gly})]^-$ and $Cu(H_{-1}\text{-Gly-Gly})AA$ (AA = Gly, Ala, Aba, or Val),[318] calorimetric studies on copper(II) complexes containing Ala-Ala, Leu-Leu, and Leu-Tyr,[319] and the stabilities and conformations of copper(II) complexes of cyclo(-L-Met-L-His-),[282] cyclo{-(*S*-Acm-L-Cys)-D-Leu-L-His-(*S*-Acm-L-Cys)-D-Leu-L-His-},[320] and (Boc-*S*-Acm-L-Cys)-D-Leu-L-His-(*S*-Acm-L-Cys)-D-Leu-L-His-OMe[320] in aqueous solution.

Complexes of Palladium(II) and Platinum(II). – In aqueous solution at pH > 4 and with equimolar concentrations of reactants, palladium(II) forms 1:1 complexes (40) with alkyl glycylglycinates.[321] The co-ordinated ester groups in these complexes undergo base-catalysed hydrolysis some 10^5 times faster than the free peptide esters at 25 °C. Activation parameters for nucleophilic attack by OH and H_2O on the ester groups in these complexes are reported.

For the reaction of $Pd(H_{-1}\text{-Gly-Gly})H_2O$ with aliphatic amines (Scheme 5) log*K* was found to increase linearly with the pK_a of RNH_3^+.[322] The reaction of

$$Pd(H_{-1}\text{-Gly-Gly})H_2O + RNH_2 \overset{K}{\rightleftharpoons} Pd(H_{-1}\text{-Gly-Gly})NH_2R + H_2O$$

Scheme 5

315 T. S. Srivastava and S. V. Deshpande, *Inorg. Chim. Acta*, 1983, **78**, 37.

316 A. M. El-Naggar, M. R. Zaher, and F. A. Kora, *Egypt J. Chem.*, 1982, **25**, 269.

317 A. M. El-Naggar, M. R. Zaher, and S. A. A. El-Ghaffar, *Glas. Hem. Drus., Beograd*, 1982, **47**, 253.

318 E. L. Gogolashvili, A. V. Zakharov, and V. G. Shtyrlin, *Russ. J. Inorg. Chem. (Engl. Transl.)*, 1983, 28, 1461.

319 R. P. Bonomo, R. Cali, V. Cucinotta, G. Impellizzeri, and E. Rizzarelli, *Chem. Abstr.*, 1984, **100**, 127 657.

320 M. Kodaka, T. Shimizu, and M. Hatano, *Bull. Chem. Soc. Jpn.*, 1983, **56**, 1181.

321 R. W. Hay and M. P. Pujari, *J. Chem. Soc., Dalton Trans.*, 1984, 1083.

322 S.-H. Kim and R. B. Martin, *J. Am. Chem. Soc.*, 1984, **106**, 1707.

$Pd(H_{-1}$-Gly-L-Phe)H_2O with amines and the reaction of both peptide complexes with Cl^- have also been investigated. The enhanced binding of aromatic amines to these complexes is attributed to an aromatic-ring–metal-ion interaction. Rotamer mole percentages have been estimated for the ternary Pd(Gly-L-Phe) amine complexes.

The addition of cytidine (cyt) to an aqueous solution containing equimolar amounts of Gly-L-Tyr and K_2PdCl_4 at 60 °C leads to the mixed-ligand complex $Pd(H_{-1}$-Gly-L-Tyr)cyt, the structure of which has been confirmed by *X*-ray crystallography.[323] Binding of purines to $Pd(H_{-1}$-Gly-AA) (AA = Tyr or Phe) has opposing effects on the chemical shifts of the CH protons adjacent to the purine co-ordination sites.[324] While the proximity of an aromatic side chain causes an upfield ring-current shift, co-ordination to the metal ion produces a shift in the opposite direction. Controlled potential electrolysis of the complex $[Pd(H_{-1}\text{-}Aib_3)]^-$ gives either palladium(III) or palladium(IV) complexes depending on the anions present during oxidation.[325]

The peptide complexes *trans*-$PtCl_2$(Gly-AA-OMe)$_2$ (AA = Val, Leu, or Glu), *cis*-$PtCl_2$(Gly-AA′-OMe)$_2$ (AA′ = Leu or Glu), and *trans*-$PdCl_2$(Gly-Val-OMe)$_2$ may be synthesized from the corresponding Gly complexes using a carbodi-imide-containing polymer as the coupling agent.[326] The complexes of *cis*-$PtCl_2$AA (AA = Gly, Phe, Tyr, Met-OMe, Leu-Met-OMe, or Leu-Met-NH) have been analysed by f.a.b., f.d., and d.c.i. mass spectrometry.[327]

Complexes of Iron(II/III) and Cobalt(III). – Iron(III) complexes of the cysteinyl peptides Boc-Cys-Thr-Val-Cys-OMe (A) and Boc-Cys-Pro-Leu-Cys-OMe (B) have been prepared (Scheme 6) as possible models for the metal-binding sites of oxidized rubredoxin, which according to *X*-ray analysis contains iron(III) in a distorted tetrahedral field of thiolate ligands from the sequences -Cys(6)-Pro-Leu-Cys(9)- and -Cys(39)-Thr-Val-Cys(42)-.[328] A comparison of the absorption, c.d., and m.c.d. spectra of these and the iron(III) complexes of Boc-Cys-Ala_2-Cys-OMe (C), Boc-Cys-Ala-Cys-OMe (D), and Boc-Ala-Cys-OMe (E) shows that

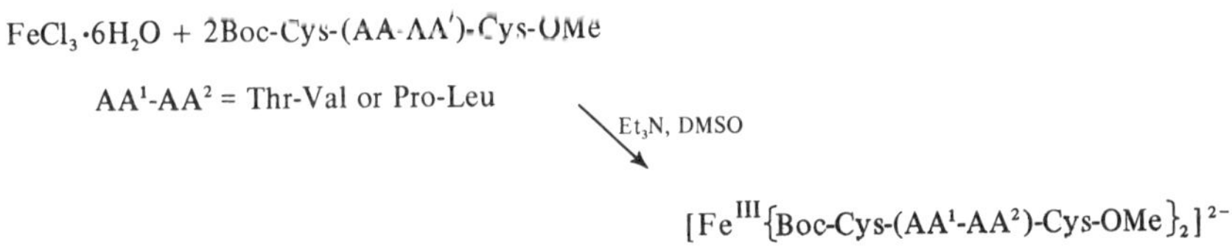

Scheme 6

[323] M. Sabat, K. A. Satyshur, and M. Sundaralingam, *J. Am. Chem. Soc.*, 1983, **105**, 976.
[324] B. Jezowska-Trzebiatowska and S. Wolowiec, *Bull. Pol. Acad. Sci., Chem.*, 1983, **31**, 33.
[325] C. Ho and A. Hamburg, Int. Conf. Coord. Chem. Abstr., Boulder, Colorado, 1984.
[326] M. Castillo, A. Romero, and E. Ramirez, *Transition Met. Chem.*, 1983, **8**, 262.
[327] D. Dalietos, A. Furst, D. Theodoropoulos, and T. D. Lee, *Int. J. Mass Spectrom. Ion Processes*, 1984, **61**, 141.
[328] M. Nakata, N. Ueyama, T. Terakawa, and A. Nakamura, *Bull. Chem. Soc. Jpn.*, 1983, **56**, 3647.

complexes of peptides B and C most closely resemble the Fe^{III}–S core electronic structure of the oxidized protein. Under anaerobic conditions the iron(III)-peptide complexes are unstable and decompose by a redox mechanism to iron(II) species and oxidized peptide. This behaviour contrasts with that of the oxidized protein, which is stable at ambient temperatures in aqueous solution. The iron(II) complexes FeA_2, FeB_2, FeC_2, and FeE_2 were prepared in aqueous 10% Triton X-100 solution.[329] Although the spectra of these complexes suggest that they have similar core structures to that in native rubredoxin, only FeB_2 shows redox behaviour similar to that of the protein.

Low-molecular-weight analogues of the 2-Fe and 4-Fe ferridoxins are discussed in a number of papers. The reaction of $[Fe_2S_2Cl_4]^{2-}$ with the peptides Ac-Gly_2-Cys-Gly_2-Cys-Gly-Gly-NH_2,[330] Boc-Ala-Cys-OMe, and Boc-Cys-Ala_2-Cys-OMe[331] in the presence of Et_3N gives cluster analogues of oxidized native 2Fe–2S ferridoxin. The incorporation of inorganic sulphide into $[Fe(\text{Boc-Cys-Ala}_2\text{-Cys-OMe})_2]^-$ also results in a 2Fe–2S cluster. Chemical ($Na_2S_2O_4$) or electrochemical reduction of these 2Fe–2S complexes results in binuclear to tetranuclear conversions, giving products with Fe_4S_4 cores (Scheme 7).

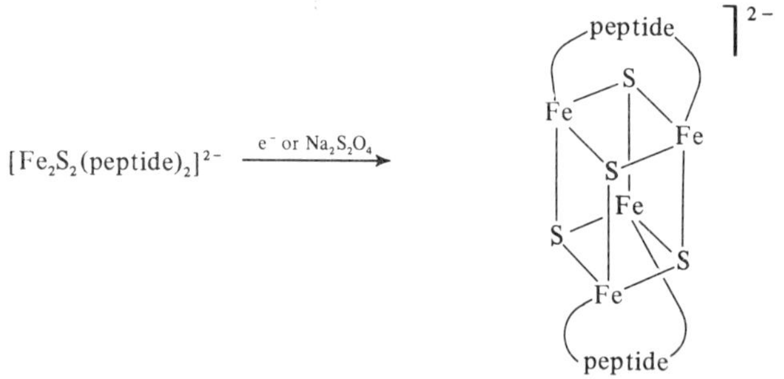

Scheme 7

The tetranuclear clusters $(Me_4N)_2[Fe_4S_4(\text{Boc-Cys-Gly-OMe})_4]$ and $(Me_4N)_2$-$[Fe_4S_4(\text{Boc-Cys-Gly-Ala-OMe})_4]$ have been investigated as models for 4Fe ferridoxins.[332] The calculated redox potential of the latter complex at 233 K in H_2O (−0.6 V) is similar to that of *C. pasteurianum* ferridoxin (−0.67 V) at ambient temperature. Effects of NH–S hydrogen bonding, which induces a folding conformation of the peptide, on the redox potential and spectra of this complex are discussed. A series of complexes of the type $[Fe_4S_4(SC_6H_4\text{-}o\text{-}X)_4]^{2-}$ ($X = NH_2$, OH, OMe, and SMe) have been investigated to examine the possible

[329] M. Nakata, N. Ueyama, M. Fuji, A. Nakamura, K. Wada, and H. Matsubara, *Biochem. Biophys. Acta*, 1984, **788**, 306.

[330] A. Balasubramaniam and D. Cocouvanis, *Inorg. Chim. Acta*, 1983, **78**, L35.

[331] N. Ueyama, S. Ueno, M. Nakata, and A. Nakamura, *Bull. Chem. Soc. Jpn.*, 1984, **57**, 984.

[332] N. Ueyama, T. Terekawa, M. Nakata, and A. Nakamura, *J. Am. Chem. Soc.*, 1983, **105**, 7098.

occurrence of five-co-ordinate iron sites in biologically important cubanes.[333] Evidence for chelation of -Cys-AA-Cys- sequences (as are present in metallothioneins) to Fe_4S_4 clusters is obtained in the reaction of $[Fe_4S_4(SBu^t)_4]^{2-}$ with *N*-PhAc-Cys-Gly-Cys-NH_2.[334] The electrochemical behaviour of the complexes $(NEt_4)_2[Fe_4S_4$(Boc-Cys-Ile-Ala-OMe$)_4]$, $(NBu^n{}_4)_2[Fe_4S_4(SC_6H_2$-2,4,6-$Me_3)_4]$, and $(NMe_4)_2[Fe_4S_4(SC_6H_2$-2,4,6-$Pr^i{}_3)_4]$ shows that bulky hydrophobic thiolate ligands stabilize $[Fe_4S_4(SR)_4]^-$ complexes but destabilize the corresponding $[Fe_4S_4(SR)_4]^{3-}$ species.[335] Inclusion of the water-soluble cluster $[Fe_4S_4$-$(SCH_2CH_2CO_2)_4]^{6-}$ into bovine insulin and bovine serum albumin protects the cluster against aerobic oxidation.[336] The possible interactions in the protein–cluster complex are discussed, and the pH dependences of the redox potentials are compared with those for naturally occurring iron–sulphur proteins. A model for hydrogenase activity at pH < 7 is proposed.

The anaerobic interaction of iron(III) salts with glutathione (HSG) and other sulphydryl-containing compounds proceeds *via* blue-coloured intermediates that are thought to be iron(II)–thiolate radical complexes.[337] A proposed mechanism for the oxidation of glutathione (pH 3) is outlined in Scheme 8. Potentiometric, Mössbauer, and magnetic-susceptibility measurements have been used to establish structures for iron(II/III) complexes with HSG and GSSG in the pH range 3–7 and in alkaline solutions.[338] The reactions of these complexes with O_2 have also been investigated.[339] The kinetics of oxidation of HSG by 12-tungstocobaltate-(III) indicate that the anion SG^- is the reactive species.[340]

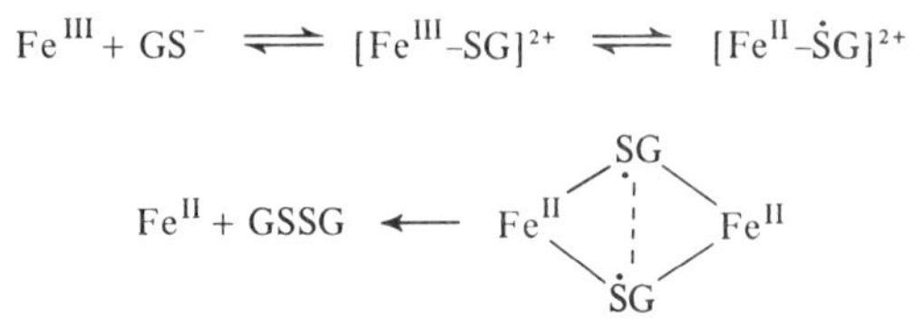

Scheme 8

Complexes of iron(III) with nine tetra- and penta-peptides containing Cys at the N-terminus and Ser, His, Thr, Tyr, or Cys at the C-terminus have been prepared in solution.[341] While these complexes resemble cytochrome P450 both

[333] R. E. Johnson, G. C. Papaefthymiou, R. B. Frankel, and R. H. Holm, *J. Am. Chem. Soc.*, 1983, **105**, 7280.
[334] J. Nieman, A. J. Naaktgeboren, and J. Reedijk, *Inorg. Chim. Acta*, 1984, **93**, L9.
[335] N. Ueyama, T. Terakawa, T. Sugawara, M. Fuji, and A. Nakamura, *Chem. Lett.*, 1984, 1287.
[336] B. Odell and P. J. Geary, *J. Chem. Soc., Dalton Trans.*, 1984, 29.
[337] M. Y. Hamed, J. Silver, and M. T. Wilson, *Inorg. Chim. Acta*, 1983, **78**, 1.
[338] M. Y. Hamed and J. Silver, *Inorg. Chim. Acta*, 1983, **80**, 115.
[339] M. Y. Hamed, J. Silver, and M. T. Wilson, *Inorg. Chim. Acta*, 1983, **80**, 237.
[340] G. A. Ayoko and M. A. Olatunji, *Inorg. Chim. Acta*, 1983, **80**, L15.
[341] H. Sakurai, E. Hatayama, T. Yoshimura, M. Maeda, H. Tamura, and K. Kawasaki, *Biochem. Biophys. Res. Commun.*, 1983, **115**, 590.

in their ability to hydroxylate acetanilide and in their e.s.r. spectra (with pyridine present), their optical spectra differ somewhat from that of the enzyme. The results form a basis for discussing the nature of the sixth haem co-ordination site in cytochrome P-450.

A number of papers deal with the effects of peptide-bridging ligands on rates of intramolecular redox processes (see below), and this work has been elegantly extended to proteins using $[Ru(NH_3)_5]^{2/3+}$ bound to the His-33 site of cytochrome *c*.[342,343] In the reduction of the ruthenium(III)-modified protein by the radicals $CO^{\bullet -}$, $Me_2\dot{C}OH$, $(CH_2OH)_3C\dot{C}HOH$, and $^-O_2CCH(OH)\dot{C}(OH)CO_2^-$, all of which were generated by pulse radiolysis, two reactions on the ms time-scale were observed (Scheme 9). A series of cobalt(III)–ruthenium(III) binuclear complexes containing a structural variety of amino acid- and peptide-bridging ligands have been synthesized and reduced to the corresponding Co^{III}–Ru^{II} species (41).[344] Rates of intramolecular electron transfer in the reduced complexes containing isonicotinylamino acid-bridging ligands ($n = 1$) are 100–300 times slower than that of the parent isonicotinate complex ($n = 0$), a retardation that is almost completely reflected in $\Delta S^{\#}$. For the isonicotinyl-peptide ($n = 2$) complexes comparable rate reductions occur, although in this series variations in both $\Delta H^{\#}$ and $\Delta S^{\#}$, attributed to differences in peptide conformations and hydration properties, are observed. Similar binuclear complexes containing monomeric ($n = 1$) and oligomeric ($n = 2–4$) proline-bridging ligands (42) have been prepared and then reduced.[345] Because of their rigidity these oligomers act as spacers, separating the metal ions at distances determined by the peptide conformation and structure. The 2000-fold decrease in rate observed on increasing n from 0 to 2 is due to increasing distance between donor and acceptor. On increasing the peptide chain from $n = 2$ to $n = 4$, however, a 20-fold increase in rate is observed, and this is attributed to *trans–cis* isomerization of the oligoproline chain, which allows the metal ions to move closer to each other.

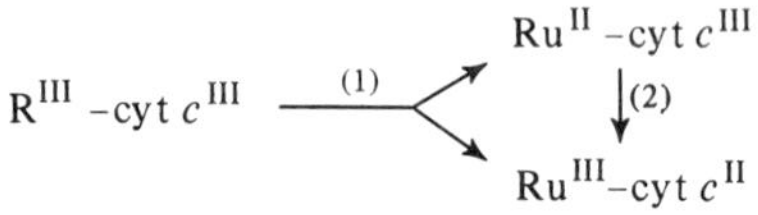

Scheme 9

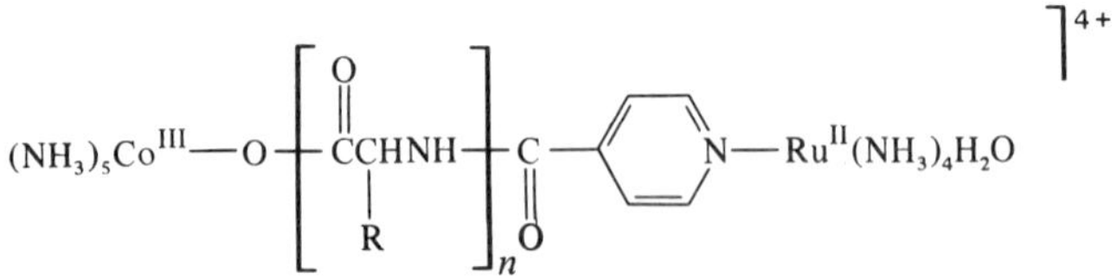

(41)

[342] S. S. Isied, G. Worosila, and S. J. Atherton, *J. Am. Chem. Soc.*, 1982, **104**, 7659.
[343] S. S. Isied, C. Kuehn, and G. Worosila, *J. Am. Chem. Soc.*, 1984, **106**, 1722.
[344] S. S. Isied and A. Vassilian, *J. Am. Chem. Soc.*, 1984, **106**, 1726.
[345] S. S. Isied and A. Vassilian, *J. Am. Chem. Soc.*, 1984, **106**, 1732.

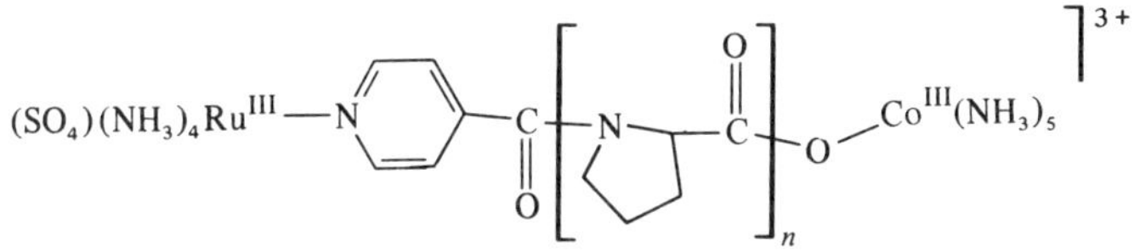

(42)

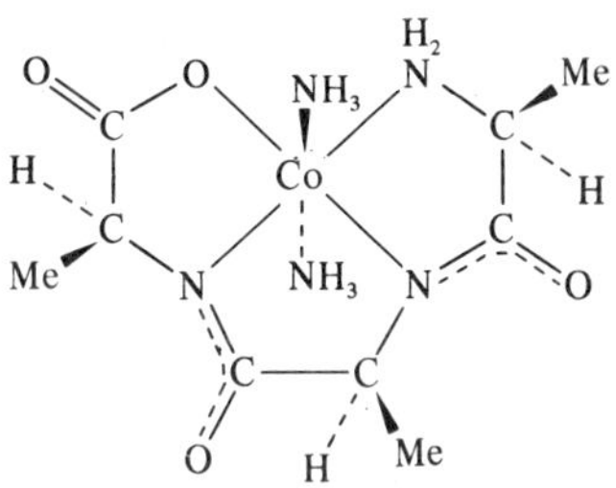

(43)

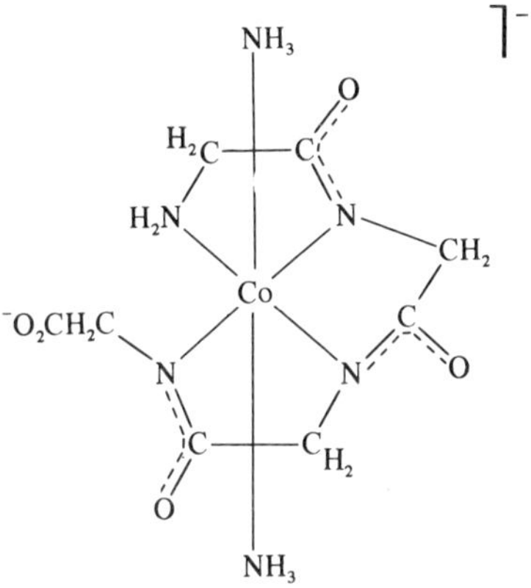

(44)

Seventeen cobalt(III) complexes of the type Co(peptide)$(NH_3)_2 \cdot xH_2O$ containing tri- and tetra-peptide ligands (43) and (44) with various combinations of Gly, L-Ala, and β-Ala residues have been prepared and characterized by n.m.r., electronic absorption, and c.d. spectroscopy.[346] In aqueous solution the complexes undergo aquation with loss of the ammine ligands. An *X*-ray crystallographic study of Co(H_{-2}-L-Ala-Gly_2)$(NH_3)_2 \cdot 2H_2O$ shows that the peptide behaves as a 3N,O donor and that the N-terminal chelate ring is non-planar with the methyl groups adopting an equatorial orientation.[347] In the complex [Co(H_{-2}-Gly-Gly-L-His)$(NH_3)_2$]$\cdot 2\frac{1}{2}H_2O$ the peptide behaves as a 4N donor; the six-membered chelate ring is puckered and the CO_2^- group adopts an axial orientation.[348] The pK_a values at 298 K are 4.06 ± 0.03 (CO_2H) and 9.81 ± 0.03 (imidazole NH).

Complexes of Nickel(III), Copper(III), and Silver(III). – Investigations by the Margerum group continue to contribute significantly to this area of metal–peptide chemistry. A number of these studies deal with copper(II/III) and nickel(II/III) redox chemistry. The reported reduction potentials for various copper(III/II)–peptide complexes span the range 0.37–1.2 V (*vs.* SHE).[349] Factors found to stabilize copper(III) include (*a*) increasing the number of

346 H. Kawaguchi, M. Ishii, T. Ama, and T. Yasui, *Bull. Chem. Soc. Jpn.*, 1982, **55**, 3750.
347 E. J. Evans, C. J. Hawkins, J. Rodgers, and M. R. Snow, *Inorg. Chem.*, 1983, **22**, 34.
348 C. J. Hawkins and J. Martin, *Inorg. Chem.*, 1983, **22**, 3879.

deprotonated amide or peptide groups, (*b*) increasing alkyl substitution on the α-carbon, and (*c*) non-aqueous solvents. An *X*-ray structure determination on $[Cu(H_{-2}\text{-}Aib_3)]\cdot 2H_2O\cdot 1\frac{1}{2}NaClO_4$, the first copper(III)-peptide structure to be reported, confirms the preference of this ion for square-planar co-ordination and allows the structural changes accompanying reduction to be predicted.[349] The copper(III)-ligand bonds are 0.12–0.17 Å shorter than those in copper(II)-peptide and 0.02–0.04 Å shorter than those in nickel(III)-peptide complexes.

Cross-exchange reactions between Cu^{III}-(H_{-2}-Aib_3) and a series of Cu^{II}-(H_{-3}-L) complexes (L = Gly_2-Aib-Gly, Gly_4, Val_4, Ala_3-Gly, or Aib_3-NH_2), all of which take place by outer-sphere mechanisms, have been investigated, and Marcus correlations have been used to determine self-exchange rate constants for the copper(III/II)-H_{-3}-L series.[350] The copper(III)-peptide complexes are rapidly reduced by $[Ru(NH_3)_5X]^{2+}$ (X = NH_3, pyridine, or picoline) at rates that are accurately predicted from the Marcus theory, using the self-exchange rate constants for the copper(III/II)-peptides and for the ruthenium complexes. Rates of electron transfer between copper(III/II) and nickel(III/II) (Scheme 10) have been measured for sixteen peptide complexes.[351] The reactions are relatively rapid and, with some exceptions, give excellent Marcus correlations. However, the nickel(III/II) self-exchange rate constants calculated from these data do not agree with those calculated from nickel(III/II) cross-reactions.[352] Reaction mechanisms involving formation of an aquo-bridged binuclear species or interaction between the nickel d_{z^2}-orbital and a copper-co-ordinated peptide N are proposed to account for these discrepancies.

$$Cu^{III}(H_{-n}L^1) + Ni^{II}(H_{-m}L^2) + 4H_2O \rightleftharpoons Cu^{II}(H_{-n}L^1)(H_2O)_2 + Ni^{III}(H_{-m}L^2)(H_2O)_2$$

Scheme 10

The photochemical decomposition of $Cu(H_{-2}\text{-}Aib_3)$ and other copper(III)-peptide complexes has been investigated, and the primary photodecomposition products are thought to be σ and π copper(II)-amidyl radicals.[353]

Chemical or electrochemical oxidation of the blue complex $[Ni(H_{-1}\text{-}Gly\text{-}Gly)_2]^{2-}$ gives a violet-black paramagnetic nickel(III) complex of proposed formula $[Ni(H_{-1}\text{-}Gly\text{-}Gly)_2]^-$ (45).[354] Similar behaviour is observed for complexes containing Ala-Gly, Ala-Ala, Gly-Ala, and Aib-Gly. Acidification of these solutions (Scheme 11) irreversibly produces transient yellow species (46), which undergo self-redox reactions to give colourless diamagnetic products.

It has recently been shown that nickel(III) plays an important role in the activation of hydrogen by hydrogenases from various sources. By comparing the

349 L. L. Diaddario, W. P. Robinson, and D. W. Margerum, *Inorg. Chem.*, 1983, **22**, 1021.
350 J. M. Anast, A. W. Hamburg, and D. W. Margerum, *Inorg. Chem.*, 1983, **22**, 2139.
351 G. D. Owens, D. A. Phillips, J. J. Czarnecki, J. M. T. Raycheba, and D. W. Margerum, *Inorg. Chem.*, 1984, **23**, 1345.
352 C. K. Murray and D. W. Margerum, *Inorg. Chem.*, 1983, **22**, 463.
353 A. W. Hamburg and D. W. Margerum, *Inorg. Chem.*, 1983, **22**, 3884.
354 S. A. Jacobs and D. W. Margerum, *Inorg. Chem.*, 1984, **23**, 1195.

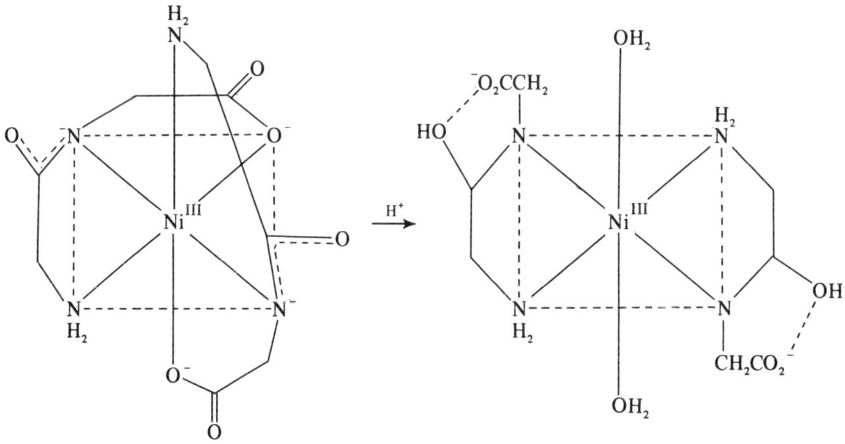

(45) (46)

Scheme 11

e.s.r. spectra of hydrogenases with those of nickel(III) complexes containing $HSCH_2CONH$-Gly-L-His, $HSCH_2CONH$-Gly_3, and other ligands, it is concluded that the metal ion in the enzyme is in a tetragonal ligand field with a cysteine sulphur occupying an equatorial site.[355] The results also eliminate the possibility of a sulphur-rich (*e.g.* four S donors) ligand field or of axial nitrogen ligation to the metal.

The wealth of knowledge of copper(III)-peptide complexes has recently been complemented by equilibrium studies on similar silver(III) systems.[356] Reaction of $Ag(OH)_4^-$ with triglycine (H-Gly_3) and tetraglycine (H-Gly_4) in aqueous alkali affords the complexes $[Ag(H_{-3}\text{-}Gly_3)]^-$ and $[Ag(H_{-4}\text{-}Gly_4)]^{2-}$, respectively. Acidification of $[Ag(H_{-4}\text{-}Gly_4)]^{2-}$ (47) gives the conjugate acid $[Ag(H_{-3}\text{-}Gly_4)]^-$ (48) (Scheme 12). The electronic spectrum of $[Ag(H_{-3}\text{-}Gly_4)]^-$ is similar to that of its copper(III) analogue except that the charge-transfer bands are blue-shifted by about 50 nm. The spectrum of $[Ag(H_{-3}\text{-}Gly_3)]^-$, which is similar to that of $[Ag(H_{-4}\text{-}Gly_4)]^{2-}$, is unaffected by the presence of phosphate even at high concentrations, confirming the reluctance of this d^8 ion to alter from square-planar to octahedral co-ordination. At pH $>$ 10 both the tri- and tetra-glycine complexes decompose by a mechanism involving proton abstraction from a methylene group followed by reduction of silver(III) to silver(I).

Complexes of Other Metal Ions. – Results of ^{13}C n.m.r. studies of manganese-(III) and gadolinium(III) complexes with trigalactosylated Ac-Gly-Ser_2-Thr_2-Gly-

[355] Y. Sugiura, J. Kuwahara, and T. Suzuki, *Biochem. Biophys. Res. Commun.*, 1983, **115**, 878.

[356] L. J. Kirschenbaum and J. D. Rush, *J. Am. Chem. Soc.*, 1984, **106**, 1003.

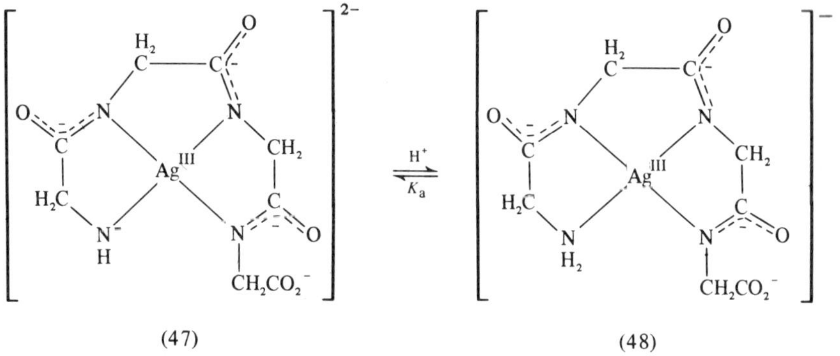

$pK_a = 12.05 \pm 0.2$, 25 °C, $I = 1.2$ mol dm^{-3}

Scheme 12

NHMe indicate that the metal ions interact weakly, although differently, with the carbohydrate part of the glycopeptide.[357] A manganese(II) complex, containing the ligand cyclo(-L-Met-L-His-) (L), of stoicheiometry $MnL_5(ClO_4)_2 \cdot H_2O$ has been reported.

The reactions of the molybdenum(IV) complexes Mo(Ac-Cys)$_4$, Mo(Sbu^t)$_2$-(Boc-Ala-Cys-OMe)$_2$, and Mo(Boc-Cys-Ala$_2$-Cys-OMe)$_2$ with the 4Fe ferridoxin models $[Fe_4S_4(SPr^i)_4]^{2-}$ and $[Fe_4S_4(Ac\text{-}Cys)_4]^{2-}$ to give Mo–Fe clusters have been investigated.[358] Molybdenum(IV) complexes of the sequential polypeptides (Gly$_2$-Cys)$_n$ and (Gly$_3$-Cys)$_m$ have been found to show high catalytic activity in the reduction of acetylene by borohydride.[359] A number of chromium(III) complexes of the type Cr(Gly-Gly)(AA)$_2$ (AA = Gly or Ala), Cr(Gly-Ala)(Gly), and Cr(Gly-Asp)Gly have been prepared and investigated as model systems for chrome tanning.[360]

[357] K. Dill, M. E. Daman, R. L. Batstone-Cunningham, M. Denarie, and A. A. Pavia, *Carbohydrate Res.*, 1983, **123**, 137.
[358] N. Ueyama, M. Nakata, A. Nakamura, M. Kamata, and S. Otsuka, *Inorg. Chim. Acta*, 1983, **80**, 207.
[359] N. Oguni, S. Shimazu, and A. Nakamura, *J. Mol. Catal.*, 1983, **22**, 1.
[360] K. Govindaraju, N. C. Kumar, C. N. Krishnam, and D. Ramaswamy, *Leather Sci. (Madras)*, 1983, **30**, 158.